DESIGN GUIDE
FOR
INSTITUTIONAL
COMMERCIAL
AND INDUSTRIAL
BUILDINGS
IN
COLD CLIMATES

High Performance Enclosures

JOHN
STRAUBE

Building Science Press

30 Forest Street
Somerville, MA 02143
www.buildingscience.com

ISBN: 978-0-9837953-9-1

First edition printed July 2011.

Revised second edition printed January 2012

Limits of Liability and Disclaimer of Warranty:

The authors and the publisher of this book have used their best efforts to provide accurate and authoritative information in regard to the subject matter covered. The author and publisher make no warranty of any kind, expressed or implied, with regard to the information contained in this book. The information presented in this book must be used with care by professionals who understand the implications of what they are doing. If professional advice or other expert assistance is required, the services of a competent professional shall be sought. The authors and publisher shall not be liable in the event of incidental or consequential damages in connection with, or arising from, the use of the information contained within this book.

Printed on recycled paper.

Illustrations and book production: Stephanie Finegan
Building Science Corporation

Contents

Contents

Preface and Acknowlegements

The creation of this book was sparked by a number of diverse factors. The course notes from my building science classes at the University of Waterloo School of Architecture formed the outline, but seminars given to many groups (especially architectural associations, facilities managers, and engineers) by Building Science Corporation developed the ideas and provided many of the details, designs, and graphics. However, it was Architecture Canada (Royal Architectural Institute Canada) and Graham Murfitt who suggested the need for a course, with course manual, on high-performance enclosures in cold climates. I, along with my legendary co-presenter Dr Ray Cole of the University of British Columbia, developed such a course. I collected the technical information required to produce a manual and together we presented it to architects across Canada.

Once the course manual was created the idea of a book became an obvious next step. With the support of Building Science Corporation, and part of a sabbatical leave from the University of Waterloo, the book before you was completed.

Any rigour and completeness is largely due to the lasting influence of my graduate supervisor, Dr Eric Burnett, while the errors and sloppiness are my own contribution. Ideas, commentary, and material that improved the content were provided, often without realizing it, by the many fine people I work with at BSC, especially Peter Baker, Robert Lepage, Alex Lukachko, Joe Lstiburek, Chris Schumacher, Jon Smegal, Aaron Grin, and Kohta Ueno. Individuals from other firms, such as, Steve Kemp (Enermodal Engineering), Edward Takahashi, FAIA and Greg Leslie (Garland) were also very helpful. Particularly detailed reviews were provided by three people: Jackson Clark, AIA, (Gould Evans Associates), Graham Finch, P.Eng. (RDH Building Engineering), and Richard Keleher AIA .

Stephanie Finegan, who produced the stunning graphics and designed the book layout, was a very patient collaborator every step of the way, and was a tremendous help in guiding the project to completion.

Of course my life partner, Professor Victoria Lamont, deserves much credit for tolerating the mumbling over my laptop while lost in thought during the many moments this project stole in the two years, most especially during episodes of Coronation Street.

John Straube
St Agatha, Ontario
January 19, 2012

Chapter 1 Introduction

The environmental crisis, and hence green building design, revolves around a wide range of issues: habitat destruction, stormwater run-off, air pollution, climate change, and resource use. However, the on-going consumption of energy to operate, condition, and light buildings, as well as the energy embodied in on-going maintenance is the largest single source of environmental damage and resource consumption. Concerns regarding energy security and carbon emissions have signalled an even stronger focus on energy in green buildings, particularly as the energy consumption of rapidly-industrializing countries such as China, Russia, India and Brazil increases at alarming rates.

Reducing the operational energy use and increasing durability should be the prime concern of those who wish to design and build "green" buildings. This position is based on the author's professional experience as well as the scientific literature. Scientific life-cycle energy and resource analyses have repeatedly found that the energy used in the operation and maintenance of buildings dwarf the so-called "embodied" energy of the materials. Cole and Kernan (1996) and Reepe and Blanchard (1998), for example, found that the energy of operation was between 83 to 94% of the 50-year life cycle energy use.

The environmental damage caused by a building's energy consumption and its production is by far the largest impact, usually dwarfing the impact of the materials used to construct a building. If buildings become much more energy efficient (i.e, use $^1/_5$ or $^1/_{10}$ as much energy as average buildings in 2010), embodied energy and pollution will become almost as important as operational energy. The industry is a long way from accomplishing this goal, even for the most exemplary of buildings.

The energy used in institutional, commercial, industrial (ICI) and multi-unit residential buildings (MURBs) in cold climates is substantial. The building enclosure and its performance almost always has a greater impact on the energy used in a building than any other building component: more than lighting, more than mechanical systems, more than office equipment. Hence, to reduce the operational energy consumption of buildings high-performance building enclosures are required. In fact, it is almost impossible to conceive of a low-energy consumption building (and all net-zero energy buildings must first be low-energy consumption buildings) in cold climates that does not have an exceptional building enclosure. The enclosure is also implicated in the majority of building durability problems, legal claims against design and construction teams, and a host of comfort and health problems.

There are surprisingly few technical references (i.e., based on building science) to explain how one should design high performance building enclosures. Ronald Brand's *Architectural Details for Insulated Buildings* (1990) is still surprisingly relevant, although building technology and science has improved in the last 20 years. The Canada Mortgage and Housing Corporation series of *Best Practise Guides* is also highly recommended. Neither of these references are placed in the context of low-energy buildings although the information they provide will certainly improve over current practice in many buildings even today.

This book aims to provide designers with the information they needed to help dramatically reduce operational energy consumption of their buildings by improving building enclosure design. The energy needed to operate equipment and processes within the building, and the efficiency with which heating, cooling, ventilation, and hot water are provided are also important, but are not directly part of this book's scope. Renewable energy will almost certainly play some role in the future low-energy, net-zero energy, or net-zero-carbon building, whether that energy is generated on site, or off. Regardless of how and where it is generated, reducing the demand for energy via good building enclosures is always far less expensive, and far less resource intensive, than photovoltaic's, wind turbines, combined

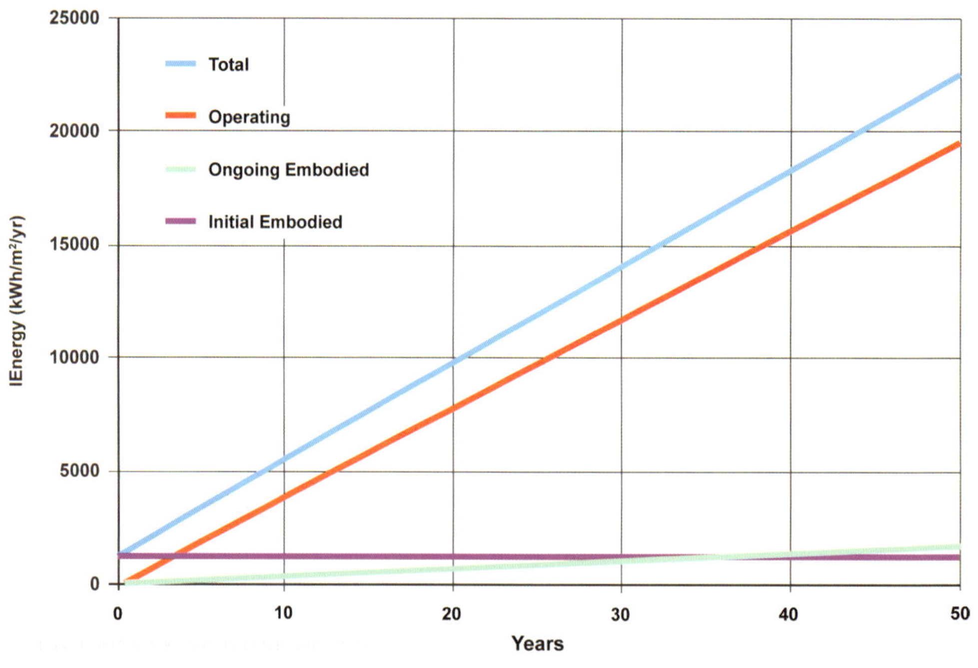

Figure 1.1: Operational energy consumption typically is ten times more significant than embodied energy for typical commercial buildings. [Cole & Kernan 1996]

Figure 1.2: Heating degree days (Base 18 °C) in North America. To convert to Base 65 °F multiply the values on the map by 1.8.

Figure 1.3: North American climate zone map (check detailed climate data for individual locations)

heat and power or most other renewable energy sources until and unless extreme energy conservation measures are implemented.

High-performance enclosures should reliably reduce operational energy needs, provide the durability necessary for the desired service life, ensure the comfort, health, and safety of the occupants. Much of the book discusses energy performance, but durability, comfort, and health are the basis of the enclosure principles presented in Chapter 3, and embedded throughout the drawings in Chapters 4 and 5.

The focus is on new exterior enclosure systems for institutional, commercial, industrial and multi-unit residential buildings. Although these building types have some similarities with low-rise, single-family housing, there are significant differences due to the different materials, systems, energy uses, and interior environments. These differences should be considered when reviewing other sources of information intended for low-rise wood-frame single-family housing.

Cold climate buildings have different concerns and priorities than hot-humid, hot-dry, and temperate marine climates. Hence, the scope of this book is intended for buildings in climates with more than about 5400 HDD (base 65 °F, 3000 Heating Degree Days, base 18 °C) that is, ASHRAE/ DOE climate zones 5 and higher. Figure 1.1 plots HDD (base 18 °C) for North America. Figure 1.3 plots the ASHRAE climates zones: ASHRAE 90.1 standard can be referenced to identify the climate zone for every county in the US. Most of the discussion and drawings are relevant for climate zone 4, although more emphasis on cooling and less emphasis on thermal control would be required.

Although cold climates are typified by cold winters, cities on the east half of the continent also have rather hot-humid summer weather, which increases the need for cooling and dehumidification. Much of the literature on low-energy and net-zero energy buildings is written in the context of climates with less annual heating needs, less extreme cold (e.g. central Europe and even most of Scandinavia are warmer than Minneapolis and Edmonton). Many parts of eastern cold climate regions have similar cooling demands as locations in the southern US or south-central Europe (e.g., Toronto, Chicago, and Boston have significant cooling and humidity control challenges). Hence, high-performance buildings in these climates need to manage hot and humid weather as well as cold dry weather.

Enclosure performance has a significant, often dominant, impact on the energy consumption of cold-climate commercial building. When such buildings are constructed according to current industry, norms, heating consumes the largest portion of site energy, despite the commonly held belief that cooling dominates the energy use in such buildings.

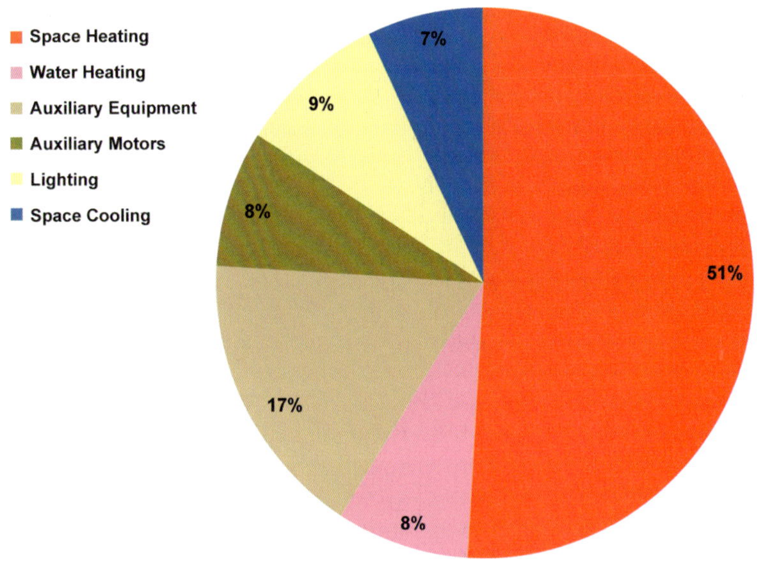

Figure 1.4: Energy consumption of Canadian office buildings [NRCan 2007]

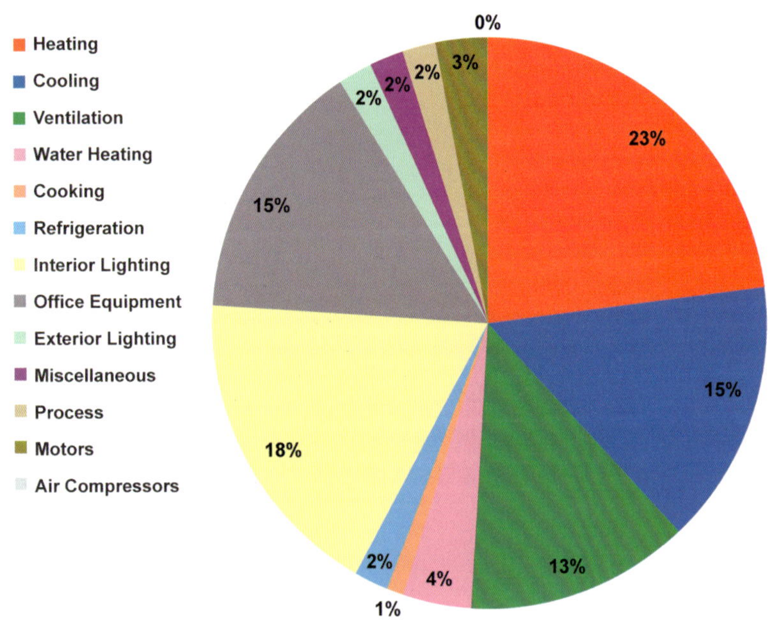

Figure 1.5: Energy consumption of large California office buildings [CEC 2006]

The total energy demand of Canadian office buildings, which are mostly cold climate buildings (only Vancouver is in zone 4), surveyed in 2007 is plotted in Figure 1.4. It can be seen that the largest energy use is clearly heating. Even large office buildings in California, a state and building type that one would assume would have very low heating demands, consume almost $^1/_4$ of their total site demand (23%) as heating.

Lighting is also a major use in commercial and institutional buildings, although far less important than heating in cold climates (and of equal importance to large California office buildings). The distribution, type, location relative to the space, and area of windows (an enclosure design choice) are all important to the ability to harvest daylight, collect free solar heat, and reject unwanted solar gains.

The approach to cooling dominates the design choices for most commercial HVAC systems, but is not usually as significant an energy consumer as lighting or heating. Although interior heat sources generate significant cooling loads, the contribution of solar gain through glazing often defines the peak cooling load, and usually is the largest single cooling load under the control of the building designer.

This chapter has identified the importance of building energy use and the need for high-performance enclosures in reducing this. Chapter 2 explores the role of the building enclosure, building shape and massing, window area, and enclosure R-values on energy consumption. Chapter 3 describes the functions of the building enclosure, and introduces the four most important control layers, and their characteristics, required to deliver occupant comfort, health and safety, low energy demands, and durability. Chapters 4 and 5 present specific approaches and details for different types of high performance enclosure systems. Finally, the appendices provide a simplified approach to assessing heat loss and gain through a building enclosure, and concise and specific technical information on over a dozen topics.

Chapter 2 Enclosures for Low-Energy Buildings

Low-energy buildings are more than buildings with high-performance enclosure assemblies. Many other design choices are required, especially mechanical system design, lighting, and controls. These items are, however, not for the most part tightly coupled to enclosure decisions. Enclosure design choices will largely define the load, and hence choices regarding the performance of the enclosure are inter-twined with some other important building decisions: building shape, massing, orientation, and window area.

For a typical mid-size office in DOE climate zones 5 and 6, especially one with an "exemplary" enclosure, other energy uses, office equipment, ventilation, and lighting comprise about half the total energy use (Figure 2.1). The role of the enclosure is, of course, less in milder climates such as Seattle (Figure 2.2). The enclosure, mostly airtightness and solar control, grow in importance in the hot-arid and hot-humid climates of Phoenix and Miami. Hence, it should be clear that different climates often require different strategies (especially with regard to low-energy design). Although many of the concepts in this book apply to all climates, and reference is

MD40SQ-C

- 4-storey, square floor plate
- 50,000 ft² GFA
- 40% WWR (N, S, E & W)
- Enclosure "C-Institutional"

MD40SQ-C

- 4-storey, square floor plate
- 50,000 ft² GFA
- 40% WWR (N, S, E & W)
- Enclosure "A-Exemplary"

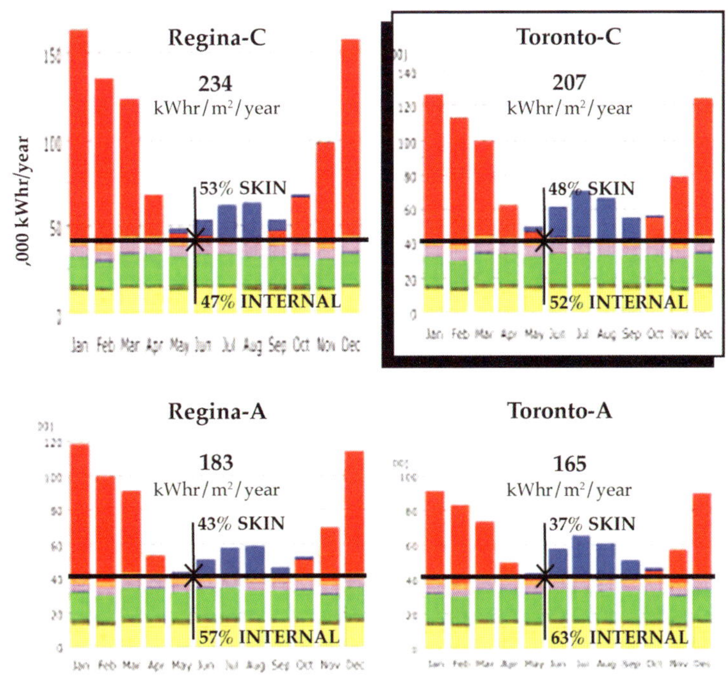

Figure 2.1: Energy use intensity and distribution of energy use of an office building in two cold climates with good (C) or very good (A) building enclosure specifications. [Ross 2009]

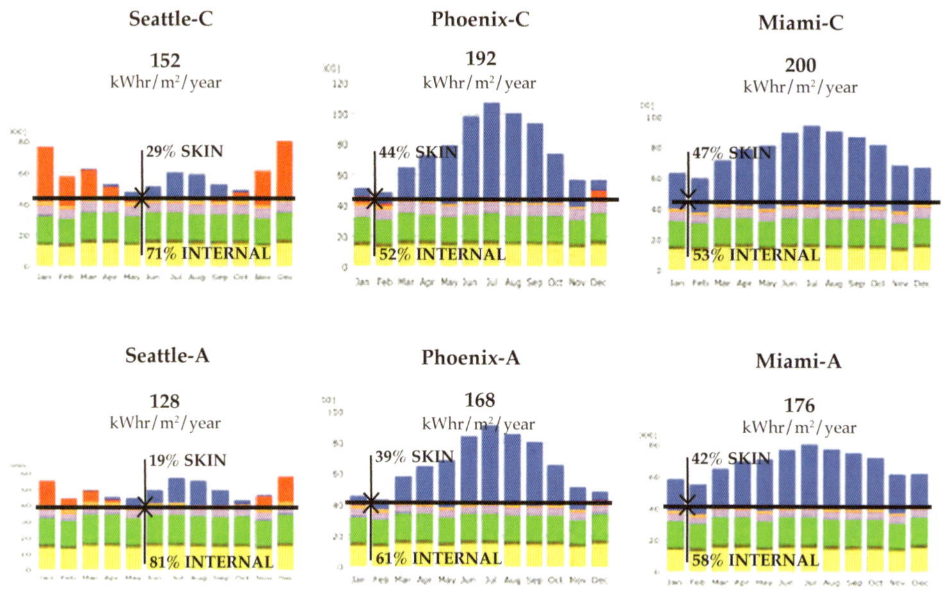

Figure 2.2: Energy use intensity and skin influences in mixed (zone 4) and hot climate zones for a mid-size office building. [Ross 2009]

occasionally made to various climate zones, the primary focus is on cold climates.

2.1 PRIORITIES FOR LOW-ENERGY COMMERCIAL BUILDINGS

To achieve low-energy performance in most commercial and institutional mid-size buildings for cold climates, the following is a general list of approximate priorities based on this author's experience.

1. Limit window-to-wall ratio (WWR) to the range of 30-40%. Up to 50% may be possible with ultra-high-performance windows (U<0.15, operable exterior shade).
2. Increase window performance (select the lowest U-value affordable, always including frame thermal bridging effects, and select the SHGC carefully).
3. Increase wall/roof insulation levels (especially by controlling thermal bridging) and increase airtightness.
4. Separate ventilation air supply from heating and cooling systems. Typically this means the use of a Dedicated Outside Air System (DOAS).
5. Use occupancy and daylighting controls for lights and equipment but avoid unnecessary building automation systems (which often are too complex to be commissioned, operated, or maintained properly).

6. Reduce equipment/plug & lighting power densities.
7. Use demand-controlled (e.g. CO_2 controlled) ventilation and heat or energy recovery. Don't over ventilate; follow ASHRAE Standard 62.
8. Improve boiler and chiller efficiency & recover waste heat (e.g., from server rooms or other sources of heat).
9. Use variable speed controls for all large pumps and fans and implement low temperature hydronic heating and cooling wherever practical.
10. Use a simple and compact building form, oriented to the sun, with a depth and layout that allows daylight harvesting.

In most situations, only when this list has been fully explored and the practical options implemented, should renewable energy from on site sources (photovoltaic, combined heat and power, etc.) be added to reach lower energy targets such as Net-Zero Energy or Net-Zero Carbon targets. Of course, access to low-cost renewable energy (e.g., waste biomass or local geothermal steam for heating) may change this recommendation.

Residential and industrial buildings will usually have slightly different priorities because of their different occupancy. The R-value/airtightness of the entire enclosure will be more important for Multiple Unit Residential Buildings (MURBs) because of their lower internal heat gains and the production of domestic hot water becomes significant, but the importance of WWR, appliance energy use, and ventilation remain.

2.2 WINDOW AREA AND PERFORMANCE

In terms of importance to energy consumption, the window-wall ratio (WWR)[1] and window performance are likely the most significant decisions for a low-energy commercial or institutional building. As opaque assemblies with very high levels of thermal control become more common (i.e., true R-values of 15 to 40), the impact of poor windows and/or large window areas begins to dominate enclosure performance (Figure 2.5).

Contrary to the belief of some, highly-glazed buildings (WWR>40%) in cold climates do not save more daylighting energy than they lose in heat.[2] Large swaths of typical south-facing windows also rarely collect more useful free heat during the day than they lose at night. Numerous studies of commercial buildings (Johnson et al., 1984, Carmody et al 2004, Bodart & De Herde 2001, Love et al 2008, Poirazis et al 2008, ASHRAE 2009 p. 15.51, Ross 2009, Lee 2010) have shown that the optimal WWR for energy performance centers around 25-35% in cold climates (Figure 2.4). This consensus of evidence from many scientific studies has resulted in standards such as ASHRAE 90.1-2007 limiting WWR to under 40% when

1
The area of glazed components (e.g., this usually includes translucent panels and may or may not include opaque spandrels framed with metal) divided by the area of the vertical enclosure is the Window-to-Wall Ratio (WWR).

2
Window *area* is not very important for achieving either daylighting or energy savings. Daylight quality is more directly related to the head height of windows, and lighting energy savings are more directly related to the presence of occupancy and daylight controls. Too much glass too close to the floor results in glare and comfort problems as well as wasted energy with little or no daylighting benefits.

applying prescriptive rules: if a WWR of more than 40% is desired, detailed hourly computer simulations need to be undertaken to show how energy can be saved in other parts of the building to offset the increased energy used by larger window areas. The optimum WWR tends to be higher if superwindows (e.g., overall window U<0.8 W/m^{2}K, R-value over 7, with SHGC<0.5) are used but even in such cases, the WWR will likely be less than about 50%-60% for cold sunny climates. For windows with U<0.5 W/m^{2}K (R-value>10), the window area can be even higher but dynamic solar control (e.g., operable shading) becomes necessary to achieve both comfort and efficiency.

Sections 3.3.4 Windows and Curtainwalls and 3.3.5 Solar Control provide more information regarding window performance and performance metrics.

Recommended targets for window U-value that will result in approximately the same R-value for the total vertical enclosure regardless of the WWR are shown in Figure 2.3. The target used to develop these recommendations is about twice the R-value (and hence half the heat loss) of ASHRAE 90.1-2007, assuming a true R-value of 20 for the opaque wall. It can be seen that buildings with high WWR in cold climates demand exceptionally low window U-values if low-energy consumption is desired.

Zone	**5**	**6**	**7**	**8**	**5**	**6**	**7**	**8**
	Heating Degree Day (18 °C)				Heating Degree Day (65 °F)			
WWR	**3000**	**4000**	**5000**	**6000**	**5400**	**7200**	**9000**	**10800**
25	2.5	2.4	2.2	1.9	0.44	0.42	0.39	0.33
30	2.3	2.2	1.9	1.7	0.41	0.39	0.33	0.30
35	2.3	1.9	1.7	1.4	0.41	0.33	0.30	0.25
40	2.1	1.65	1.45	1.3	0.37	0.29	0.26	0.23
50	1.8	1.5	1.2	1.1	0.32	0.26	0.21	0.19
60	1.5	1.2	1.1	0.95	0.26	0.21	0.19	0.17
70	1.3	1.1	1	0.85	0.23	0.19	0.18	0.15
80	1.2	1	0.9	0.8	0.21	0.18	0.16	0.14

Figure 2.3: Recommended full-system U-value (SI and IP units) for cold-climate glazing systems as a function of WWR for high-performance commercial building enclosures.

Note that these values are based on an assumed true R-value for the opaque wall of R-20 (RSI3.5). For buildings with mid-range WWR (30-50%), higher opaque wall R-values can be used to trade off window U-values that are higher than tabulated. For example, in zone 7, a window U-value target of 0.21 (1.20 W/m^{2}K) can be increased to 0.23 (1.30 W/m^{2}K) by increasing the opaque wall R-value from 20 to 28. Any number of combinations can be assessed by using the Overall Average R-value equation presented in Section 2.3

However, R- or U-value is primarily a metric for heat loss. To control heat gain, the Solar Heat Gain Coefficient should decrease as the WWR increases. While a SHGC of about 0.30 to 0.40 for the fenestration is recommended at a WWR of 40%, either exterior shading, lower SHGC, or both, should be implemented if a WWR of 50 to 60% is chosen.

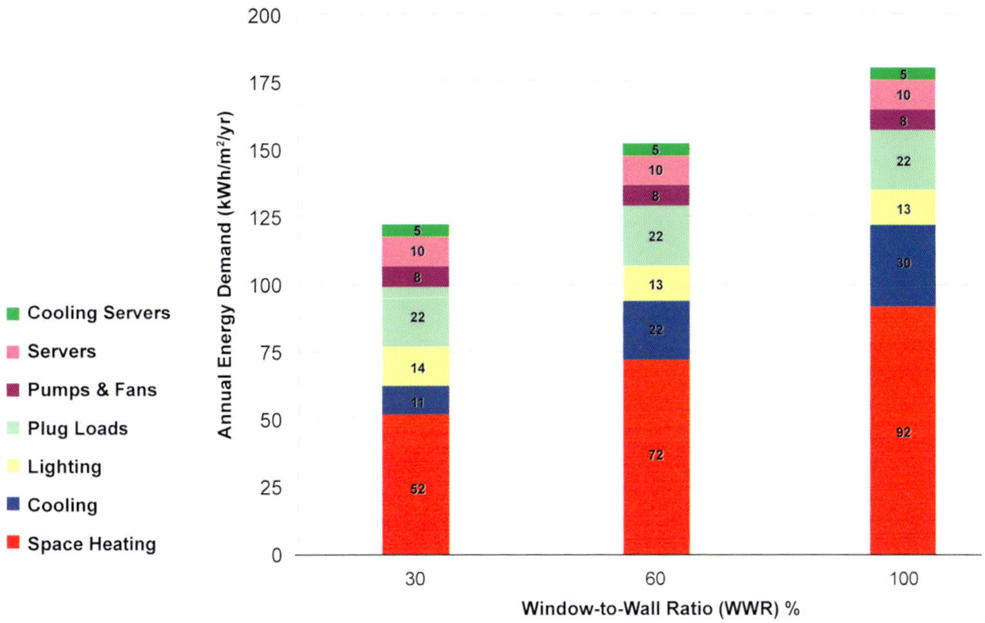

Figure 2.4: Impact of WWR on a mid-size cold-climate (Sweden) low-energy office building (after Poirazis et 2008). Note: true R-20 walls, R-3.5 windows, daylighting controls, and demand-controlled ventilation.

2.3 WALL, ROOF AND SLAB R-VALUES

There is a wide range of opaque enclosure R-values that can result in a low-energy, high-performance building. The climate, surface-to-floor area ratio, interior heat generation rate, window-to-wall ratio and the performance target (e.g., Net-Zero Energy at one extreme, LEED at the other) will influence the selection of an R-value target.

In small buildings with few interior heat gains, R-values approaching those of high performance houses (R-40 walls, R-60 roofs, R-20 basements) will generally be in the optimal range. In larger buildings (or more precisely, buildings with a low surface-to-volume ratio) and those with higher interior gains, lower wall and roof R-values (in the range of true R-15 to R-25) can often be optimal.

The R-value targets for roofs depend to some extent on the nature of their construction: low-slope roofs with waterproof membranes must usually be insulated with rigid board products (foam or dense rockwool) to control both thermal bridging and the potential for air leakage condensation. The higher cost per R-value of these products results in lower recommended R-values; typically in the R-30 to R-40 range. For pitched and easily ventilated roofs, roof R-values of 60 and higher are quite economical as much more economical low-density air- and vapor-permeable insulations can be used. For smaller buildings in cold climates choosing a pitched ventilated roof can often be desirable just to allow for economical high R-values (air-tightness is, of course, a practical challenge).

Slab insulation has not received much focus in the past, but this is slowly changing. For buildings with a narrow dimension (for example 30-40 ft / 9-12 m), continuous sub-slab insulation of R-7 to 12 should be considered,[3] with twice these levels for the vertical sub-grade foundation component. For wider buildings with significant interior gains (e.g. high intensity server cores, some retail malls), insulating only the perimeter 6-12 ft (2-4 m) to R-6 to R-12 is recommended with twice these levels for the vertical sub-grade foundation component.

3
There are significant comfort and moisture benefits to continuous sub-slab insulation. For smaller buildings, these benefits will often justify continuous insulation regardless of, or in spite of, the energy savings.

A useful metric that can be used to guide initial design choices of window, wall, and roof insulation levels is the overall average R-value. For a vertical enclosure comprised of opaque assemblies with one R-value, and windows with a different R-value, one can calculate the overall average R-value as:

$$\textit{Overall Average } R = \frac{1}{WWR \bullet U_{window} + (1-WWR)/R_{wall}}$$

Increasing the *overall average* R-value of the vertical enclosure to over R-7 to 10 (see also Figure 2.5), and roofs to over R-30, will result in dramatically reduced demands for heating. In milder cold climates (e.g., zone 5), compact mid-size buildings with high interior gains may not require R-values this high: investments in better controls, better mechanical systems, and/or renewable energy generation may be superior choices. In zone 7 or 8 climates, higher R-values are often justified, and for small buildings, which have a large enclosure-to-floor-area ratio, average vertical enclosure R-values over 12 and over R-60 for roofs can be justified.

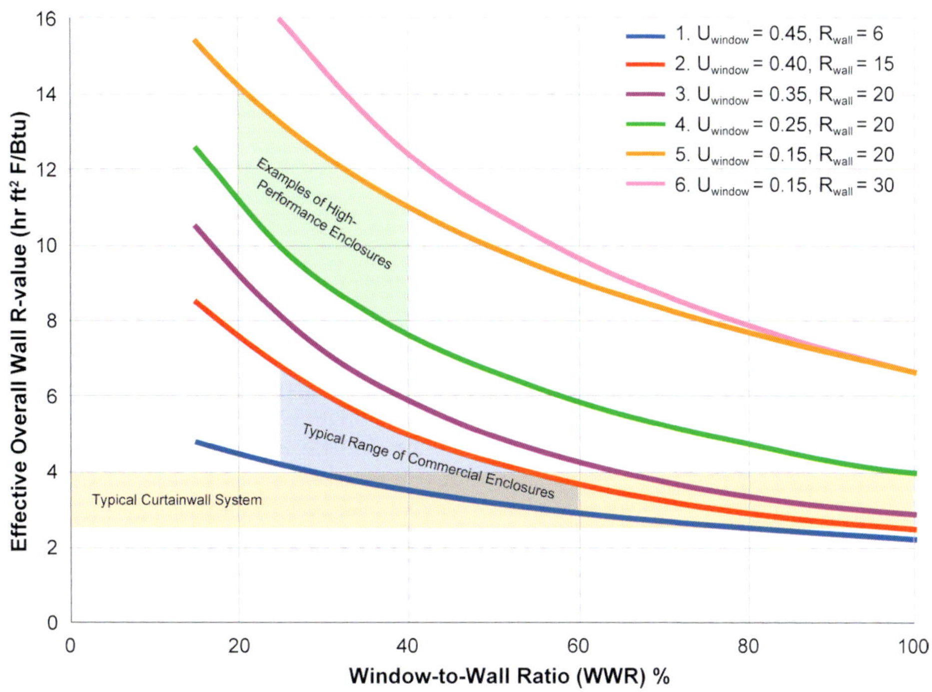

Figure 2.5: The proportion and area of windows to wall dramatically influences the thermal control performance of the overall vertical enclosure.

2.4 INTERNAL AND SOLAR HEAT GAINS

In a well-insulated building, the interior heat generated by occupants and activities can be quite important. In cold weather, this interior heat offsets the heat required to warm the space. In warm weather, the internal gains add to the cooling load.

In smaller buildings (or buildings with a large enclosure surface area to interior floor area ratio) with residential occupancy (e.g., single-family housing) interior heat gains do not play a large role in most cases. Only in very well insulated homes or mild heating weather (i.e., around 10 °C or 50 °F) do interior heat gains form a significant proportion of heat flows in a small building.

Large boxy buildings (that is, those with a small ratio of enclosure surface area to floor area) are often dominated by internal heat gain. Thermal flow in well-insulated commercial office buildings is generally dominated by heat gain and loss through windows at the perimeter (that is, within about 3–9 m / 10–30 ft of the perimeter) and interior heat gains in the core area require cooling all year. By employing moderate areas of high performance (over R-4 or metric U less than 1.25 W/m^2 °C) windows in a well-insulated opaque enclosure, many commercial buildings will require little or no heating in below freezing weather when occupied. Most of the annual space heating energy will be required at night when far fewer internal gains are present. This lack of coincidence of available solar gain and heating demand means that passive solar heating is far less important for large office towers than it can be for houses, and that using heat generated in the core (e.g., by servers and "always on" lighting) is a far more important and reliable source of heating the perimeter than solar gains are.

2.5 BUILDING FORM AND ORIENTATION

Building form and orientation do not have as large an impact on energy consumption as sometimes thought, especially for mid-size or large buildings. In all buildings, the ratio of enclosure area to floor area is important, and hence simple shapes are preferred (as well as being less expensive to build and maintain). In Europe, the ratio of volume, V, to surface area, S, is a typical metric, labeled Compactness C:

$$\textit{Compactness } C = \frac{\textit{Volume}}{\textit{Surface Area}}$$

The German energy code goes as far as prescribing higher R-values for buildings that are less compact than others.

The heating load of small buildings (e.g., houses) can vary by around 25% (Gratia and De Herde 2003) from the most compact (high C) to the most sprawling (low C) designs (Figure 2.6). Most ultra-low energy single-family houses have V/S ratios of around 1.0 or larger.

Another metric, preferred by this author for commercial buildings, is the ratio of the usable floor area, F, to above-grade enclosure area E. The more compact the form, the higher the ratio F/E. By explicitly removing volume from the assessment, this metric rewards buildings that require less floor-to-floor height. Maximizing ceiling height relative to floor-to-floor height is also always desirable from a resource use and first-cost perspective. Most building uses (such as office, education, retail)[4] do not require volume for their function, and are not sold or leased on a volume basis: it is floor area that matters. This metric also does not include the ground contact area, but does include the roof. The ground is always at a more moderate temperature difference than walls or roofs, and is not affected by solar radiation, therefore the slab and basement influence on decisions should be de-rated. The roof is also different in that it is usually easier to insulate to a high level. Counting the full roof area and not counting the slab area is intended to approximately balance the impacts.

4
Warehouses are one specific exception, and special purpose manufacturing is another.

Figure 2.7 depicts the impact of size and form on the floor: enclosure (F/E) area ratio for an office with a 14,000 ft^2 floor plate. As can be seen, the

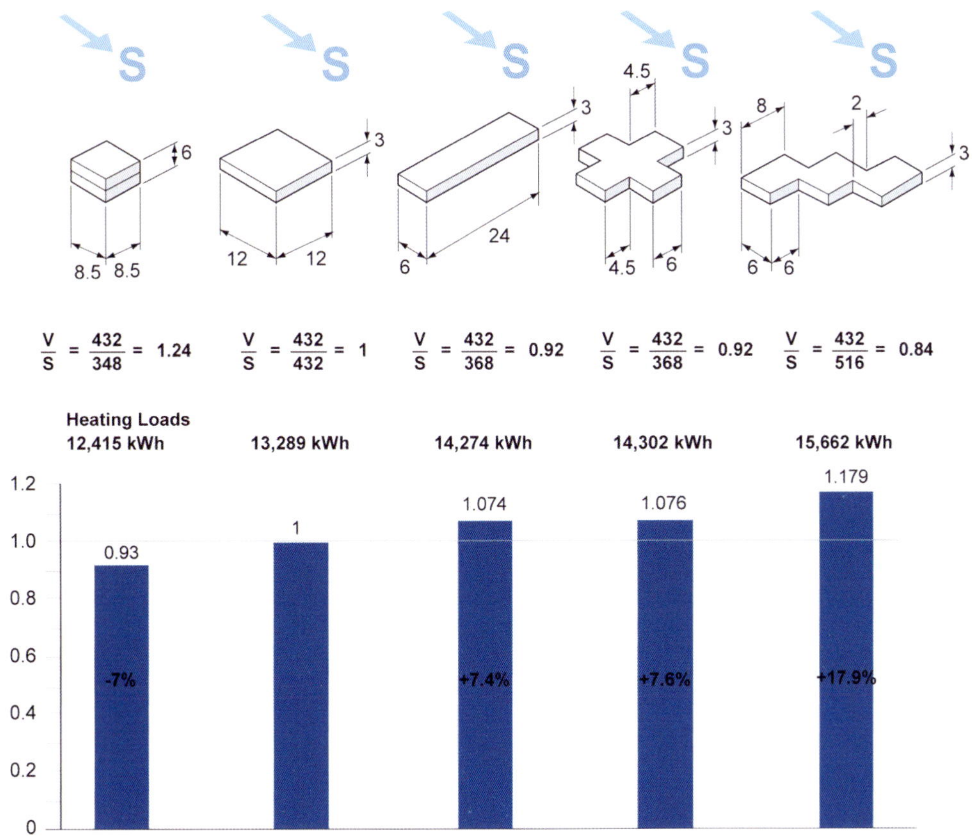

Figure 2.6: Impact of building shape on annual heating energy for a small 144 m^2 (1500 ft^2) building in a cold climate.[Gratia & De Herde 2003]

more compact the form (square is close to the perfect optimum, the circle), the higher the ratio. Large buildings (e.g., 172,800 ft^2 over 12 stories) have a much more efficient form than small buildings (houses) or large high-bay buildings (e.g., malls). It follows, that for the same heat loss through the enclosure (i.e., the same overall U-value or R-value) on a winter night, the heating energy intensity (W/m^2 or Btu/hr/ft^2) will be higher for low F/E ratio (e.g., the house or mall) and lower for high F/E ratios. It is this logic that leads to much higher recommended overall R-values for enclosures of houses than large offices.

Figure 2.8 provides a range of F/E ratios for two sizes of office floor plan area (10,000 ft^2 per floor). For the small office of 20,000 ft^2 (1800 m^2) a narrow two-storey form, ideal for natural ventilation and daylighting, has an F/E of 0.88, whereas a deep square plan has an F/E of 1.02. For the long

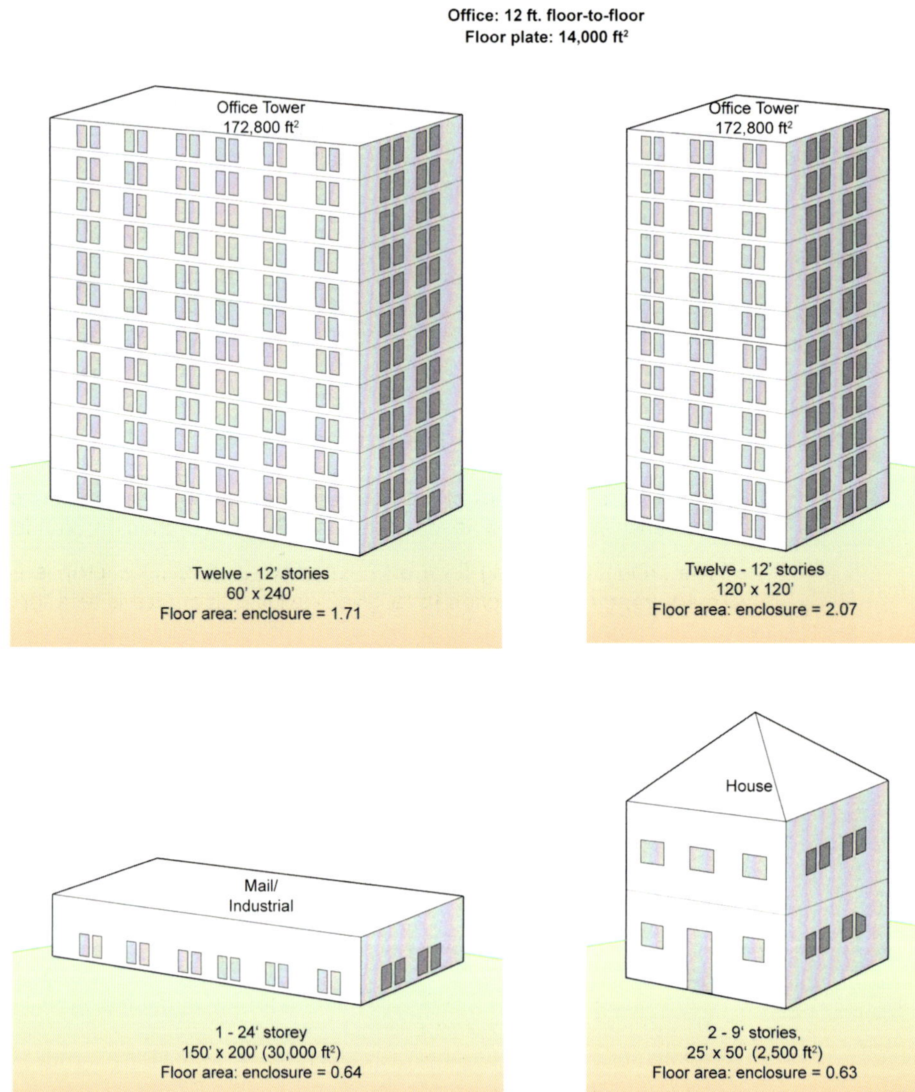

Figure 2.7: Impact of form on floor-to-enclosure (F/E) area ratio of different building types

narrow building to have the same enclosure heat loss coefficient, its overall average enclosure R-value would need to be 1.02/0.88 = 16% higher. In practise, this might be achieved by increasing the average R-value from 7.5 to 9.0. This small increase in overall average R-value appears deceptively easy to achieve: as shown in Section 2.3, it would require a significant increase in the opaque wall area R-value, a reduction in window area, or a much more expensive window.

Nevertheless, it is relatively easily to achieve technologically.

The larger building of 60,000 ft^2 (5500 m^2) has a much higher F/E ratio. For the same floor plan form, the F/E reduction is around one third (0.88/1.33 and 1.02/1.55) and hence the larger building would have notable reduction in heat loss relative to the smaller building. The impact of plan form is similar, at around 15%, as the smaller building.

The size of the building in floor area is a better indicator of energy gain/loss through the enclosure than plan shape form for most common buildings. Unfortunately, in practise, total floor size, floor plate and number of stories are constrained by the needs of the project far more than the plan form.

The impact of building form on *total* (of which heating energy is around half) energy consumption for a given building floor size is less for larger

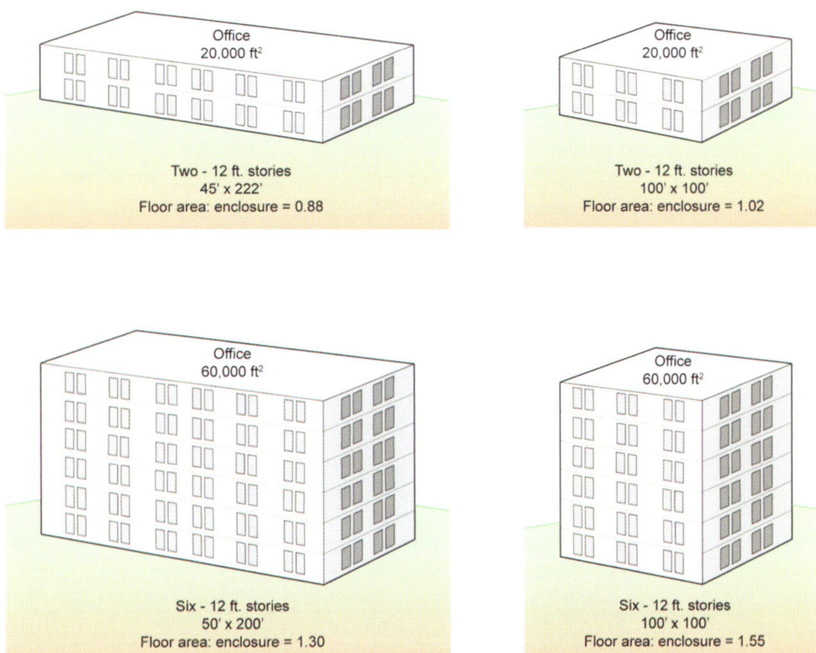

Figure 2.8: Floor area-to-enclosure area ratios for different building forms, each with 10,000 ft^2 (930 m^2) floor plate

buildings than small buildings: research suggests that around 10% separates the energy use of a compact square building to a long, narrow "bar" building (Ross 2009). This is an important observation for two reasons:

1. it allows designers of larger buildings considerable freedom of form to fulfill program requirements with site constraints and
2. daylighting and natural ventilation cooling can be important energy-saving strategies, and both require one dimension of the building to be relatively narrow, in the order of 45 to 60 ft (14 to 18 m).[5]

5 The depth of useful daylight harvesting is limited to from 2.0 to at most 2.5 times the head height of the windows serving the space. As the finished ceiling height is the highest head height possible, and ceilings are often 9 to 10 ft (2.7-3 m) high, offices around a double loaded corridor can be daylit if the building is about 2 x 2-2.5 x 9-10 = 36 – 50 ft plus the corridor / core width.

These observations lead many low-energy commercial-occupancy building designs (and almost all ultra-low energy designs) to choose a simple, compact form with the short dimension of around 45-60 ft (14 to 18 m). Such buildings can reduce lighting loads (which occur mostly during the daytime occupancy) to a minimum using daylight controls and daylight harvesting. The small increase in heat loss that a non-square floor plate form incurs can be eliminated by increasing the enclosure performance at little cost. If at all possible, the building should be oriented towards the south (for useful winter solar gain while easily rejecting summer gain and minimizing exposure to hot west summer sun).

It should also be noted that the higher the F/E, the lower the ratio of enclosure area to floor area, and hence the lower the cost of the building enclosure proportional to the usable or rentable floor area. More expensive building enclosures on a per area basis can actually be less expensive on a per rentable/saleable square footage basis if the form is kept simple and compact. Numerous very low-energy buildings have been constructed at market cost simply by choosing a more economical to build and energy-saving form for the building. In fact, the F/E ratio often has a bigger impact on first cost than it does on energy consumption.

Figure 2.9: A large modern office tower with highly efficient shape and window-wall area (note the number of windows with shades drawn)

Chapter 3 Fundamentals of Building Enclosure Design

The building enclosure is defined as the physical component of a building that separates the interior from the exterior: it is an environmental separator. In practice, the building enclosure has to provide the "skin" to the building, i.e., not just separation, but also the visible façade. Unlike the superstructure or the service systems of buildings, the enclosure is always seen and is therefore of critical importance to owners, the occupants, and the public. The appearance and operation of the enclosure have a major influence on the interior environment and on factors such as comfort, energy efficiency, durability, and occupant productivity and satisfaction (see also Appendix B–*BSD-018: The Building Enclosure* or online at www.buildingscience.com).

In general, the physical function of separation required of the building enclosure may be further grouped into three sub-categories as follows:

1. **Support** functions, i.e., to support, resist, transfer and otherwise accommodate all the structural forms of loading imposed by the interior and exterior environments, by the enclosure, and by the building itself. The enclosure or portions of it can be an integral part of the building superstructure, usually by design, but sometimes not.
2. **Control** functions, i.e., to control, block, regulate and/or moderate all the loadings due to the separation of the interior and exterior environments. This largely means the flow of mass (air, moisture, etc.) and energy (heat, sound, fire, light, etc.).
3. **Finish** functions, i.e., to finish the surfaces at the interface of the enclosure with the interior and exterior environments. Each of the two interfaces must meet the relevant visual, aesthetic, wear and tear and other performance requirements.

A fourth building-related category of functions can also be imposed on the enclosure, namely:

4. **Distribution** functions, i.e., to distribute services or utilities such as power, communication, water in its various forms, gas, and conditioned air, to, from, and within the enclosure itself.

Figure 3.1 is a conceptual drawing of part of a building enclosure showing its functional layers.[6]

Almost all enclosures must satisfy support, control, finish and distribution functions. However, only the support and control functions are needed everywhere: control and support functions must continue across every pene-

6
The enclosure, of course, often serves other non-physical purposes such as advertising, as a symbol, or as a cultural message but these are not considered here.

tration, every interface and every assembly. The lack of this required continuity is the cause of the vast majority of enclosure performance problems. The need for finish and distribution varies over the extent of the enclosure. It is rather unlikely to find an enclosure that requires a finish on the interior and exterior everywhere. It is even more unlikely to find a building that imposes the distribution function on the enclosure over its entire surface.

The **support function** is of primary importance. Without structural integrity, the remaining functions are useless. However, the industry has reached a high level of understanding and accomplishment in this area. Support systems have evolved from massive elements pierced at a few locations, to efficient primary structural systems with light-weight framed infill and sheathing. The trend to lightweight enclosures is likely to continue as the demand for more resource-efficient buildings grows.

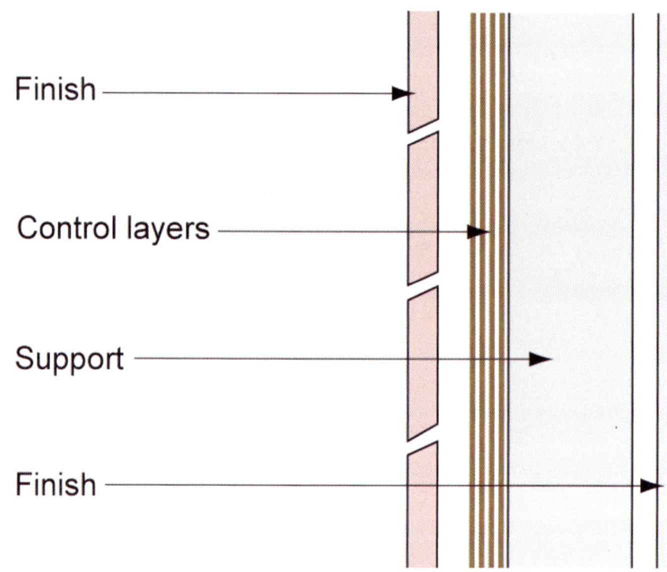

Figure 3.1: An idealized enclosure and its functional layers

For physical performance, the most common required enclosure **control functions** include rain, air, heat, vapor, fire & smoke, sound, light (including view, solar heat, and daylight), insects, particulates, and access. Again, as these functions are required everywhere, continuity of these control functions, especially at penetrations and connections, is critical to a successful enclosure. The most important control function with respect to durability is rain control followed by airflow control, thermal control, and vapor control. The level of fire and sound control required varies with code requirements and owner requirements.

Unlike the control and support functions, which rely on continuity to achieve performance, the **finish function** is optional, and may not be needed in some areas. For example, above suspended ceilings or in service or industrial spaces where the finish is often deemed unimportant. Exterior finish is often termed "cladding," but the term is imprecise, since cladding systems and materials often includes some control functions (such as UV control, solar control, impact resistance, etc.) while also delivering the finish function.

The **distribution function**, a building function often imposed on the enclosure, largely services the adjacent interior spaces and only needs to be met where there is a service or utility to be distributed – large proportions of most enclosures do not need to fulfill this building function. The distribution function of the building however usually impacts the control-related functions, for example, service entrances penetrate the entire enclosure, and pipes, ducts and wires that run through insulated stud spaces can seriously reduce the performance of insulation installed here.

Confusion about the classification of the functional roles of enclosure components and materials is far too common, and this confusion can cause serious performance and durability problems. For example,

- Vinyl wallpaper is often applied as a finish, but in fact fulfills the control-related function of vapor diffusion control and acts as a Class I vapor control layer. This unintentional control of diffusion can, and too often does, create serious mould problems in air-conditioned buildings.
- Gypsum board is often seen as filling a finish function, but in fact, the paint is more often the finish and the gypsum board often serves as a control layer for fire, sound, and air flow. If a designer or builder stops the gypsum board above a suspended ceiling because a finish function is not needed here, the required fire, sound, and airflow control will be missing.
- A thick self-adhered bituminous membrane is often used to drain water and control airflow in high-performance assemblies. However, this membrane is also a very low permeance Class I vapor control layer, and locating it on the outside of insulation in a cold climate can lead to damaging cold weather condensation.
- Polymeric housewraps are often specified for use as "water resistant barriers," but may be installed as "air barriers." If the joints between sheets are not gravity lapped, then the WRB function is compromised. If the joints are lapped properly but not sealed with tape and the sheet attached with staples, rather than cap nails or furring strips, the air control functionality is compromised. Often both conditions exist on the same building!

Function	Sub-categories	Examples
Interior Finish	Color, pattern, reflectance, gloss	Paint, wallpaper, veneers, laminates, knock-down textured coatings
Support	Gravity	Poured concrete, load-bearing block, wood stud, load-bearing steel studs
	Lateral out-of-plane (wind, EQ, window washing equipment)	Concrete, wood studs, infill wood and steel studs
	Lateral in-plane (imposed wind and EQ from diaphragm)	Concrete, masonry, wood sheathing, metal bracing
Control	Thermal Control Layer	Insulation: on the inside, within structure, or continuous on the exterior, radiant barriers, gas-fills (glazing)
	Solar (surface solar absorption and reflectance, thermal emissivity)	Paints, coatings, finishes, cool roofs, overhangs, plantings
	Air Control Layer (air barrier system)	Airtight gypsum board, taped and supported housewrap, peel-and-stick membrane
	Vapor (Vapor Control Layer)	Latex paint, closed-cell foam insulation, polyethylene sheet, kraft-facing, FSX membrane
	Rain Control Layer	Waterproofing, or water resistant barrier (WRB), drainage plane, drainage gap, roof membrane
	Fire and smoke	Gypsum board, concrete, masonry, fire caulking, fire-stopping, intumescent paint
	Sound	Gypsum board, resilient channels, fibrous insulation, perforated sheets
Exterior Finish	Color, texture, pattern, gloss, reflectance	Metal panels, concrete panels, stucco, brick, fiber-cement siding, concrete, wood

Table 3.1: Building enclosure functions

From a more practical point of view it is useful to divide the functions of the enclosure into various sub-categories to which we can assign actual products and sub-system assemblies. With respect to the control function, these are termed *control layers*.

The most important and commonly defined enclosure control function layers are, in approximate order of importance:

1. water/rain control layer (e.g., drainage plane & gap or waterproofing),
2. airflow control layer (e.g., an air barrier system),
3. thermal control layer (e.g., insulation, radiant barriers, etc.), and
4. vapor control layer (e.g., vapor retarder or vapor barrier as required).

Each control "layer" may be a single material, or a sub-assembly of materials that together provide the control desired. In many cases, two or more of the control functions are provided by a single material or layer (e.g., a membrane may provide water and air control, or spray foam insulation may provide vapor, air, and thermal control).

Figure 3.2 shows an idealized enclosure with the four control layers labelled, along with cladding, support and interior finish function layers. The water, air and vapor control layers are shown as lines to indicate that they can, in reality, be quite thin (e.g. less than $^1/_{16}$" or 1 mm) and still perform their functions very well. The support and the thermal control layer are shown as thicker components, because in practice these layers need to

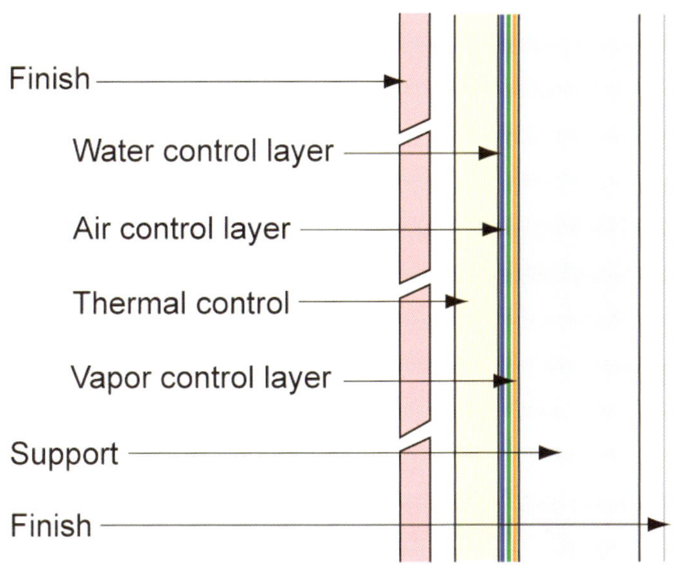

Figure 3.2: Idealized enclosure showing the four primary control layers

be thicker (e.g., well over 1" or 25 mm) to perform their function. Depending on loads and spans, the vertical support structure will usually be in the range of 3" to 8" (75 to 200 mm). Depending on its properties, the climate, and building design, the thermal control layer will usually be 2" to 10" thick (50-250 mm).

Figure 3.3 depicts the special, but common and practical, case of an assembly that collapses the water, air, and vapor control layers into a single physical material (such as a peel-and-stick membrane), and provides an optional service distribution space and interior finish.

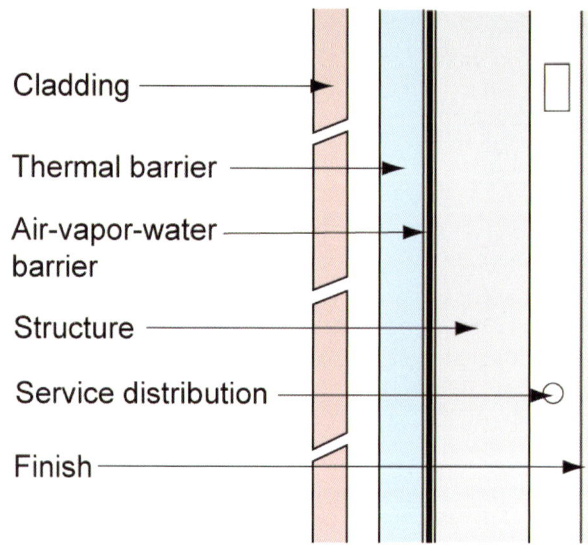

Figure 3.3: An idealized enclosure and its functional layers

The key to high performance is that the four control layers be provided and that they be as continuous and unbroken as possible across penetrations and transitions. The design and construction team must be able to identify which materials/sub-assemblies are providing each of the four control functions. Once the control layers in each enclosure component (e.g., window, roof, wall) are identified or specified, continuity analysis is conducted by drawing a line for each control layer around the entire enclosure through all penetrations and transitions. Any interruption in a line is a defect that must be rectified.

The need for distinct and clearly defined functional layers has grown from the changes in building construction in the last 50 to 100 years (see Appendix B–*BSI-042: Historical Development of the Building Enclosure* or online at www.buildingscience.com). In the past, a single solid material was used to provide structure, control, fire protection and even finish (e.g., log cabins, adobe, masonry). As the demand for better comfort, energy efficiency, economy, and speed of construction grew, enclosures became lighter, more specialised, and comprised of a range of different layers in complex arrangements (Figure 3.4). Although modern enclosures may appear at first glance (from the inside and outside) to be the same as those built 50 or 100 years ago, they are assembled in completely different ways of very different materials, and often behave entirely differently than past enclosures.

High-performance enclosures are those that require little operational energy, provide high levels of occupant health, safety and comfort, while ensuring the desired level of durability. The design of high-performance building enclosures for low-energy buildings must of course be highly focused on both the thermal and airflow control. However, the interaction of these two functions with the other primary functions cannot be avoided in many designs. For durability it is often water control that is the most important. Hence, all four of the primary control functions are briefly reviewed in the following sections.

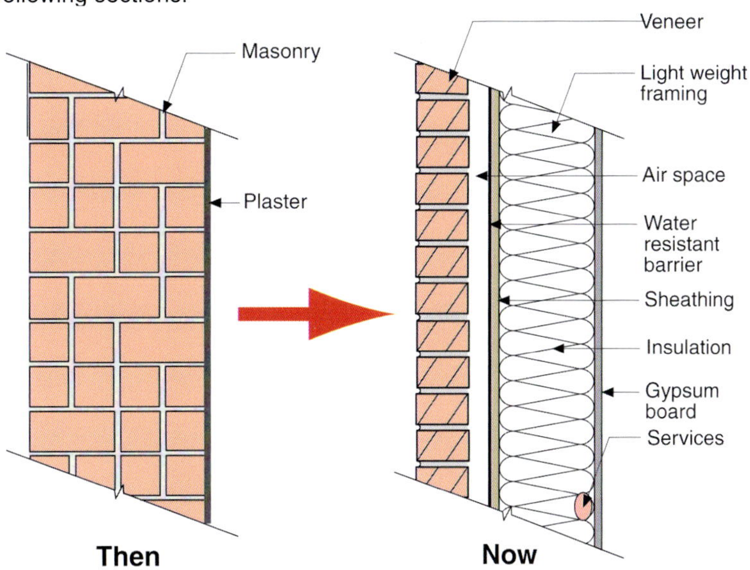

Figure 3.4: Changes in enclosure complexity

3.1 WATER CONTROL

Water (used to denote liquid moisture here) control for enclosures can include rainwater, meltwater, groundwater, and surface water. Rainwater penetration control and rainwater management is one of the most important and most difficult to provide enclosure functions. The significance of rainwater management cannot be over-emphasized: along with the structural support function it is usually this functional requirement that defines an enclosure design approach. Highly-insulated and airtight enclosures require a higher level of rain control because large energy flows are not available to provide drying of small rainwater leaks that occurred in older buildings without resulting problems.

The holistic state-of-the-art approach to rain control can be described by the three-D's: **D**eflection, **D**rainage/Storage/Exclusion, and **D**rying (see Appendix B: *BSI-028: Energy Flow Across Enclosures* or online at www.buildingscience.com).

3.1.1 Deflection

The climate and the site play a large role in defining the rain exposure that a building is exposed to. Tall buildings (over 3 stories) and buildings in the open landscape (not protected by trees and other buildings) are exposed to much higher wind speeds, and hence, much higher rain deposition rates. The rain deposited on walls can be significantly reduced by massing and overhangs. Rain exposure can be further reduced by employing surface rainwater shedding details such as projecting drips, belt courses, window

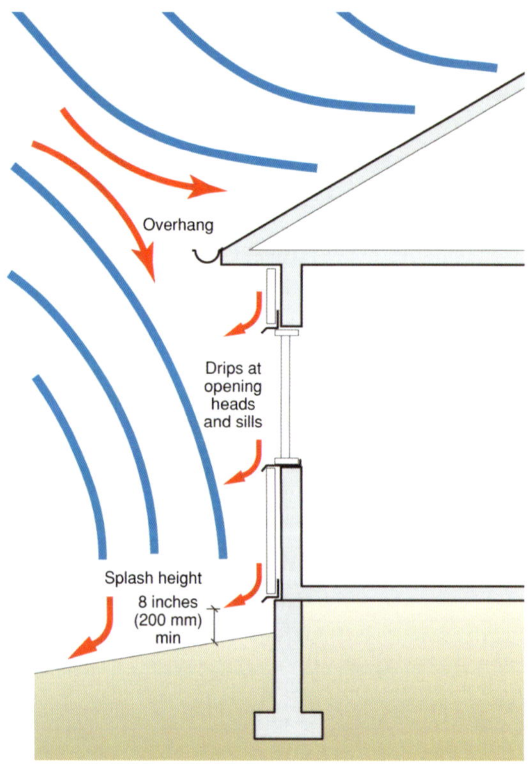

Figure 3.5: Deflection of rainwater from the building and the enclosure by slopes, drips, and surface features is the important first step to controlling rain

"eyebrows," etc. A tall building with no overhangs or projecting belt courses rising up from a suburban surroundings can be exposed to ten times as much rain as a two-storey suburban home with overhangs.

3.1.2 Drainage/Storage/Exclusion

Deflection cannot reduce all water on walls or roofs in most cases. Hence, the enclosure itself needs to be designed to control or manage rainwater penetration. There are three recognized design strategies to control rain penetration within and through the enclosure (Straube & Burnett 1999):

Storage: **Mass** or **storage** is the oldest strategy. This approach requires the use of an assembly of materials with enough storage mass and moisture tolerance to absorb all rainwater that is not drained or otherwise removed from the outer surface. In a functional mass or storage wall this moisture is eventually removed by evaporative drying from both the inside and outside before it reaches the inner surface of the wall as a liquid. Although enclosures employing this strategy might be best termed "moisture storage" systems, "mass" is often used because a large quantity of material is required to provide sufficient storage. The maximum quantity of rain that can be controlled is limited by the storage capacity available relative to drying conditions. Some examples of mass systems include adobe walls, thatched roofs, solid multi-wythe brick masonry, and single-wythe block masonry still employed for some modern buildings.

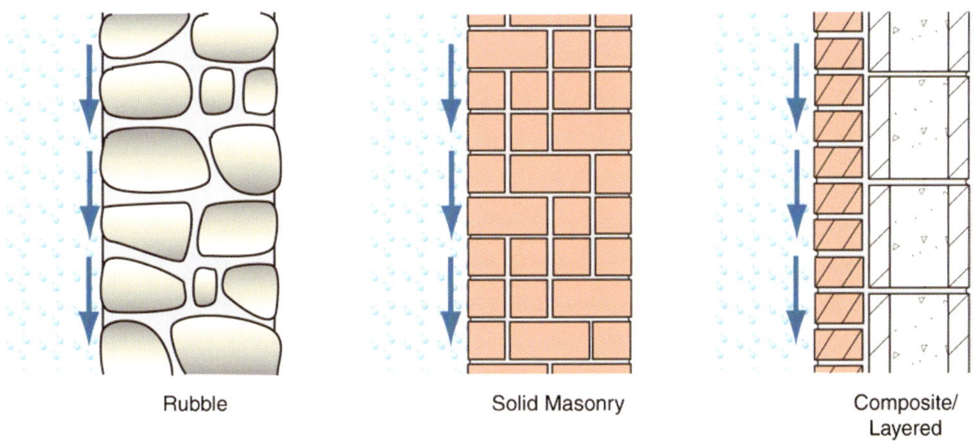

Figure 3.6: Mass (storage) enclosures

Drainage: **Drained** and screened enclosures assume some rain water will penetrate the outer surface (hence the cladding "screens" rain) and remove this water by designing an assembly that provides drainage (a capillary breaking drainage plane, a drainage gap, flashing, and weep hole/

drain). As most cladding systems, such as brick veneers, stucco, and the joints between cladding panels of shakes, terra-cotta, metal or natural stone leak[7] in the field, this is the most practical and successful system of enclosure wall rain control. Even EIFS systems (which leak mostly at the joints) and architectural precast (only at the joints) are now usually designed as drained joint systems.

7
It is sometimes possible to build such claddings with special chemical treatments or polymer additives so that they do not leak in lab tests or mock-ups, but experience has shown that such systems inevitably leak rain water in an unacceptably high proportion of buildings after only a few years service.

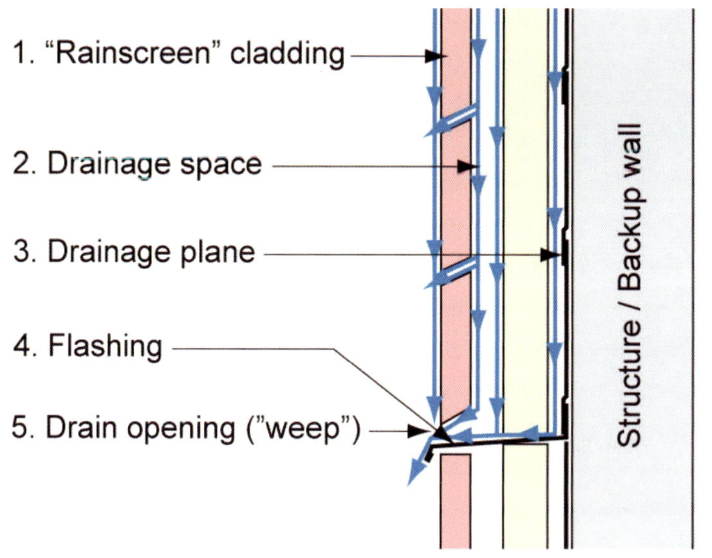

Figure 3.7: Drained systems, sometimes called "rainscreen" systems, are comprised of five components

Drained, or "rainscreen"[8] systems have five components (Figure 3.7):

1. cladding, or rainscreen,
2. one or more drainage gaps, or drainage spaces
3. one or more drainage planes, a capillary inactive material which acts as the water control layer together with the drainage gap
4. flashing, which collects water flowing vertically on the drainage plane and directs it to the exterior, and
5. weep holes or drain openings to allow water collected by flashing to exit.

8
It should be noted that the screen is much more than a rainscreen; it must also resist solar radiation, wind, snow, impact, flame spread, graffiti, etc.

As a drainage gap behind the cladding is easily provided in walls with exterior insulation, drained systems complement walls with continuous insulation.

Exclusion: **Perfect barrier** systems stop all water penetration at a single plane. Such perfect control required the advent of modern materials. Some examples of perfect barrier walls are some window frames, low-slope membrane roofs, and some metal and glass curtain wall systems. Because it is difficult to build and maintain a perfect barrier, most walls are designed as, or perform as, imperfect barrier wall systems of either the mass type or the screened type. However, some systems, usually factory built, provide wall elements that are practical perfect barriers. For example, architectural precast concrete can be considered watertight, as can glazing, and roof membranes. The joints between perfect barrier elements should be almost always be drained joints, in the form of two-stage sealant joints or similar.

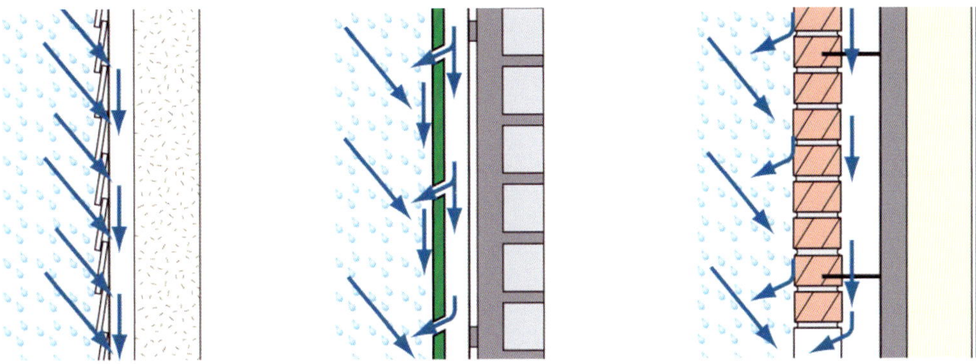

Figure 3.8: Drained enclosures

Perfect barriers may be face-sealed – in which the perfect barrier is at the exterior face – or concealed barrier, which protects the barrier behind some layers. Face-sealed sealant joints (e.g. a single line of caulking) have a poor record of performance and cannot be recommended for controlling rain entry.

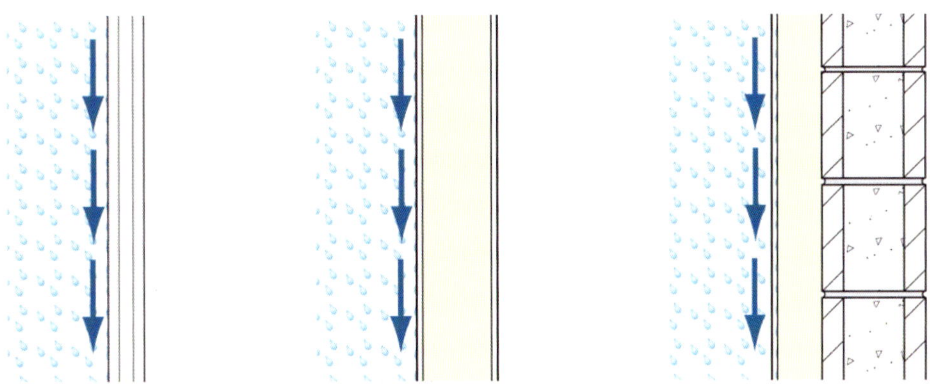

Figure 3.9: Perfect barrier enclosures

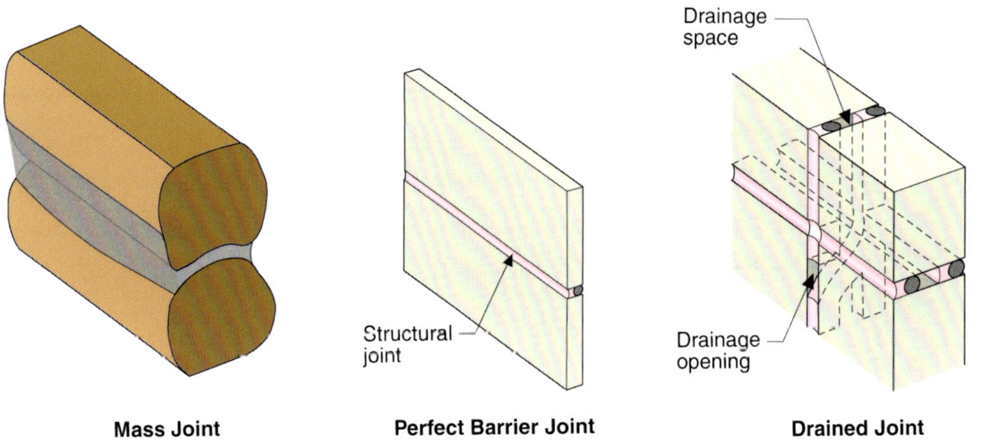

Figure 3.10: Joints are classified in the same manner as enclosure elements, although the joint and the element may use different approaches.

Perhaps the most common source of rainwater control failure is window penetrations. Regardless of which enclosure rain control strategy is used, window and door penetrations through enclosure walls should be drained. Sub-sill flashing (Figure 3.11) of various types is widely available for this purpose. For drained systems, the flashing can drain into the drainage gap. For perfect barrier and mass systems, the flashing must direct drain water to the exterior face of the assembly.

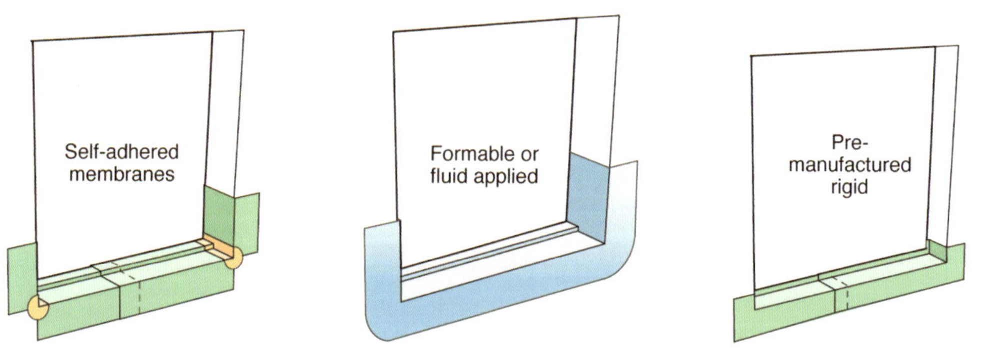

Figure 3.11: Sub-sill flashing below window and door openings is a critical requirement for all high-performance buildings if rain penetration is to be reliably controlled.

3.1.3 Drying

Despite all attempts to resist penetration and drain water, field experience has shown that some water may still penetrate or be built in during construction. Drying of this moisture must be provided for by allowing diffusion, capillary wicking to the surface followed by evaporation, and/or ventilation with outside air.

Diffusive drying of water that is not removed by drainage can dry in either direction, depending on the wall system and the climate. In colder climates the vapor flow is often outwards, and inward drying only occurs during warm weather or when the sun shines on a wall. In warmer climates, drying will often be to the interior.

Ventilation can be important as a means to encourage drying of cladding (unimportant for some cladding materials but critical for the durability of some such as wood) and drying of sheathing (very important for walls with exposed wood-based sheathing, not very important for walls with insulating sheathing outside of water proof membranes).

3.1.4 Water Control Layers: Materials and Systems

The water control layer takes several different forms depending on the enclosure rain penetration control strategy taken. Three common layers are: a waterproof membrane, a capillary break drainage plane combined with a drainage gap, and an assembly of capillary active and moisture tolerant materials that store significant quantities of water. These three water control layer types are all used in modern buildings.

For the common drained system, the water control layer can be formed from a drainage gap outside of peel-and-stick membranes, polymeric housewraps, asphalt-infused building papers, fluid-applied barriers or even spray-applied closed-cell foam. Provided a drainage gap exists, there is no need for the material to restrict hydrostatic pressures. The materials rarely fail, but layers often leak at reverse laps, penetrations and joints of all types.

Fully-adhered water (and air) control layers are always preferred, as they limit a failure, rip, tear, or penetration to a local area.[9] Loose-applied sheets allow water that penetrates a water control layer to flow laterally for long distances, significantly increasing the consequential damages.

9
Fully-adhered membranes come in numerous forms, but are adhered over all of almost all of their surface. Examples include roof membranes, "peel-and-sticks", self-adhered membrane flashing, fluid-applied coatings, and even rigid boards with inherent water (and air) resistance.

In perfect barrier systems (by definition, undrained), a perfectly watertight system able to resist backed up hydrostatic pressures becomes necessary. The material and installation quality required for this are much higher, failures are more common, and durability is usually more limited. Roof membranes comprised of high quality materials, with great attention at joints and penetrations, are the most common example. High-quality, reinforced concrete such as precast panels, are perfect barrier water control layers that often have finish and support functions.

In mass or storage systems, materials such as brick or concrete masonry, adobe, or poor quality concrete are used. The water control layer is often the same as the support or finish layer, but need not be.

3.2 AIRFLOW CONTROL

To control the flow of air, an air barrier system (air control layer) is required in all building enclosure components that separate two different environments (see also Appendix B–*BSD-014: Airflow Control in Buildings* or online at www.buildingscience.com). All modern buildings in all climates require air barriers and they are a critical component for high performance enclosures. In multi-unit/townhouse/apartment construction the air barrier system also separates the conditioned air from any given unit and adjacent units. Air barrier systems also define the location of the pressure boundary of the building enclosure. In multi-occupancy construction the air barrier system is also almost always part of the fire barrier and smoke barrier and sound control of the inter-unit separations. In such assemblies, the air barrier system must also meet the specific fire resistance and sound-resistance rating requirement for the given separation.

An air barrier system[10] is provided to control airflow across the enclosure. Airflow should be controlled for a number of reasons, including the following, in approximate order of importance:

1. to ensure good indoor air quality (accidental paths of airflow rarely provide sufficient quality or quantity of air to guarantee indoor air quality, and flow paths through enclosure elements often contain pollutants such as mould spores, off-gassing, etc.),
2. to reduce inward airflow and its attendant comfort problems due to cold drafts, and increased humidity problems in hot-humid climates,
3. to avoid condensation of water vapor as warm humid air flowing from the inside (in cold weather) or outside (in hot weather) contacts cold surfaces within the enclosure,
4. to reduce heat loss/gain, of which air leakage is an important component, particularly in well-insulated buildings, and
5. to reduce sound and odor transmission, which is typically dominated by small penetrations through the enclosure.

3.2.1 Airtightness Metrics

A recommended (used by ASHRAE 189, and the US General Services Administration) performance specification for the air control layer is:[11]

- air permeance of materials comprising part of the air barrier system shall not exceed 0.004 cfm/ft^2 at 0.3" w.g. [0.02 L/s•m^2 @ 75 Pa]
- the air leakage rate of opaque assemblies that comprise the air barrier system shall not exceed 0.04 cfm/ft^2 at 0.3" w.g. [0.2 L/s•m^2 @ 75 Pa]

10
System is used here to emphasize that materials are not enough: most leakage occurs at joints and laps, and hence it is the system that matters far more to performance than the materials used. *Air flow retarder* is another term used but is falling out of favor as it becomes clear that while a perfect air *barrier* is unlikely to be achieved, it is the desired outcome of an enclosure design.

11
ASTM E2183 can be used to measure the air permeance of a material. This is mostly an academic exercise for solid materials, as air leakage in practice almost always occurs through cracks, joints, and penetrations. Hence, ASTM E2357 *Standard Test Method for Determining Air Leakage of Air Barrier Assemblies* was developed to test the air permeance of specific assemblies. This is a rigorous test method for commercial building enclosures that can be used to verify a new design: many standard designs have passed the test of time and do not justify the need for such testing. ASTM E1677 is a less stringent standard intended for low-rise buildings, especially wood-frame low-rise buildings. Windows and doors often leak more than opaque wall assemblies: ASTM E283 and E783 can be used to measure the air permeance of windows and doors.

- the whole building shall not have an air leakage rate of more than 0.4 cfm/ft^2 (2.0 L/s/m^2) at a pressure differential of 0.3" w.g.(75 Pa).

Despite the best efforts of designers and builders, some leakage of air will still occur. The quantity of this leakage can be measured using a blower door apparatus (ASTM E-1827 *Standard Test Methods for Determining Airtightness of Buildings Using an Orifice Blower Door*) or by using the building's own air handling equipment (CGSB 149.15 *Determination of the Overall Envelope Airtightness of Buildings by the Fan Pressurization Method Using the Building's Air Handling Systems*). A building can be declared reasonably airtight if the leakage rate is below 0.4 cfm/ft^2 of enclosure area at a pressure difference of 0.3" wc (2.0 l/s/m^2 @75 Pa). Exceptionally airtight buildings will have $^1/_4$ this leakage rate and many current new buildings leak at two to ten times this rate.

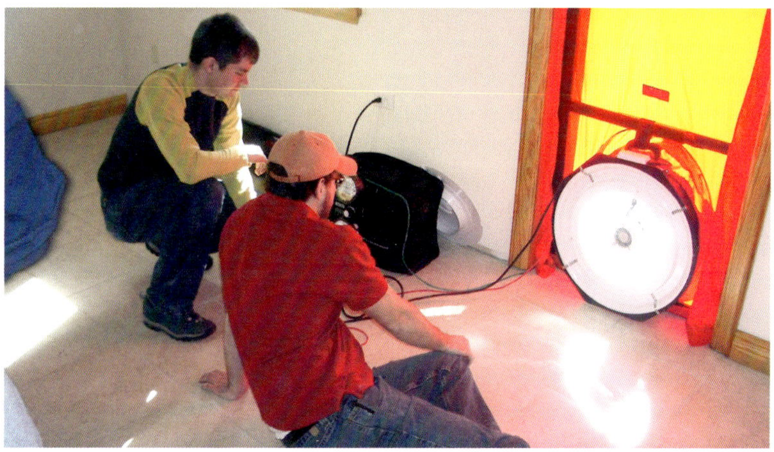

Figure 3.12: A blower door test being conducted from the 3rd floor of a five-storey residential building of 25,000 ft^2. Large buildings can be airtightness tested with simple equipment if airtight.

Although air leakage can have negative impacts on comfort and energy efficiency, it is often condensation (during cold weather or warm humid weather) due to uncontrolled air leakage through enclosures that causes the most serious and dramatic problems.

Condensation occurs because the capacity for air to store water vapor drops rapidly as the temperature drops. If the water vapor content in a parcel of air remains the same (i.e. none is added or removed) but the parcel is cooled, the relative humidity (RH) will rise. If the parcel is cooled enough, the RH will climb to 100% and the dew point, or point at which condensation can occur, is reached. If air is heated, the relative humidity will drop and condensation cannot occur. Condensation will preferentially occur on solid surfaces, and so, in most enclosures, condensation will usually occur on the backside of sheathings, membranes, etc., not within the body of insulation as is often thought.

Condensation within enclosure assemblies (termed *interstitial* condensation) can occur if warm humid air leaks out during cold weather (Figure 3.13) or, if hot humid outdoor air leaks inward while the interior is kept cool. An important function of an air barrier system is to control the flow of air that feeds this condensation. To assess whether moisture in air will condense on surfaces within enclosures, one compares the moisture content of the air (one measure of which is its dew point temperature) and the maximum moisture content of air at the temperature of the surface in question (see also Section 3.5 Relationship of Air, Vapor and Thermal Control later in this chapter). If the dew point temperature of the leaking air is higher than the surface temperature, condensation will likely occur.[12] However, air must be transported to the surface: stop airflow and condensation is controlled.

To avoid cold-weather condensation problems due to small amounts of leakage into the enclosure, the interior relative humidity should be controlled, that is, not allowed to rise too high. This can be accomplished by ensuring sufficient ventilation with outdoor air (between 5 to 15 cfm per person plus some baseline ventilation, often 0.02–0.06 cfm/ft^2) in cold weather. In essence, following the current ASHRAE 62 standard will avoid high (e.g., over 40%) relative humidities in cold weather, and will often result in an RH of 25 to 30% during very cold weather.[13] This ventilation must be accomplished using mechanical systems, as operable windows are not used by occupants in cold weather due to comfort issues.

12
This method of analysis is not valid for vapor diffusion control, as the dew point of the air in the interstitial spaces of an enclosure varies with the relative permeance of each layer. See *Building Science for Building Enclosure Design* (Straube and Burnett, 2005) for more details of how to calculate the dew point due to diffusion, and how to quantify air leakage and diffusion condensation.

13
Seattle, Portland, and Vancouver often have high wintertime interior relative humidities in multi-unit residential buildings. This is not as much because the exterior climate is humid, it is because of poor mechanical ventilation and high occupancy. Such buildings often require continuous ventilation of 50 cfm/unit to control indoor winter RH to 40%RH in these climates. Maintaining indoor relative humidity levels of 50%RH or more during freezing weather requires special enclosure designs in all climates.

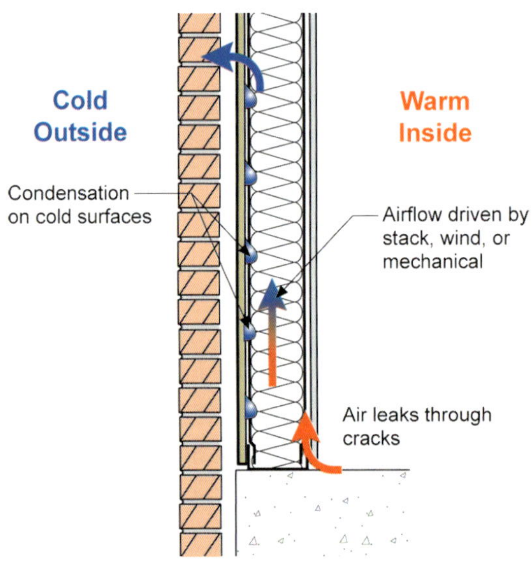

Figure 3.13: Cold-weather air leakage condensation

Excessive moisture release in the building (e.g., wet slabs, productions processes that release moisture, etc.) can over whelm ventilation's ability to remove humidity and these moisture sources should be controlled. In almost all cases, humidification, over and above that provided by human occupancy, is only needed in buildings with excessive ventilation, excessive air leakage, or both. Mechanical humidification should be limited in capacity if even installed, and its use restricted to safe interior RH levels.

Special use buildings may require higher RH. Although the need for higher winter RH in hospitals, museums, and rooms with fine wood work are usually based on tradition, and not usually required based on science, they must still sometimes be designed and built. In these cases, very high levels of airtightness and pressure control are required, and special analysis during design and testing during construction should be undertaken to avoid condensation problems.

Damaging air leakage condensation can be managed in hot humid weather, by controlling the average direction of airflow: this is achieved in practice by pressurizing the building by no more than around 2 to 5 Pa (.01 to 0.02" water column) while the interior air is dehumidified (of course, all supply ventilation air must be dehumidified as well). Pressurization cannot control condensation due to large flaws in the air barrier system and will always increases energy consumption. Pressurization will cause serious cold weather condensation in humidified buildings, even those with very little air leakage through their air barrier systems, and is to be avoided in these applications.

3.2.2 Wind Washing, Convective Loops, and Pumping

Controlling airflow through the building enclosure is the role of the air barrier system, measured using the metrics described earlier. Airflow can reduce or bypass thermal insulation in other ways than by just flowing across the enclosure.

Convective loops (Figure 3.14) can potentially form within highly air permeable insulation (low-density fibrous insulations) or, more likely, around insulation through small gaps (easily possible with improperly installed rigid board insulation or batts). The effect is enhanced by large temperature differences (over 40 °F/ 20 °C) and large gaps (e.g., over $^1/_8$" or 3 mm). If the temperature across the layer is over 40 °C (60 °F) and an interconnected air gap of $^1/_8$" (3 mm) exists on *both sides* of the insulation layer, then heat flow across the insulation can be increased by as much as 25% (Brown et al 1993) to 45% (Trethown 1991, LeCompte 1991).

To avoid the risk of convective loops, some combination of the following can be used:

1. insulation should be installed in tight contact with surfaces on at least one side (this is ensured almost by default with spray-applied insulation, whether fibrous or foam),
2. lower air permeance fibrous insulations (products with higher density or finer fibers) or airtight foams can be used, and/or
3. insulation can be installed with multiple layers of insulation with staggered joints to reduce the temperature difference across each layer

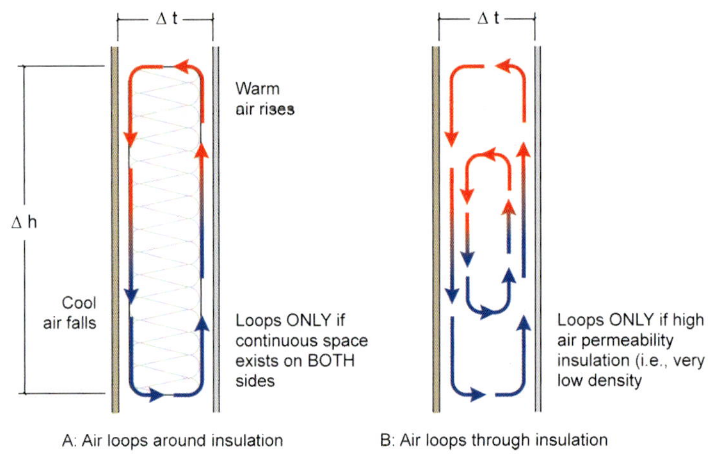

Figure 3.14: Convective air loops that reduce thermal control of insulation generally only affect insulation installed with gaps around them

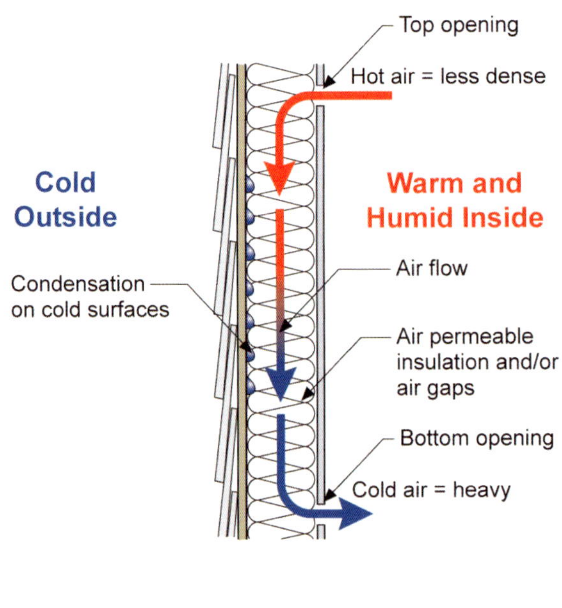

Figure 3.15: Convection can move air, and its heat and moisture content, if the interior finish is highly air permeable, and small gaps allow flow within or around the insulation. Although the interior finish need not be an air barrier, large openings and voids should be sealed to avoid convection, or an air-impermeable insulation (such as spray foam) should be used.

and to interrupt airflow pathway connections between the warm and cold sides.

Wind-washing is the term applied to the phenomenon of airflow driven by wind pressures moving air through air permeable insulations. Sometimes described as the "knitted sweater effect," it is a serious concern at the corners of building where wind pressure gradients are large, although it can be a problem in the field of the wall as well. Enclosures that use interior air barriers with air permeable insulation outboard of them are susceptible. Wind-washing can reduce the thermal control by anywhere from 25 to 90% depending on the amount of airflow involved (Uvsløkk, 1996).

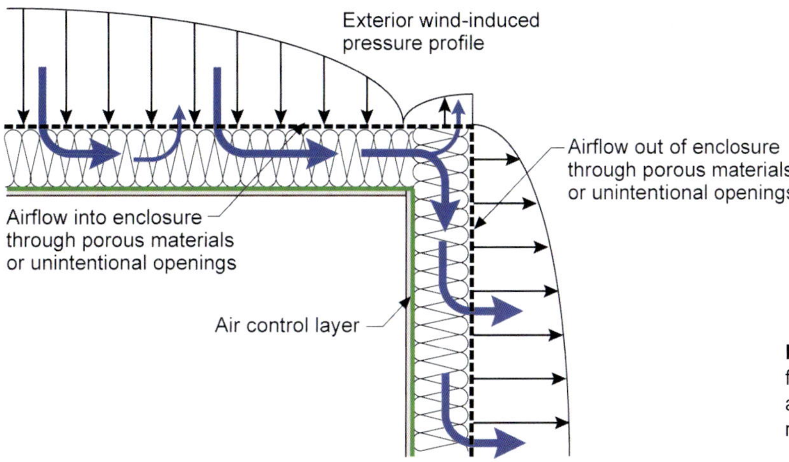

Figure 3.16: Wind-washing, the flow of wind through air permeable insulations, can significantly reduce the thermal control

Wind-washing can be controlled by applying some limited amount of airflow resistance outboard of air permeable insulations (e.g., housewrap, glass scrim facing, sealed sheathing), increasing the density of fibrous insulation (to well over about 50 kg/m^3/3 pcf), and ensuring that air-impermeable foam insulations are pressed tight (i.e., less than 1-2 mm $^1/_{16}$" gap) to their substrates to restrict lateral air flow behind them.

Finally, pumping is another concern. If the air control layer is flexible and not constrained by adhesion or clamping, it may billow under the action of the wind (Figure 3.17). Although a housewrap or roof membrane may be perfectly airtight when tested, when exposed to the gusting wind in service air can be drawn into an assembly through the insulation, resulting in energy loss, condensation, and even comfort problems.

The control of pumping is a primary reason why high performance air control layers should be either rigid, fully-adhered, or clamped between rigid materials.

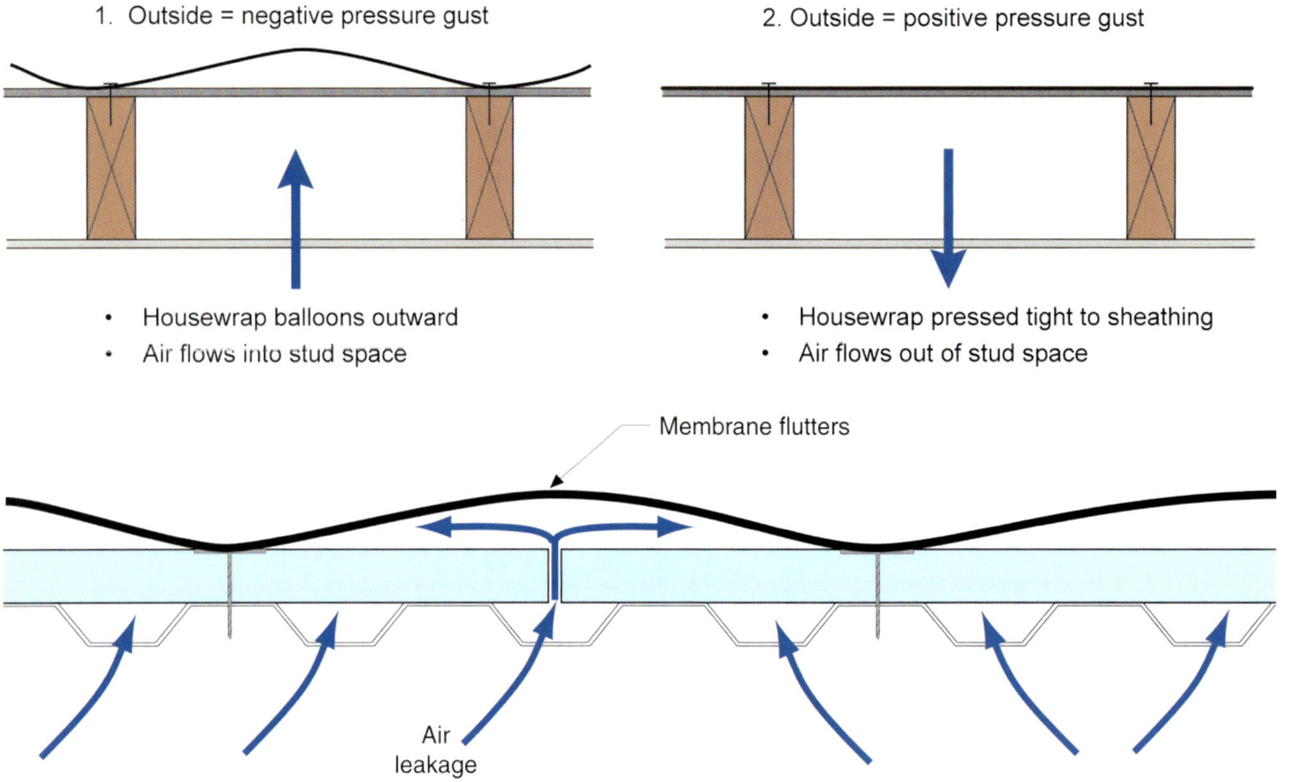

Figure 3.17: Pumping due to flexible unsupported air barrier membrane (housewraps or roof membranes) can result in damaging airflow even when airtight

3.2.3 Airflow Control Layers: Materials and Systems

The airflow control layer (i.e., air barrier) in different enclosure assemblies must be connected together to form a complete air barrier system that resists air flowing from the inside to the outside. Air control layers are comprised of assemblies of materials and components. Airflow control layers can be formed of many different materials and systems, including:

- rigid sheets such as gypsum board, exterior gypsum or OSB sheathing with the joints sealed (with sealant, special tapes, or gaskets),
- supported or adhered membranes, polyethylene sheet with the joints sealed and sheet supported by other materials to resist deflection and wind load,

- spray polyurethane foam, (provided all joints and penetrations are sealed with membranes and sealant),
- concrete, provided cracks and form tie holes are sealed,
- concrete block coated with a heavy block filler and sealed at joints,
- glass or metal sheets,
- roof membranes.

In all cases, it is the continuity of the joints and structural support that distinguish a successful application from a failed one.

Continuity is the key, and hence an air barrier system requires sealed joints at all penetrations and junctions, such as at floor penetrations, window frames, etc. Small cracks and openings can often be tolerated, however, even a small hole or long thin crack may allow too much air leakage. Unsealed cracks, openings, and joints should have a total air leakage area of less than about 2.5 in^2 per 100 ft^2 of enclosure area (170 mm^2/m^2).[14] High performance enclosures will have a fraction of this leakage area.

The air control layer must be:

1. Continuous: this it the key requirement and the one that is most difficult to achieve in practice
2. Strong: as installed the system must be able to resist most of the design wind gust load (i.e., something in the order of 1 to 2 kPa / 20 to 40 psf). Adhesion of membranes must be ensured, and loose membranes often need to be sandwiched between stiff layers (e.g., between gypsum board and studs, or between board foam and sheathing). Fastenings can be critical.
3. Stiff: to reduce stresses that cause stretching and fatigue and prevent billowing and pumping that allow air movement (Figure 3.17). Fully-adhered membranes, sheet or liquid-applied, are always preferred.
4. Durable: as the air control layer is fundamental to performance, it must remain durable for the service life of the enclosure, or be replaceable.
5. Air-impermeable: this is the simplest requirement and the least important. Although a value of 0.02 lps/m^2@75 Pa was quoted above, this material property is not critical provided the assembly and component airtightness targets are achieved.

It is important to note that an air barrier system/airflow control layer is not a product that can be purchased: the details of design and construction quality dominate the resulting performance not the materials despite the best marketing efforts of manufacturers.

14
To put this in perspective, a 10-foot wide by 10-foot high area of enclosure (100 ft^2) with a $^1/_{16}$" crack along one side would have an unsealed area of 7.5 in^2.

In practice, an air control layer located outside the primary structure and service distribution spaces has the highest likelihood of being designed and built in an airtight manner because this location dramatically reduces the number of penetrations. Of course, if insulation is installed to the inside of this location, the air barrier material must be vapor permeable (to avoid a cold side vapor barrier). However, locating the insulation to the exterior also has the major advantage of continuity. Hence the ideal location of the air control layer is to the exterior of the structure, but inboard of the insulation. In this location the air control layer is protected from wide temperature ranges and exposure to solar-radiation which should ensure much greater durability.

In some assemblies, the material layers used for rain control (e.g., fluid-applied or sheet membranes) can also often also be used for airflow control. However, not all materials that can be used as part of an air barrier system (such as sealed gypsum board) can control rain, and the installation techniques for rain control (relying on gravity via shingle laps to provide continuity to gravity drainage) are not effective for air flow control. Building paper such as #15 felt or Grade D, does not have the strength to resist the imposed wind loads.

3.3 THERMAL (HEAT FLOW) CONTROL

As society demand(s) that buildings consume less energy and generate less pollution, minimizing the flow of heat through the enclosure has become an increasingly important function for the enclosure to perform. The control of heat flow is also important for the control of interior surface temperatures, and hence ensuring human comfort and avoiding cold weather condensation. Controlling the temperature of various elements and layers within an enclosure assembly can be used to avoid condensation or enhance drying, both of which influence durability.

Thermal control is typically seen as being provided by insulation. However, radiant barriers, solar control coatings, thermal mass, and thermal breaks are also part of the thermal control system.

See also Appendix B–*BSD-011: Thermal Control in Buildings* or online at www.buildingscience.com for more information.

3.3.1 Thermal Control Metrics

Heat transfer occurs by three primary mechanisms, acting alone or in some combination (Figure 3.18):

- conduction,
- convection, and
- radiation.

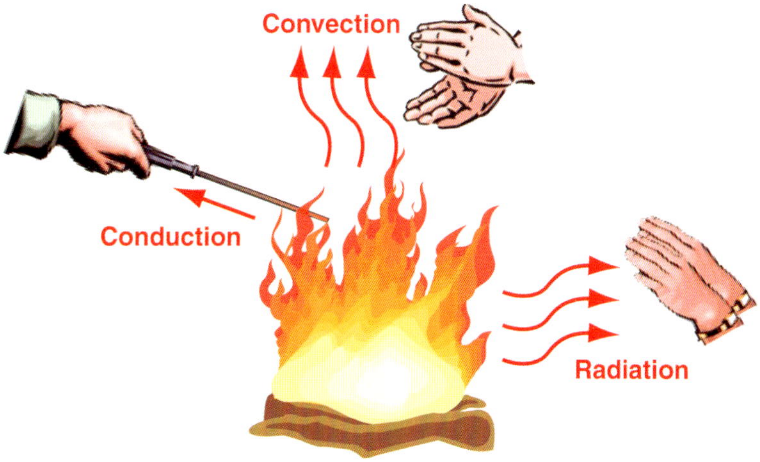

Figure 3.18: Heat flows in three different modes (from the *EEBA Insulating Guide,* 2007)

Changes in moisture state, although not strictly an energy transfer mechanism, must also be considered since these state changes absorb and release heat energy, i.e., latent heat.

Conduction is the flow of heat through a material by direct molecular contact. This contact occurs within a material or through two materials in contact. It is the most important heat transport mode for solids, it is sometimes important for liquids, and it is occasionally important for gases.

Convection is the transfer of heat by the movement or flow of molecules (liquid or gas) with a change in their heat content. This is an important heat transfer mode between fluids and solids, or within fluids.

Radiation is the transfer of heat by electromagnetic waves through a gas or vacuum. Heat transfer by this mode therefore requires a line of sight connection between the surfaces involved. All objects above absolute zero radiate heat energy; it is the net radiative heat transfer that is the heat transfer of interest. Radiation is mostly of importance for heat transfer between solids separated by gases and within highly porous solids.

State change, sometimes called phase change, occurs at a constant temperature but still entails the movement of energy. For example, evaporation absorbs energy and condensation releases energy. The energy required to create a phase change is often called latent heat.

The mode of heat transfer often changes during the process of heat flow through and within building enclosures. For example, the sun transmits heat by radiation to the earth, where it can be absorbed, for example, by a brick wall. The heat is then transferred through the brick by conduction and to the indoor air by convection, and to the indoor surfaces by radiation.

Heat flow control is measured in different ways for each of the different mechanisms. A few of the more common terms are explained below.

Conductivity (k or lamda): a material property, heat flow per unit area per unit thickness per unit temperature, (W/m K)

Conductance (C): a property of a material layer, heat flow per unit area and temperature,

$$C \text{ units } (W/m^2\ K) = \text{conductivity } k\ /\ \text{thickness } l \text{ (in meters)}$$

Resistance (R or RSI): a property of a material layer or enclosure assembly measured from surface to surface.

$$R = 1/C$$

$$RSI \text{ units } (m^2\ K\ /W)$$

$$\text{Imperial } R \text{ units } (ft^2{\bullet}F{\bullet}hr/Btu) = 5.678\ RSI$$

Thermal Transmittance[15] (U): a property of an enclosure assembly, often measured from the air on one side to the air on the other side.

15
Sometimes called the Overall Heat Transfer Coefficient and sometimes measured only from the surface to surface of an assembly of layers.

$$U = 1\ /\ R_{total}$$

where R_{total} is the sum of the R-values of each of the layers and the airfilms on either side.

The ubiquitous "R-value" is thus a measure of the resistance to heat flow by conduction. Strictly speaking, R-value cannot be used to measure the resistance to heat flow by radiation or convection. In practice, however, the three modes cannot be easily separated. Instead, the term "apparent conductance" or "apparent resistance" is used in the scientific literature.

We often speak of an enclosure's R-value, but what this actually means is the overall apparent R-value. While such shorthand is inevitable, the difference between the R-value of the materials when tested for conduction in a standard test apparatus can vary by factors of 3 or 5 from how much thermal control it offers in an assembly in the field (largely because of thermal bridging, convective loops and wind washing reductions).

Numerous enclosure systems (especially those with thermal mass, or good airtightness) and products (especially radiant barriers) attempt to quantify their benefits in marketing literature with the term "effective R-value." There is no generally accepted scientific definition of "effective R-value," and hence this label can be misused. Specifiers should demand that overall effective R-value (including all thermal bridges) be provided, and that the energy impacts of thermal mass and air leakage be accounted for via whole-building energy simulations and testing.

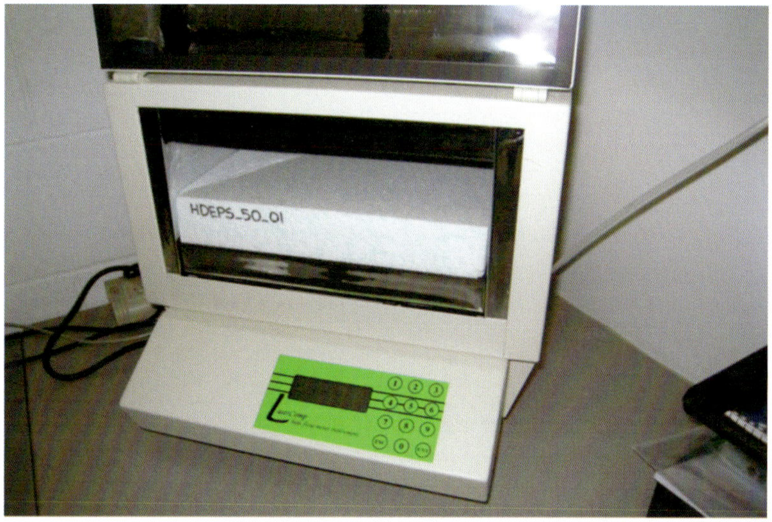

Figure 3.19: An ASTM C518 thermal conductivity testing machine. A 12" square sample up to 4" thick is inserted, and a thermal conductivity is reported within a few hours.

The traditional measure of thermal control of windows, doors and curtainwalls is the U-value, usually measuring using some standard test or modelling approach (such as NFRC 100, NFRC 500, ASTM C1199). Most window tests are conducted with outdoor temperatures of 0 °F/-18 °C and indoor temperatures of around 68 °F/20°C. European standards do not test at such cold outdoor temperatures, and hence usually report better performance than would be achieved in the more challenging North American tests.

The U-value is the reciprocal of the R-value, but includes the thermal resistance of the interior and exterior heat transfer coefficients (the fictitious "surface film" of "air films"). Surface films comprise a significant proportion of the total heat transfer resistance in most glazing products, often 25 to 40% of the total. In contrast, the R-value of wall assemblies may be reported in terms of thermal resistance from surface-to-surface (i.e., not including surface films) or from air-to-air (including surface films). Surface films comprise a small part of any high-performance wall or roof systems thermal resistance. Wall systems may also report their U-values and some codes define opaque systems' thermal control requirements in the form of U-values.

The use of two different set of values for glazed and opaque enclosure elements has, in this author's opinion, resulted in confusion and obscured the significantly lower thermal control performance of most glazing systems. As U-value, technically speaking, is the reciprocal of R-value, this book will often use R-values, which are always air-to-air, when speaking about glazing.

3.3.2 Thermal Bridging

Continuity of the thermal control layer is important and becomes critical in high-performance enclosures. Heat flow, however, deviates from simple one-dimensional flow at corners, parapets, penetrations, and intersections between different assemblies. When heat flows at a much higher rate through one part of an assembly than another, the term *thermal bridge* is used to reflect the fact that the heat has bridged over/around the thermal insulation. Thermal bridges become important when:

- they cause cold spots within an assembly that might cause performance (e.g., surface condensation), durability or comfort problems
- they are either large enough or intense enough (highly conductive) that they affect the total heat loss through the enclosure

All enclosures should be designed to avoid numerous and large thermal bridges. Unlike rain control and airflow control, small local interruptions and reductions in the efficacy of the heat flow control are widely tolerated and accepted in current practice. Even a rather small amount of thermal control, e.g. R-2 to R-4 (the thermal resistance of double-glazed windows) is tolerated over significant areas of many enclosures, although the energy and comfort penalty can be significant if the area covered is large.

Penetrations such as nails, screws, and masonry wire ties rarely affect the energy[16] or comfort-related performance of the enclosure, although they may still cause some condensation on the interior of the fastener in cold weather for buildings with high interior humidity (e.g., museums and swimming pools).

16
Closely-spaced screws that pass through the thermal bridge of metal windows and curtainwalls are one exception: heat flow through a very high performance, thermally-broken frame can increase 25% due to screws at 8" spacing.

The most effective solution to most thermal bridging, exterior layers of continuous insulation (i.e., insulating wall sheathings), "blunt" thermal bridges and also offer improved resistance to exfiltration condensation in cold weather.

Thermal bridging, especially by steel framing, or at the intersection of wall corners with roofs and floors, projecting structural elements like balconies and perimeter slabs often causes cold interior surface temperatures and thus condensation. Figure 3.20 is a schematic of how thermal bridging at studs and near corners can cause low surface temperatures. In the case of the steel framing shown, an exterior temperature of –10 °C (15 °F) can result in interior surface temperatures of 5 to 10 °C (40 to 50 °F) at studs, and below freezing temperatures at floor-to-wall corners.

Heat flow through steel stud walls and metal curtainwalls / windows is dominated by heat flow through the metal components. Failure to break these thermal bridges can reduce the R-value of the insulating components by 50 to 80% (for the insulated glazing unit or cavity-fill insulation respective-

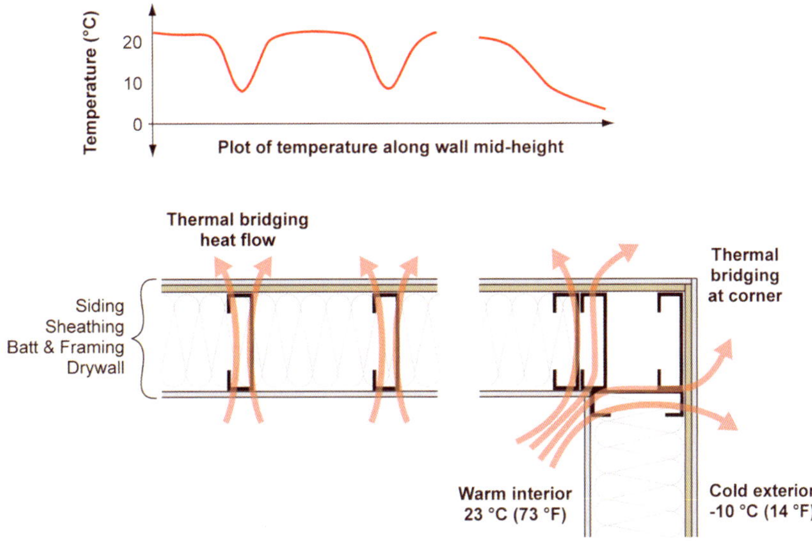

Figure 3.20: Thermal bridging can cause local temperature depressions during cold weather, resulting in condensation, mold growth, and staining.

ly) (Figure 3.21). Steel stud enclosures insulated solely between the studs typically have true R-values of under R-5 when the floor slab penetrations and typical framing details are also accounted for. As the ASHRAE 90.1-2010 commercial building energy code and USGBC-sponsored Standard 189.1-2009 (the new LEED requirement) become more widely adopted and enforced, continuous insulation, or "ci," will become the standard approach. The prescriptive method[17] of both standards requires continuous insulation[18] outside steel stud framing in almost every application.

Filling the voids in concrete block masonry with insulation is even less effective than insulating between steel studs: adding poured foam or perlite insulation to the cores of a standard density 8" concrete masonry unit (CMU) will increase the R-value of the wall assembly by only about R-2 (Van Geem 1986, ASHRAE 90.1).

Wood-framed walls and windows are not as seriously affected by thermal bridging because wood is a material with both quite good structural properties and not insignificant thermal properties (an R-value of better than 1/inch, k<0.12 W/mK). Insulating between wood framing results in thermal bridges that regularly reduce the R-value by 20-30% but 40% reductions can occur for enclosures with a large amount of wood structure (e.g., framing factors of over 25%). Commercial enclosures with 2x6 wood framing on the exterior of all conductive structural components (e.g., concrete or steel primary structure) can provide good thermal control even without exterior insulation layers. True high performance, however, requires exterior insulation.

17
Modeling of the wall assembly and the whole building can be used in the trade-off method. However much, although not all, of the thermal bridging effect of steel stud and CMU walls is considered in the prescriptive tables of 90.1. Detailed modeling should include the additional thermal bridging effect of floor slabs and partition wall intersections and framing around openings.

18
Continuous insulation means penetration only by intermittent fastenings / connections.

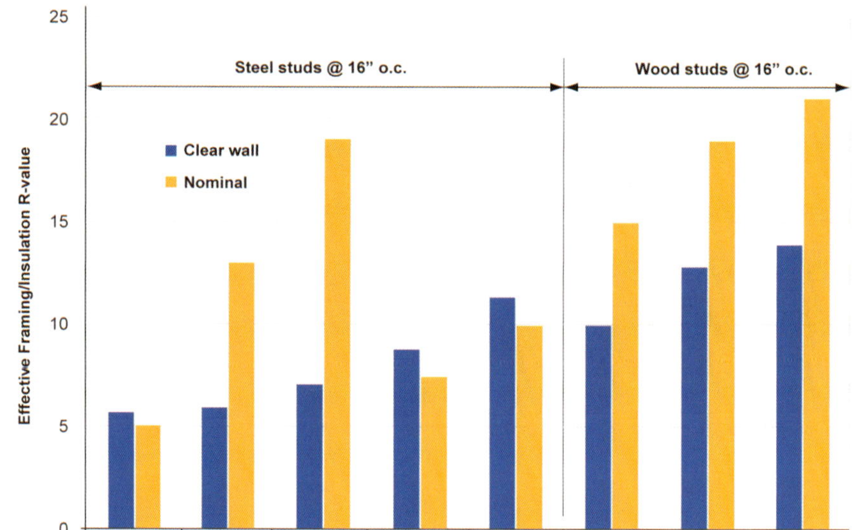

	R-5ci*/3.5 in	R-13/3.5 in	R-19/6 in	R-7.5ci*/6 in	R-10ci*/6 in	R-15/3.5 in	R-19/6 in	R-21/6 in
Clear wall	5.8	6.0	7.1	8.8	11.3	10.0	12.9	13.8
Nominal	5	13	19	7.9	10	15	19	21

* ci denotes a layer of continuous insulation with no framing penetrations

Figure 3.21: Effective R-value of the framing/insulation layer in wood and steel framed walls (*adapted from ASHRAE 90.1-2007; Table A9 2B*)

Masonry and tilt-up concrete walls can be successfully insulated on the interior, although the risk of cold weather condensation (especially due to air leakage) increases, and heat is lost through thermal bridging at the required structural connections at the roof, balcony, and floor. For buildings with few penetrations and low interior humidity applications (such as shopping malls), interior-insulated masonry may provide acceptable, although rarely excellent, thermal performance.

Serious thermal bridging is such a major problem (in that it seriously reduces heat flow resistance and results in comfort and condensation problems) for modern institutional, commercial, and industrial buildings (that is, those made with steel, concrete or masonry structures and enclosures) that it is recommended that all such enclosures include a continuous layer of insulation on the exterior of the primary and infill structure. Only small amounts of steel (under 0.1% of enclosure area for high-performance) or concrete (under 2% of enclosure area) should be allowed to penetrate the insulation, and even these guidelines allow for non-trivial heat loss. Placing all or some insulation outside the structure helps to ensure continuity of the thermal control.

In buildings that require a significantly higher level of insulation, it is feasible to include a fraction of the total insulation on the interior of or within framed assemblies. However, this placement of insulation both reduces its efficacy (thermal bridging always acts to reduce its performance) and increases the risk of damaging interstitial condensation. Assessing the type of insulation and the safe ratio (if any) of inboard to exterior insulation is further discussed in Section 3.5 Relationship of Air, Vapor, and Thermal Control.

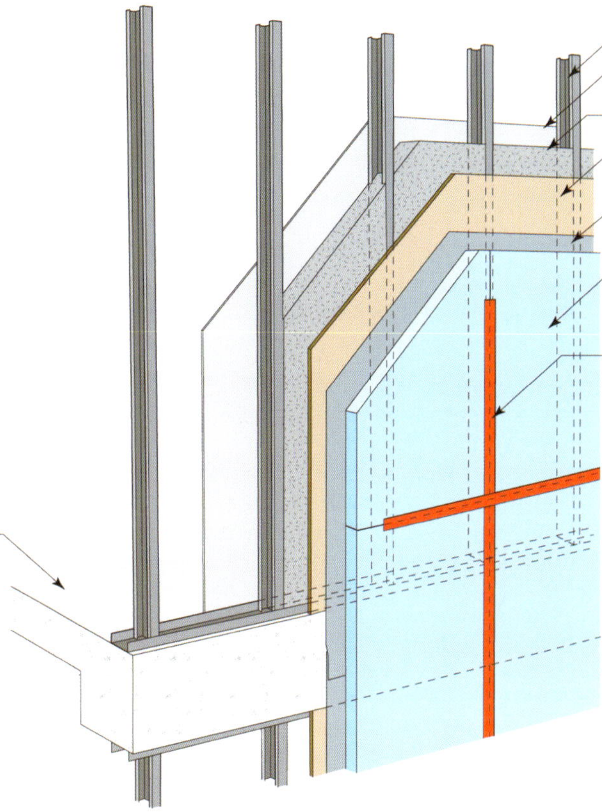

Figure 3.22: Little benefit is derived from insulation installed within the framing of a steel stud wall, where thermal bridging through the floor slabs and studs reduces the R-13 to R-19 stud bay insulation to an effective R-5 or R-6. A major advantage of continuous exterior insulation, especially in steel and concrete enclosures, is the blunting of thermal bridges at floor structures. Simply adding thickness to the exterior continuous insulation layers allows one to reach very high true R-values. In rare cases the additional R-5 gained may justify the risk and effort.

3.3.3 Quantifying Thermal Bridging

Many designers are increasingly interested in assessing the impact of thermal bridging on thermal control for reasons of energy efficiency, code compliance, comfort, or condensation control.

One common way of assessing the impact of thermal bridges is by considering the “clear wall R-value”, which is the R-value of a wall with no unique thermal bridges. This is appropriate and useful for reporting the R-value of, for example, a steel stud wall with a single stud spacing or an aluminum curtainwall with one size of repeating glass. The limitation of this approach is that it does not consider the additional heat flow of, for example, double studs at windows, floor slab intersections, cladding attachment clips, or sunshade attachment points. These project-specific changes in the clear wall heat flow often may need to be considered.

The simplest method of estimating heat flow through thermal bridges is to calculate the area-weighted impact. That is, the total effective R-value including thermal bridges can be calculated as:

$$R_{eff} = A_{total} / (A_{insulated} \bullet U_{insulated} + A_{bridge} \bullet U_{bridge})$$

where $U_{insulated}$ is the U-value of the insulated portion of the enclosure,

U_{bridge} is the U-value of the thermal bridge,

A_{bridge}, $A_{insulated}$ and A_{total} are respectively the areas of the thermal bridge, insulated enclosure, and total area projected on a plane perpendicular to the direction of heat flow.

This equation assumes parallel heat flow, that is, no heat flows in a two- or three-dimensional manner, but only along parallel one-dimensional paths. In wood-framed buildings, it is reasonable to assess the thermal bridging using the assumption of parallel heat flow. Full-scale laboratory hot-box testing at Building Science Corporation and other research facilities has shown that the parallel path method is accurate to within a few percent for wood frame walls. The same approach has been presented earlier (Section 2.3) to assess the impact of the window heat loss in an insulated wall.

In most commercial buildings, components of metal and concrete are the thermal bridges, and these highly conductive materials generate two- and three-dimensional flows that cannot be ignored or reliably predicted by simple parallel path methods. To accurately account for these types of thermal bridges computer models are commonly used. Alas, this means that simple calculations methods are not available.

To report the results of multi-dimensional heat flow models, laboratory tests, and field measurements in a useful and general way requires a standard means of reporting. The most common approach, embodied in standards, such as ISO standard 10211:2007 *Thermal bridges in building construction - Heat flows and surface temperatures :Detailed calculations,* distinguish between linear and point thermal bridges. This approach is recommended and will likely become more common over time as it allows typical details to be cataloged.

Linear thermal bridges are elements that can be accurately assessed in a two-dimensional section (such as steel studs, Z-girts, relieving angles, and slab edges) and point thermal bridges (such as those that occur in beam penetrations, clips, and complex shapes) are useful for assessing three-dimensional flows. The additional heat flow caused by linear thermal bridges is commonly assigned the Greek letter ψ, and point thermal bridges by the Greek letter χ. Given the characteristics of each of the thermal bridges, the total impact can be calculated by summing their impact and adding the heat flow to the same enclosure with no thermal bridges:

$$U = \frac{\Sigma(\psi \cdot L) + \Sigma(\chi)}{A_{total}} + U_o$$

Where:

ψ is the linear thermal transmittance (Btu/h ft °F or W/mK)

χ is the point thermal transmittance (Btu/h °F or W/K)

L is the length of linear thermal bridges (ft or m)

Q is the assembly heat flow with the thermal bridge for A_{total} (Btu/h °F or W/K)

U is the assembly U-value with the thermal bridge (Btu/h ft^2 °F or W/K)

Q_o is the assembly heat flow without the thermal bridge for A_{total} (Btu/h °F or W/K)

U_o is the assembly U-value without the thermal bridge (Btu/h ft^2 °F or W/m^{2}K)

A_{total} is the total projected surface area (ft^2 or m^2)

Once modeling has assessed the additional heat flow caused by typical and repeating thermal bridges, this approach allows the designer to predict, with sufficient accuracy, more realistic heat flow through enclosures.

3.3.4 Thermal Control Layers: Materials and Systems

All materials and layers in a building assembly have some resistance to heat flow. However, materials with an R-value of about 2/inch or more (k-value less than 0.07 W/m·K) are deliberately used in building assemblies for their ability to retard the flow of heat. These building products are called thermal insulations. Insulations are usually solid materials, but radiant barriers that control only radiation heat transfer across air spaces are also available. Polymeric foams, mineral fiber (i.e., glass, slag, or rockwool), and organic fiber (e.g., cellulose, cotton, polyester, etc.) are commonly used to produce building insulation.

High performance building enclosures generally require continuous thermal control layers on the exterior of the structure. This layer can be board or spray foam as well as semi-rigid fiberglass or rockwool (i.e., mineral fi-

Figure 3.23: Closed-cell spray polyurethane foam (ccSPF) has been used for decades as a combined thermal insulation, air barrier material, drainage plane, and vapor control product. Note the extensive use of transition membranes at penetrations.

Figure 3.24: Rockwool semi-rigid board insulation used as continuous exterior insulation.

ber) insulation. Locating all insulation as a continuous layer on the exterior of the structure effectively eliminates low-cost fibrous batt and loose fill insulations from a designer's palette. Such low-density fibrous insulations are still important to fill interior voids and can supplement the exterior insulation layer, especially when affordable high thermal resistance enclosures are desired. Hence, some drawings in this book show stud spaces with optional cavity fill. To ensure all spaces and voids in framing cavities are filled properly (to control convective loops), spray or blown fibrous insulation is preferred and is shown in this book when cavity fills are appropriate. Whenever air and vapor permeable insulation is added to a cavity, extra care is required in the location and selection of the vapor control layer and the location of the air control layer.

Cellular foam plastics are used when air-impermeable and/or water tolerant materials are required. Products such as expanded polystyrene (EPS),

Layer	R-Value	RSI (m^2K/W)
Empty stud space (0.75"–6" / 20–140 mm)	0.85 –1.4	0.15–0.25 m^2K/W
Empty stud space (3.5" / 90 mm)	approx. 1	0.18 m^2K/W
Empty stud space (5.5" / 140 mm)	approx. 1	0.18 m^2K/W
CMU 8" / 200 mm normal weight	2.0 - 2.2	0.35 m^2K/W
Insulation material	**R-value/inch**	**k (W/mK)**
Batt (mineral fiber)	3.5 – 3.8	0.034 – 0.042
Extruded polystyrene (XPS)	5.0	0.029
Polyisocyanurate (PIC)	6.0 – 6.5	0.022 – 0.024
Expanded polystyrene (EPS)	3.6 – 4.2	0.034 – 0.040
Semi-rigid mineral fiber (MFI)	3.6 – 4.2	0.034 – 0.040
Spray fiberglass	3.7 – 4.0	0.034 – 0.038
Closed-cell spray foam (2 pcf) ccSPF	5.8 – 6.6	0.022 – 0.025
Open-cell spray foam (0.5 pcf) ocSPF	3.6	0.040
Aerogel	8 – 12	0.012 – 0.018
Vacuum insulated panels (VIP)	20 – 35	0.004 – 0.008

Table 3.2: Representative insulation material properties

extruded polystyrene (XPS), and faced polyisocyanurate ("polyiso" or PIR) have long been used behind claddings and outboard of the water control layer. Polyiso should not be used in applications where it can be immersed in water for long periods of time: this is not a concern for above-grade walls, but does limit its use below-grade. Expanded polystyrene, especially at higher densities, can be used very effectively below-grade and has long been accepted in the Canadian building code for this use. Medium-density closed-cell spray polyurethane (ccSPF) is an increasingly common product that is spray applied to appropriate substrates. It has even been used under slabs by spraying it directly to the earth. This product can act as part of the rain, air, vapor, and heat flow control layers of an assembly if care is taken to maintain continuity at transitions and shrinkage cracks.

Most rigid foam board products, especially those with facers (e.g. the aluminum facer on polyiso intended for walls, faced EPS board), can act as part of the water and/or vapor, and/or air flow control so long as other requirements, such as structural support, durability, and continuity at joints, are met.

Semi-rigid fibrous insulation boards can only act as heat flow control layers, although they can provide a drainage path (i.e., they provide a drainage gap in the large air voids between fibers) for rain water, either by design or in actuality.[19] In many applications, air- and vapor permeable mineral-fiber semi-rigid insulations (MFI), such as fiberglass and rockwool can be used as exterior continuous insulation and have a very long track record of good performance. To perform in unsupported applications, fiberglass products generally require a minimum density in the range of 2.5-4 pounds pcf (40-70 kg/m^3), whereas rockwool should have a density of over about 3 pcf (50 kg/m^3). Such MFI products tend to be less expensive, and are always more fire resistant, than foamed plastics.

19 Semi-rigid mineral fiber insulation (MFI), particularly rockwool, provides a drainage path in the fibers of the outer $^1/_8$" – $^1/_4$" (3-6 mm) without materially affecting its thermal resistance. MFI have been used in drained masonry cavity walls with great success for over 50 years in most parts of the world and have been widely deployed as below-grade drainage layers with some insulating value.

Radiant Barriers and Low-E Coatings

Some thermal control products are designed to primarily control radiation heat transfer. Radiant barriers can be provided as a sheet, be adhered to sheathing and insulation products, or be part of a roof or sheathing membrane. Transparent metallic coatings are can be applied to any of the surfaces of glazing.

In all cases the efficacy of radiation control can be measured in the infra-red region by emissivity. Low emissivity, low-e, surfaces neither emit nor absorb energy effectively in the infra-red temperature range (i.e. the temperature of building enclosures in service). Normal building materials have an emissivity of around 0.90, or 90%. Low-e surfaces, almost always based on some metallic compound, have emissivities of well below 0.20, and the best coatings have values of 0.03 or even lower. A low-e surface will reduce the radiation that transfers across an air space regardless of which side of the airspace it is applied to. Applying a low-e coating to both sides will have only a very small additional benefit. In all cases, a radiant barrier must have an air space associated with it to have any thermal control benefit.

The most powerful application of radiation control is in the form of transparent coatings on glass, the ubiquitous low-e coating. The coating is usually applied to one of the surfaces facing a sealed air gap in an insulated glazing unit (IGU). In this location, the surface is protected from corrosion, condensation, dust, and dirt, and therefore the emissivity of the coating is protected from change. Low-e coatings increase the apparent thermal resistance of a $^{1}/_{2}$" (13 mm) air gap in a sealed IGU by from R-1 to R-2. When the gap is filled with a less conductive gas such as argon, the benefit is even larger, boosting the R-value of a double-glazed IGU from around R-2 to around R-4. Some newer glazing systems are adding a low-e coating to the inside face of the glass as well: this improves R-value and occupant radiant comfort but does result in colder glass surface temperatures.[20]

20 Adding a low-e coating to surface 4 (the interior-facing surface) of a double-glazed window reduces the interior surface heat transfer coefficient substantially. The occupant (and room) radiates less heat to the cool interior window surface, thereby increasing comfort, but also reducing actual interior surface temperatures, increasing the risk of condensation. Hence, additional care is needed to reduce interior RH during cold weather.

Emittance Factor		0.05			0.2			0.82		
Airgap width (mm)		13	20	90	13	20	90	13	20	90
Vertical	Warm	0.41	0.57	0.60	0.19	0.21	0.38	0.14	0.15	0.15
Flow across	Cold	0.55	0.63	0.58	0.45	0.51	0.47	0.26	0.27	0.26
Horizontal	Warm	0.36	0.39	0.47	0.27	0.28	0.32	0.13	0.28	0.14
Flow up	Cold	0.35	0.37	0.45	0.31	0.32	0.38	0.20	0.21	0.23
Horizontal	Warm	0.41	0.58	1.44	0.29	0.37	0.60	0.14	0.15	0.18
Flow down	Cold	0.56	0.79	2.03	0.46	0.60	1.12	0.26	0.30	0.39

Emittance Factor		0.05			0.2			0.82		
Airgap width (inch)		$^{1}/_{2}$"	$^{3}/_{4}$"	$3^{1}/_{2}$"	$^{1}/_{2}$"	$^{3}/_{4}$"	$3^{1}/_{2}$"	$^{1}/_{2}$"	$^{3}/_{4}$"	$3^{1}/_{2}$"
Vertical	Warm	2.33	3.24	3.41	1.08	1.19	2.16	0.79	0.85	0.85
Flow across	Cold	3.12	3.58	3.29	2.56	2.90	2.67	1.48	1.53	1.48
Horizontal	Warm	2.04	2.21	2.67	1.53	1.59	1.82	0.74	1.59	0.79
Flow up	Cold	1.99	2.10	2.56	1.76	1.82	2.16	1.14	1.19	1.31
Horizontal	Warm	2.33	3.29	8.18	1.65	2.10	3.41	0.79	0.85	1.02
Flow down	Cold	3.18	4.49	11.53	2.61	3.41	6.36	1.48	1.70	2.21

Emittance Factor: 0.05 = very clean, shiny aluminum
0.2 = dull, slightly dusty
0.82 = normal building materials

Table 3.3: Thermal resistance (RSI at top, R-value at bottom) of air gaps of different sizes, at different temperatures, with different emissivities.

Low-emissivity coatings or films are often applied to carrier sheets, insulation boards, or sheathing. These coatings may initially have emissivities as low as 0.05, but the value tends to rise in service as dust, corrosion, and aging increases the emissivity. Some low-e paints and coatings may have emissivities as high as 0.30: the benefit of these coatings is pretty small.

As radiation emission increases with the fourth power of absolute temperature, radiation is more important at high temperatures than at cold temperatures. Hence, radiant barriers/low-e coatings are slightly more important at high temperatures than low temperatures. As heat is carried by rising air, heat flow upward across a horizontal air space is dominated by convection, whereas heat flow downward is dominated by radiation: the impact of low-e coatings will be very important in the latter case and not the former. All of these varied factors means there is no one correct apparent thermal resistance for an airspace with low-e coating, or a radiant barrier product exposed to an air space.

Table 3.3, using selected values from the *ASHRAE Handbook of Fundamentals*, demonstrates the wide range of possible thermal resistance values. A normal air gap can be assumed to have an emittance of 0.82, whereas the best radiant barrier product would have a value of around 0.05 when new and clean, or 0.20 when dusty.

In the solar spectrum (visible light), absorption of solar energy can be reduced by using light–colored and reflective coatings on surfaces exposed to the sun. Reflective roofs, also called cool roofs or white roofs, are a form of thermal control in that they limit the solar gain into a roof assembly. They do nothing to prevent heat loss at night. The solar reflectance, or albedo, is used to quantify the efficacy: high reflectance values like 90% means 90% of incoming radiation is reflected back. If the aim is to control solar gain, high reflectance (e.g. silver or white color) is obviously critical, but high emissivity is also important.[21]

21
When describing cool roofs, the reflectance measures radiation in the visible and solar spectrum (around 0.5 micrometer wavelength), whereas emissivity measures the emittance of the surface in the infra-red spectrum (about 10 micrometer wavelength).

White stone ballast or special white concrete pavers can be very effective cool roofs. The thermal mass supplements the reflectivity to reduce the rooftop temperature to, in some cases, below the temperature of a new white roof membrane or even below that of a green roof.

3.3.5 Windows and Curtainwalls

In windows, thermal control is provided by a combination of sealed airgaps filled with dry air or rare gases (argon, krypton), low-emissivity and solar reflective coatings, and thermal breaks of polyurethane, nylon, or PVC (Figure 3.25).

The heat flow through glazed systems is measured using the U-value (including surface films). However, windows have very significant three-

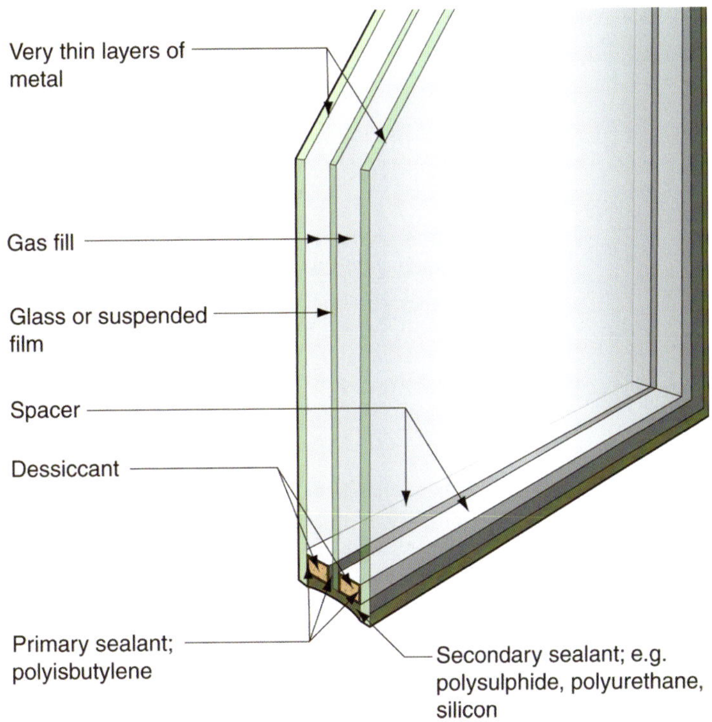

Figure 3.25: Anatomy of a higher performance insulated glazing unit (IGU)

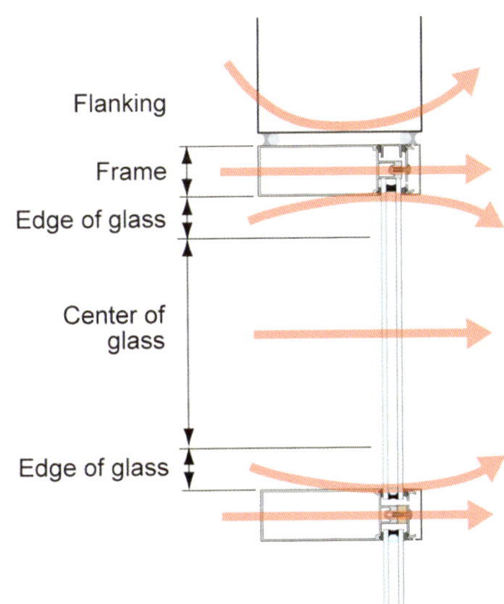

Figure 3.26: Conductive flow components through a window comprises flow through frames, edge of glass and glazing

dimensional heat flows: through the frames, through the edge-of-glass region, and through the center-of-glass region (Figure 3.26). Heat flow through each component of the three heat flow paths can be calculated separately and given its own U-value. The overall U-value comprises the area-weighted area average of these three U-values. Heat flow around an installed window, termed thermal flanking, can result in serious problems due to excess heat loss and condensation around the frame.

Today's best-available glazing units are very effective at controlling heat flow (Figure 3.28): low-e coated, gas-filled units can be readily procured with U-values of less than 1.5 W/m^2K (R-values over 4) for double glazing and less than 0.8 W/m^2K (R-values over 7) for triple-glazed units. Triple-glazing, especially with low-e coatings and gas fills, are almost always justified for low-energy buildings in climate zones 7 and 8, and as far south as zone 5. The glazing layers do not need to be glass, but could be suspended polymeric films to save weight and thickness.

The spacers used to separate the sheets of glazing create additional edge losses, especially if made of aluminum. Thermally-broken edge spacers made of stainless steel or polymers are now available to reduce this edge-of-glass loss.

The most important impact of thermal bridging occurs through frames because frames have less thermal resistance than most glazing. This is a major problem in all window systems but especially so for high-performance glazing (i.e., U_{SI} less than 1.5 W/m^2K or over R-4/under U-0.25). Large thermal breaks (e.g. over 12 mm or $^1/_2$") and thicker glazing units (over 25 mm/1") can reduce this problem although there are few widely available aluminum framing products with the required large thermal breaks.

Double-façades, or double-skin systems usually have thermal resistance no better than a medium quality triple-glazed curtainwall: the numerous connections through the ventilated cavity in these systems act as thermal bridges. Designing a façade with less than 100% glazing with high-performance triple-glazed fenestration will always outperform typical double façade designs.

Significant heat loss can result from improper installation of a window or curtainwall in a wall (Figure 3.29). Thermal flanking can be avoided by properly aligning the thermal control layer in the wall with the thermal control layer of the fenestration (Figure 3.30). If the thermal break in the window is not aligned with the thermal insulation layer, then some insulating/non-conductive transition should be provided (e.g., backer rod and sealant, wood or nylon blocking, etc. to provide some level of thermal control). The same concern needs to be dealt with at door thresholds (Figure 3.32).

Wind washing and convection within open aluminum frames can seriously compromise the thermal performance of frames (Figure 3.31) if they are allowed to occur. Airsealing on both sides of the frames can help reduce this problem, but filling the void with low-expansion foam or pre-formed insulation inserts can limit the impact and make aligning the thermal control layers in the opaque wall with that in the window easier.

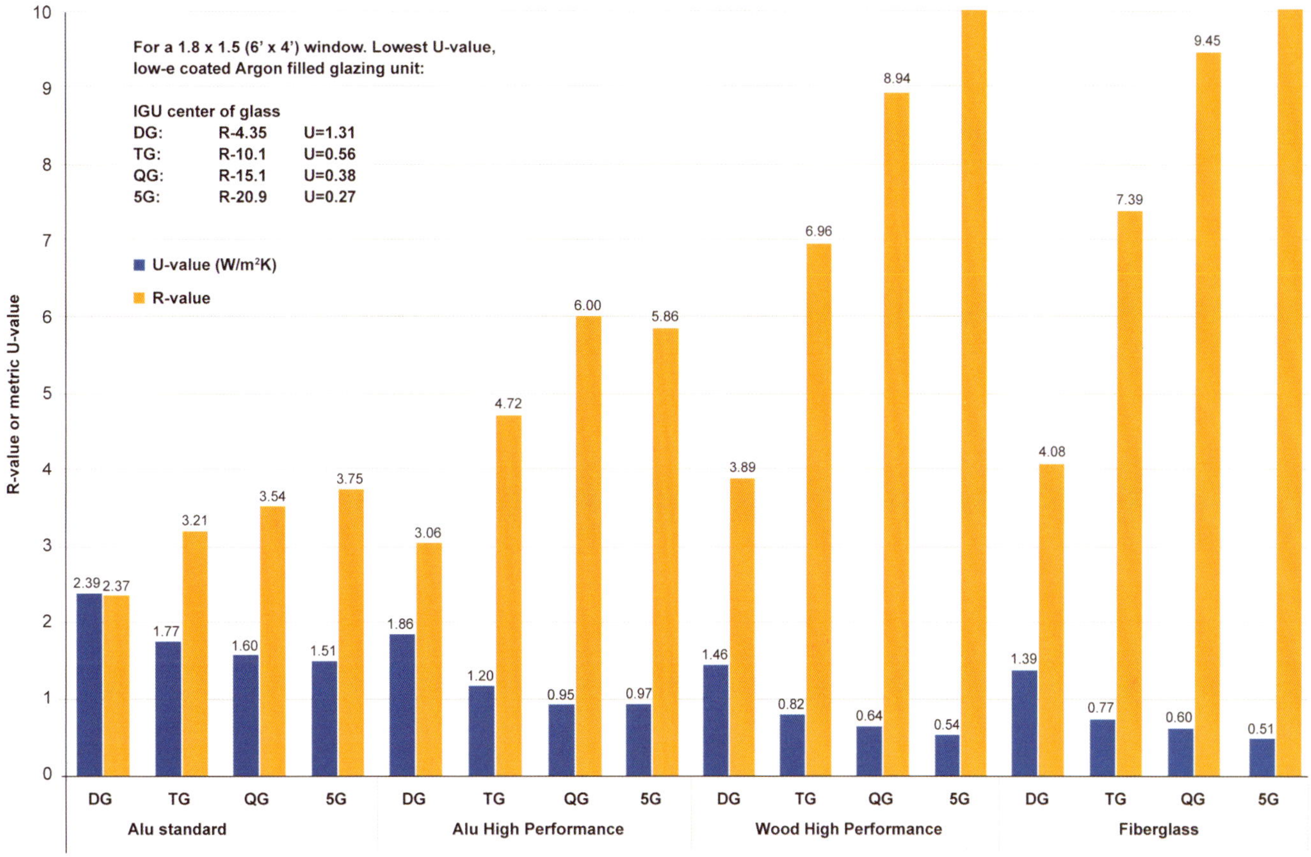

Figure 3.27: Thermal control of best-available glazing units with insulating spacers, low-e coatings and argon fill. This chart presents overall R-value (U-value) of a window depending on the frame and number of glazing layers. (Lee, 2010)

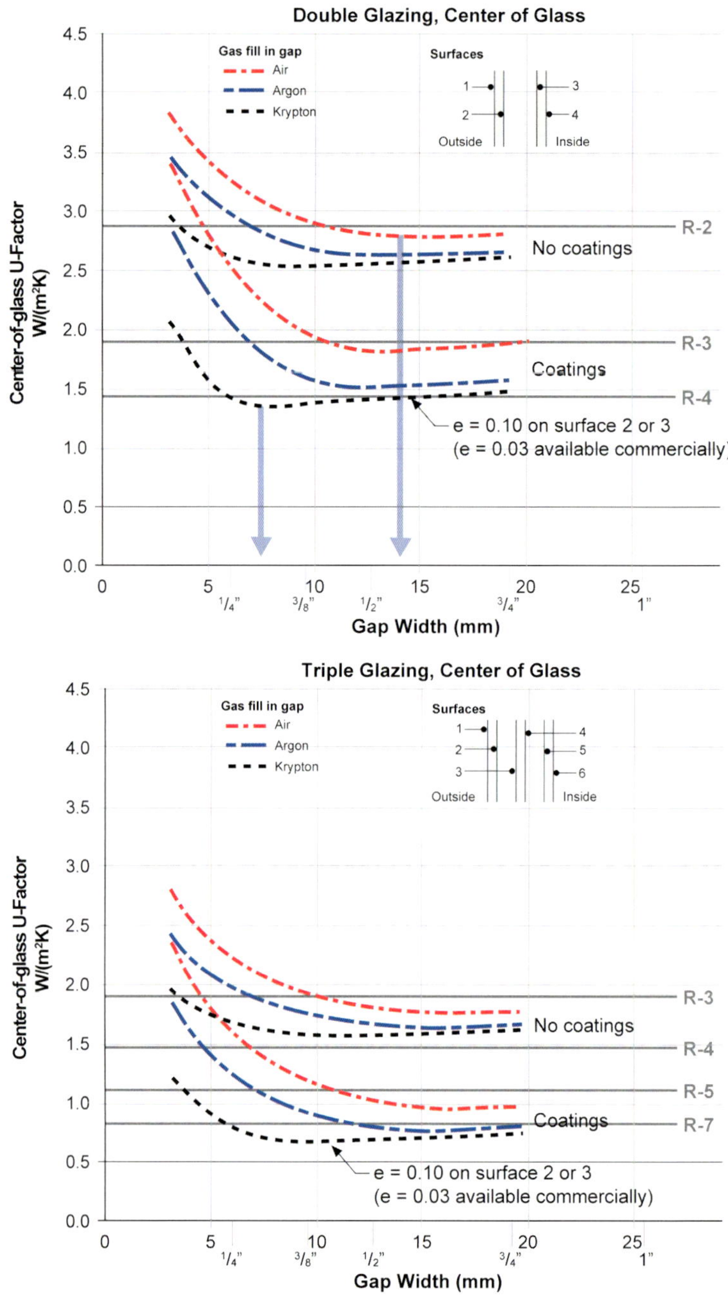

Figure 3.28: Center-of-glass glazing U-value (R-value)

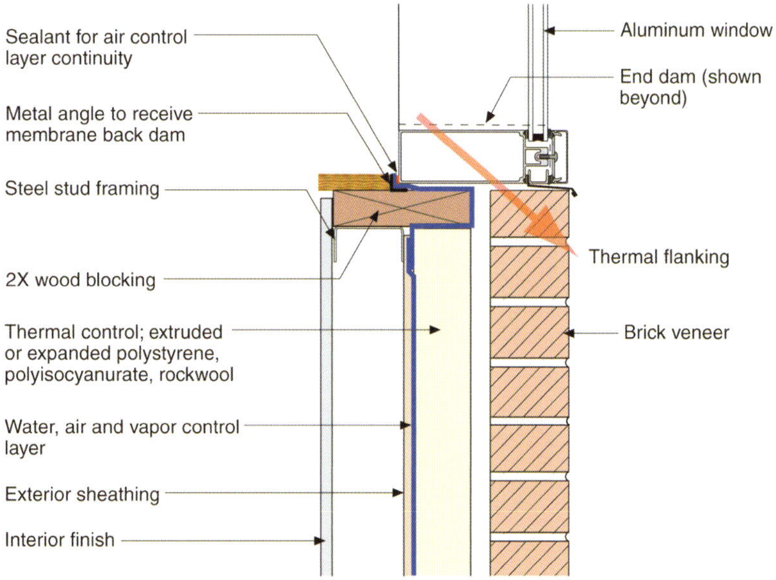

Figure 3.29: This aluminum curtainwall frame sill detail allows heat to flow uncontrolled to the exterior brick and airspace via the frame. Although heat loss through this discontinuity in the thermal control layer will be significant, condensation and discomfort are larger concerns.

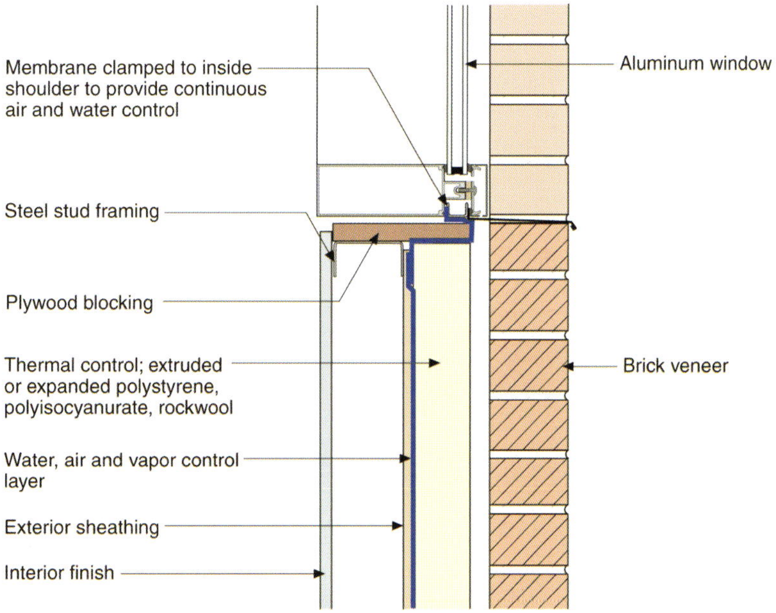

Figure 3.30: This detail aligns the thermal control of the window frame and IGU with the thermal control layer of the wall to avoid thermal flanking.

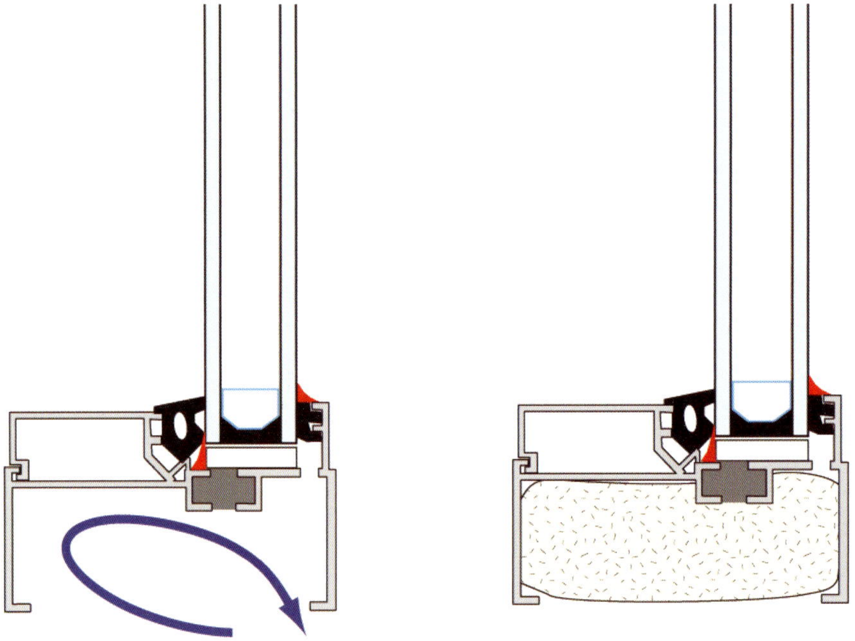

Figure 3.31: Open window frames can allow air to enter and reduce the thermal resistance of the frame. Filling the frame with low expansion foam on site, or preformed insulation in the factory, minimizes this impact.

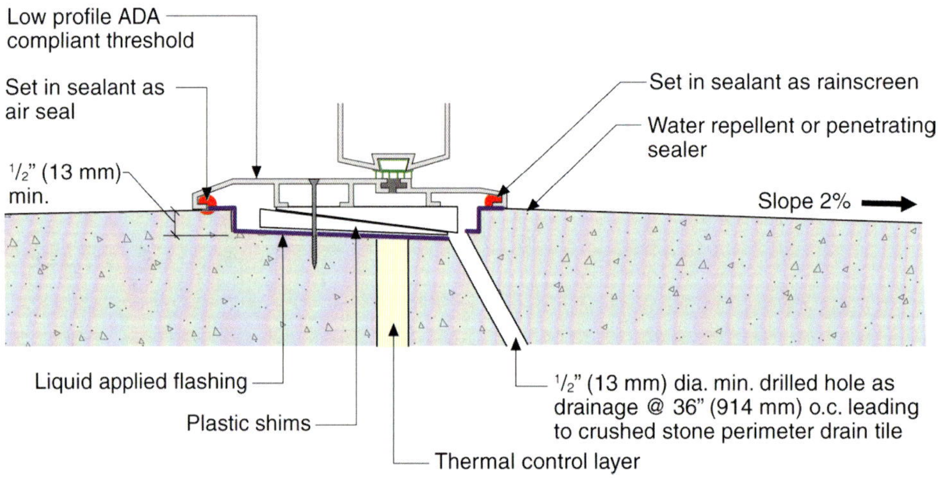

Figure 3.32: The thermal, air, and rain control do not need to be compromised at on-grade door thresholds. Aligning the thermal control layers in the door threshold and the enclosure below is important.

3.3.6 Solar Control and Windows

Solar control is an important subset of thermal control, both to allow solar gain in cold weather when beneficial, and to avoid solar gain in warm weather. In some cases, radiant barriers and high reflectance "white" or "cool" roofs (both discussed earlier) are thermal control layers that can play an important role in ensuring comfort and reducing space-conditioning energy consumption. However, in most cases, solar control relates to solar gain through windows, as this remains the highest source of solar gain in well-insulated buildings.

Solar Gain Through Windows

Solar gain in cold climate buildings can be a blessing or a curse. If heating is needed during daylight hours, solar gain is beneficial. If too much solar gain is allowed, air conditioning loads are increased in warm weather, and comfort problems (overheating and radiant asymmetry) are experienced even in cold weather. In northern latitudes (over 40° N), solar radiation will reach a peak in summer on the east and west side, and peak in winter on the south side. Hence, shading is most important on the east/west side to avoid summer over heating, and windows on the south will generate the most useful solar gains during winter.

Solar gain through windows exposed to either the direct sun or reflected sun (reflected off the particles in the sky, creating diffuse radiation, or reflected off a surface) can dramatically affect heat flow across the building enclosure. Solar radiation levels on a vertical surface of 750 W/m^2 or 250 $Btu/hr/ft^2$ are typical "sunny day" values for the peak hour,[22] whereas heat loss on a cold winter night through a window might be less than 10% of these values. Hence, any assessment of building energy flows must account for the solar gain through windows. Solar gain can dominate the performance of a modern commercial building with relatively high window coverage (i.e., more than about 30% window-to-wall ratio) and is almost always the largest component of the peak cooling load in such buildings.

22
These intensity values are essentially the same for every climate zone, but they occur on different orientations at different times of the day.

Solutions to control this form of heat flow include reducing window area, using projecting horizontal shading (effective on the south side only, and only in summer), operable shading on the exterior, and solar control coatings on windows. Interior shades have a relatively modest impact (because modern windows do not let heat flow out easily), but have the important role of controlling glare and providing privacy. A fully-drawn white (reflective) shade may reduce the SHGC of a high-performance glazing unit by around 20-30% (but will have a larger improvement for older low-performance glazing units). An important feature of interior blinds is that they provide the occupant with a measure of personal control.

Double façades usually place the exterior shading device within a ventilated cavity. This shading is less effective than exterior shading, but more effective than interior shading. Using lower areas of glass, and/or lower SHGC glazing, can generate inexpensive and simple indoor shading solutions with better solar control than a good double façade.

The level of control of solar gain through glazed components is measured using the Solar Heat Gain Coefficient[23] (SHGC), which is defined as:

23 The Shading Coefficient (SC) is an outdated metric, retired more than 20 years ago, but still seen in many specifications. The SC is equal to SHGC times 0.86.

$$SHGC = \frac{\textit{Amount of solar energy that enters building as heat}}{\textit{Solar radiation falling on glazing or window product}}$$

For example, if the SHGC of a glazing unit is 0.50, and the sun is shining on the window with an intensity of 500 W/m^2 (150 Btu/ft^2/hr) only 250 W/m^2 (75 Btu/ft^2/hr) will enter the building. The SHGC may be applied to measure the glazing only (center of glass) or the entire fenestration product. For small windows with a significant portion of the total area provided by the opaque window frame, the difference can be important.

Passive solar heating design is used to capture the heat of the sun in a beneficial manner—this requires that most of the windows face south and that window area be limited to collect only as much energy as needed for heating and to warm storage (i.e., thermal mass). Modern passive solar buildings have better control of the thermal losses in cold weather and hence should have almost normal ratios of window to wall area. Unlike houses, which have similar needs for heating during the day as they do at night, commercial buildings have reduced heating needs during the daytime because of the addition of internal gains. High-performance enclosures (with high R-value windows and walls) will have such small heat losses that internal heat gains can provide a significant part of the heating during all but the coldest weather. Thus, a limited amount of solar gain can be gainfully utilized, and large window areas will provide too much gain. Hence, in buildings with high-performance enclosures in moderately cold climates (e.g. zones 5 and 6) with summer air conditioning needs, increasing window area beyond a WWR of around 30%, it is not uncommon to consume more air conditioning energy in the summer than was saved by collecting solar heating gains passing through larger window areas (i.e., WWR over 0.30) in the winter.

Not only is there a net decrease in annual energy performance because of high glazing ratios, comfort problems (due to overheating) are common in rooms with high solar gains and low heat loss (Buelow-Huebe, 2001).

Consider the following simple example. A typical large commercial window 5 ft x 7 ft (1.5 x 2.1 m) in size will, without solar control, allow almost 6000 Btu/hr (1.75 kW) of heat to enter. If this window has a U-value of 0.4 (USI2.3), and the exterior temperature is 0°F/-18 °C, heat loss will be 0.4 *

Figure 3.33: Fixed shades, especially horizontally projecting shades on the south side, have long been used to control summer time heat gain.

Figure 3.34: Shading integrated into a building's architecture can provide both vertical and horizontal fixed shading elements.

Figure 3.35: Fixed overhangs are not effective in controlling excessive solar gain and glare during winter (occupants draw the shades in response)

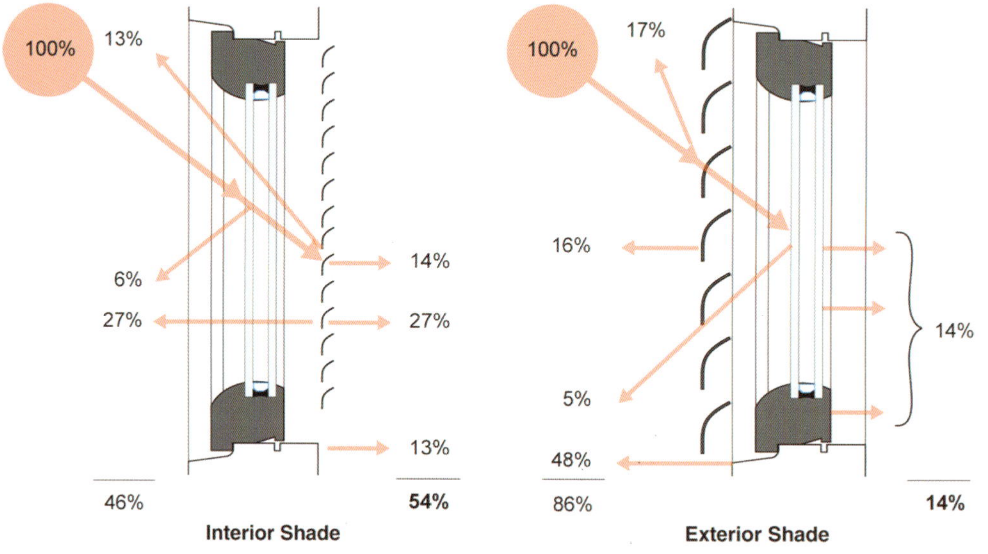

Figure 3.36: Exterior operable shades are a powerful technology that enables solar heating to be collected when desirable, and solar heat rejected when not. Interior glare control shades are still usually required.

Figure 3.37: Exterior operable shading takes several forms: roller shades, blades, and rolling shutters are common.

(70 – 0) * (3 * 5) = 420 Btu/hr. If the window is installed in a wall insulated to a true R-20 (RSI3.5) with a WWR=0.30, heat loss through the opaque wall will be an additional ($^1/_{20}$* (70 – 0) * 30 ft^2) = 105 Btu/hr. Thus, if bright sunshine strikes the south wall on a very cold winter day, the space will be heated by the net gain of over 6000 – 420 – 105 = 5475 Btu/hr (1.6 kW)! The result is a rapidly heating office behind the window. Ideally, some of this heat gain will be stored in mass to be released later, but the air temperature must rise to begin to transfer heat to the mass that is not directly exposed to the sun (in practice, solar gain does not strike massive concrete floors in offices because of furniture). If the office being served by this window has additional internal heat gains (computers, prints, people, and lights) the problem of overheating is worse. While it may sometimes be possible to open windows or ventilate away the overheating (often a challenge in 0 °F/ -18°C weather!), the rapid temperature fluctuations and high radiant temperatures are difficult to manage in a comfortable manner.

The relative need for heating and the availability of sun in northern, cold climates must also be considered. During the coldest portion of the winter, during which the large proportion of heating is needed, the number of hours of sunshine are limited. These few hours of sunshine coincide with the highest interior gains due to occupancy for many commercial applications. Most of the heating is needed after business hours when there is no sun. As high-performance enclosures are deployed, the heating seasons becomes shorter (rather than requiring heating for 6 months per year, perhaps heating is needed for only 4 months) and becomes focused on the coldest months with the least sun hours because internal gains can provide most of the daytime heating needs for many months. As the number of sun hours increases (e.g., March-April/September-October) the heating needs approach zero in a high performance building. It was exactly during these shoulder periods that passive solar heating made large useful contributions to heating the less-well-insulated buildings of the past.

All of the above reasoning leads to the conclusion that a paradigm shift in thinking is required when selecting windows for the new breed of commercial buildings with high-performance enclosures. Window area for each space should be selected based on the desire for view and daylighting, not based on assumptions that the winter heat gain is beneficial. The location of the windows in the room (high windows are always preferred for daylighting, windows below desktop height provide no benefit to daylighting), their size (larger windows allow more daylight) and visual transmittance is then selected. Finally, solar control is applied to reduce solar gains to avoid overheating (the south side is critical in the winter, and the west side is critical in the summer) and high window surface temperatures.

All standard modern glazing units will have low-e coatings that reduce the SHGC to below around 0.60, and sometimes to below 0.30. Higher performance products, usually with three or more layers, have multiple low-e coatings and will reduce the SHGC lower, usually well below 0.50. These types of Insulated Glazing Units (IGU) help avoid solar over heating comfort problems and can provide exceptionally low U-values (under 1.0 W/m^2K, R-values over 5). There is a direct technological trade-off for a given low-e coating between R-value and SHGC: to achieve the highest R-value, the SHGC will be lower. For many cold-climate ICI buildings, a WWR of under 40%, and a SHGC of 0.30 to 0.50 will balance heat loss, heat gain, occupant comfort, daylighting, view and energy consumption.

The balance of winter solar gain and occupant comfort is somewhat different for single-family homes. In general, south-facing windows in residential applications designed for passive solar heating (i.e., with horizontal projecting elements to provide shade in the summer, and window sizing based on the heat loss and thermal mass available) should have windows with a modestly-high SHGC (e.g., over 0.40) to allow in beneficial solar heat gain in the winter. Only in houses with low window area (e.g., under 10% of floor area) should a SHGC of 0.6 or higher be specified. East or West facing windows that receive large amounts of undesirable sun in mornings and afternoons and windows in houses with large south-facing windows (a window to floor area ratio of over 15%), should have a lower SHGC (0.40 or lower).

Shading Windows

Exterior shades intercept solar gain before it can impinge on the glazing product and hence can drastically reduce solar gain. The sun's energy heats the shade, which rejects its heat to the exterior by convection and radiation to the exterior. Interior shades reject most of their absorbed solar gain to the interior air, and hence are not as effective.

Exterior fixed shades reduce the direct solar incident on the glazing for specific, and predictable, periods of the day and the year (Figure 3.33). Diffuse radiation from unshaded parts of the sky will still impinge on the glass, allowing some radiation to pass. However, this diffuse light does provide high-quality daylight and usually allows for good view lines. Fixed shades have the problem that they limit the ease of access for window cleaning in multi-storey buildings, and do not control winter-time over heating unless they are combined with modest window-to-wall ratios.

Operable exterior shades that cover the glazing have dramatically better performance than conventional interior shades and blinds (Figure 3.36). The performance advantages provided by exterior operable shades are so profound to a low-energy building that they can generate quite different optimal glazing solutions. For example, given an operable shade over a south-facing window, the solar gain can be easily reduced when it is undesirable (winter or summer), and allowed to enter when it can be used. This reduces the need for the designer to choose lower SHGC windows, and will allow for better solar gain utilization (as more solar energy can be collected with a high SHGC window during the non-bright sun hours). Designers must still be careful however: over glazing with high SHGC windows can result in the shades being kept closed for many hours, obstructing the view and reducing daylighting benefits while increasing the need for space heating during non-sunny hours. Glare control is still required, and for high glazing ratios with high visual transmittance (VT), glare control may be critical for the enjoyment of a space.

Many different forms of operable exterior shading are available (Figure 3.37), although this technology is not yet common in North America.

3.3.7 Windows, Glazing, and Daylight

The potential energy saving benefits of the daylighting provided by windows relate to the reduced hours of operation and/or intensity of artificial lighting. As installed lighting power density has dropped with better lighting technology and stricter codes, the energy saving benefits of daylighting have also decreased. Hence, daylighting is not as significant an energy saving strategy as it once was, especially if occupancy sensors are used to switch lights off (Love et al, 2008). For this reasons choosing window area on the basis of achieving daylighting is rarely justified. However, for low-energy buildings almost all strategies must be used and daylight harvesting and lighting controls will almost always be justified. To achieve the energy savings possible, it is critical to include automatic daylighting controls: occupants regularly leave lights on even with sufficient daylight unless controls turn the electric lights off.

The ability of glazing to pass daylight is measured using the Visible Transmittance (VT):

$$VT = \frac{\textit{Daylight passing through glazing or window product}}{\textit{Total daylight landing on glazing or window product}}$$

VT can be applied to the center of the glazing, or to the entire product (that is, including the area of the framing).

To maintain a clear appearance and provide significant daylight benefits, it is desirable to maintain the VT of the glazing unit itself over about 0.40: even 0.50 may be desirable. Anecdotal experience suggests that glazing with VTs of about 0.50 are not perceived as any clearer by most occupants than IGUs with much higher VT's of 0.65 or higher.[24] While a higher VT is likely to be noticeable when two samples of glazing are placed side by side, when a single type of glazing is installed in a room, the benefits of very high VT (often provided at considerable extra expense by the use

24 This appears to be true for rooms with typical window areas. Rooms with only 20% window-wall area from the inside will appear slightly dull even with a VT=0.50 (Buelow-Huebe, 2001). The higher the WWR, the lower the optimal VT.

of low-iron glass) are hard to justify. In cold climates, low-e coatings that dramatically increase R-value will always have some effect on VT, and high-performance triple-glazing will not easily be able to provide VTs of over 0.60. There is little if any benefit of striving for higher levels.

Glazing units with both a high VT and a low SHGC are termed *spectrally selective*. For almost all high-performance enclosures, glazing should be spectrally selective to allow as much daylight in as possible while managing solar over-heating issues.

The enclosure design characteristics that control the amount of possible daylight harvesting are the WWR and the VT. The term *daylight aperture* is used to describe the product of the two. Numerous studies of schools and office occupancies have shown that for modern lighting power densities[25] of around 1 W/ft^2, an effective daylight aperture of around 0.20 (e.g., WWR=0.5 and VT=0.4 or WWR=0.4, VT=0.5), combined with automatic continuous dimming control of the electric lighting will result in the lowest annual energy consumption.

25
Lighting Power Density (LPD) is the metric used to report the installed lighting capacity. The LPD in office spaces has dropped in the last 30 years.

Although WWR is important to daylighting performance, for the normal range of commercial window areas, it is the window head height that dominates daylighting performance more than any other design feature. The area over which daylight intensity is distributed is dependent on the depth of daylight penetration, which is in turn dependent on the height of the window. There is no benefit to window area below desktop height, and from a daylighting point of view, strips of window just below the ceiling give the best performance as the light projects most deeply into a space. In practice, one needs to balance the desire for view (mid-height glazing) with the desire for deep daylight penetration (high glazing).

In the end, glazing is a classic design problem that requires one to balance the desire for thermal comfort, energy efficiency, and light quality (all of which require small window areas) with equally important desires for view, daylight, and connectivity with the outdoors (all of which benefit from large vision-glass areas). Less window/curtainwall area, and higher system performance is the low-cost, high performance prescription for buildings suffering from excessive glazing area. Restricting WWR to around 40% or less, carefully choosing and specifying overall glazing system U-value, SHGC, and VT is the best start to delivering high performance glazing for a high-performance building.

3.3.8 Thermal Mass

Thermal mass acts to store and release thermal energy (heat). The heat capacity of a material (in units of kJ/kg/C or Btu/lb/F) can be easily found from numerous sources. Mass located on the inside of the insulation layer is far more important than mass on the outside: as the level of thermal control increases, the difference in effectiveness between indoor and outdoor thermal mass grows. For high performance buildings, exterior mass has only a very small benefit. Thick layers of thermal mass are not as effective as thin layers widely spread through the interior. Thermal mass in the floor is often isolated from the interior by finishes and furniture, whereas as wall and ceiling mass (if there is no suspended finished ceiling) are more accessible and hence more effective.

Figure 3.38: U (W/m²K), SHGC and VT values for representative glazing products (modified from ASHRAE 2009)

			Center of IGU	**Operable Aluminum TB**	**Wood Vinyl**	**Fiberglass**	**Fixed Aluminum TB**	**Wood Vinyl**	**Fiberglass**
Low-e	2	U	1.70	2.60	2.20	1.80	2.35	2.00	1.90
Solar control		SHGC	0.41	0.37	0.31	0.32	0.38	0.36	0.36
		VT	0.76	0.69	0.68	0.69	0.69	0.68	0.69
	3	U	1.00	2.00	1.70	1.40	1.75	1.40	1.30
		SHGC	0.27	0.25	0.21	0.22	0.25	0.24	0.25
		VT	0.70	0.65	0.64	0.65	0.65	0.64	0.65
	4	U	0.50	1.75	1.10	0.95	1.65	1.00	0.85
		SHGC	0.22	0.20	0.19	0.20	0.20	0.19	0.20
		VT	0.62	0.59	0.58	0.59	0.59	0.58	0.59
Low-e	2	U	1.95	2.89	2.39	2.07	2.60	2.23	2.13
Solar gain		SHGC	0.68	0.62	0.52	0.52	0.61	0.61	0.61
		VT	0.64	0.58	0.57	0.58	0.58	0.57	0.58
	3	U	1.42	2.85	1.95	1.70	2.10	1.73	1.66
		SHGC	0.62	0.55	0.46	0.46	0.56	0.54	0.54
		VT	0.6	0.54	0.53	0.54	0.54	0.53	0.54
Reflective	2	U	2.70	3.40	2.87	2.53	3.22	2.84	2.72
		SHGC	0.29	0.27	0.22	0.22	0.27	0.26	0.26
		VT	0.27	0.25	0.24	0.25	0.25	0.24	0.25
	3	U	1.75	2.60	2.19	1.91	2.39	2.01	1.93
		SHGC	0.34	0.31	0.26	0.26	0.31	0.30	0.3
		VT	0.53	0.48	0.47	0.48	0.48	0.47	0.48

Thermal mass is almost always beneficial for buildings with near constant interior temperature. For buildings with aggressive setback temperatures or long periods of unoccupied hours (community centers, churches, cottages), thermal mass can cause difficulties with control and may even increase energy consumption.

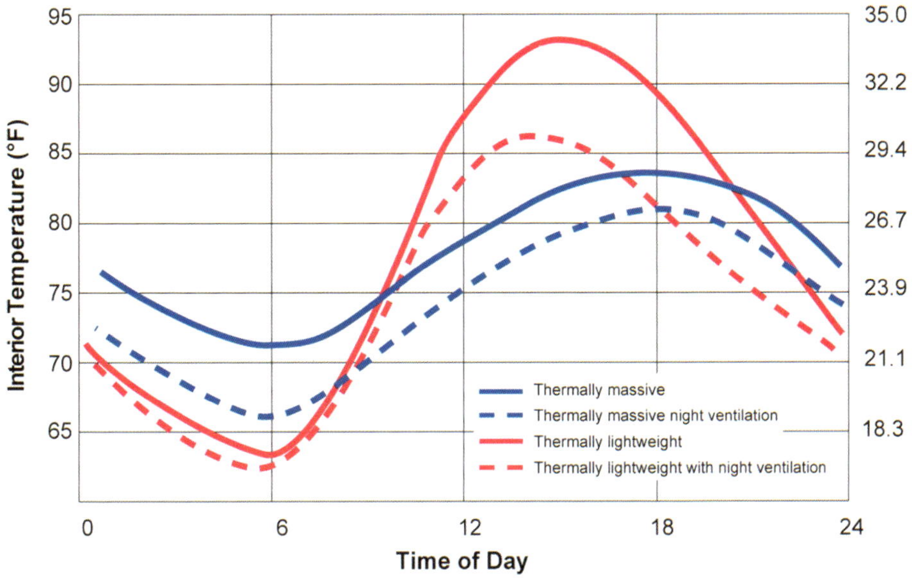

Figure 3.39: Example temperature profile of an office over a 24 hour period. Thermal mass acts to reduce the daily temperature variation and delays the peak temperatures to later in the day.

Thermal mass is often desirable to store solar heat gains and internal heat generation from occupancy during the day, and release it during the evening (Figure 3.39). This is one of the basic tenets of passive solar design. The limit to this strategy in all types of buildings is the rate at which heat gain can be transferred to the thermal mass during sunny period (without overheating) not the total amount of thermal mass. Because the heat loss of high performance buildings is so small, increasing thermal mass is often not useful unless the rate of heat transfer to the thermal mass is increased. This latter goal can be achieved by increasing the surface area of mass, by ensuring that the mass is not covered by insulating or thermally-lightweight finishes, and by encouraging airflow from the heated zone to remote thermal mass in other zones.

For high-performance buildings, the building heat loss coefficient is low enough that the thermal mass of typical construction is quite substantial: if the heating supply is interrupted during cold weather it can take several hours for the temperature to drop by even one degree. In many high-performance buildings setback schedules will not work well because the building does not cool down to the setback temperature before the temperature is set up again for the next day: this is a natural outcome of high thermal mass, low heat loss coefficients, or some combination.

As an example, consider a school with an occupancy schedule of 7 a.m. to 9 p.m. With a high-performance enclosure, the thermal mass of the CMU interior walls and insulated concrete floor slab will release sufficient heat with a 2°F/1°C temperature drop to heat the school for the 10 unoccupied hours even during quite cold weather. Hence, a more aggressive setback strategy of 10°F/5°C would have no effect as the temperature would simply not drop more than 2 °F/1 °C. Over an unoccupied weekend, however, the full setback may be achieved, just in time for the Monday morning temperature reset. A precisely "right-sized" mechanical system for the school may then require many hours to reach set point as its capacity was based on steady-state heat loss. If the heating system is based on heating air, the temperature of the CMU walls and concrete slab will remain cool and uncomfortable for many hours, even if the heating system manages to reach set point quickly by providing extra capacity.

Hence, while thermal mass is desirable, large setbacks and setup temperatures should be avoided in high-performance buildings with even modest levels of thermal with careful consideration of the dynamic heat flow response under a wide range of operating and climatic conditions.

For ultra-low energy use buildings, thermal mass can be a powerful and often necessary strategy to allow for very small heating and cooling plants, to capture night cooling, or absorb high intensity but short-term internal gains. However, it is a design strategy that is difficult to use because of the complexity of assessing actual dynamic building use, heat transfer to and from mass, the temperature swing acceptable to the occupants and other poorly characterized variables.

3.4 VAPOR DIFFUSION CONTROL

Although most condensation problems occur because of air leakage (convection) vapor diffusion can also occasionally cause damaging amounts of wetting. However, vapor diffusion is also an important drying mechanism, and hence allowing diffusion drying of wet materials within the enclosure must also be considered as part of enclosure design (especially rain control).

Vapor diffusion is the movement of water vapor molecules through the microscopic open pore structure of porous materials (glass, solid plastics, and metals are not porous, wood, gypsum, foam, and concrete are). Vapor diffusion always moves from more to less vapor concentration, but practically this tends to mean from the warm side to the cold side of an enclosure or material layer. As the process is relatively slow, it usually requires weeks or months to move significant quantities of water vapor. Vapor permeance[26] is used to describe the ease of vapor diffusion through a layer of material. (A Building Science Digest entitled *BSD-106: Understanding Vapor Barriers* is available at buildingscience.com for more specific guidance.)

26
Permeance has units of US perms. There are 57.3 metric perms (ng/Pa s m^2) in a US perm. Permeance is typically measured using ASTM E96.

Vapor control layers of some type are often required in enclosure designs. A vapor impermeable material (i.e., a barrier) can be used to block most vapor flow by diffusion. However, in most cases allowing a vapor permeable path is desirable to allow drying of construction moisture and any accidental wetting that might occur. The direction of vapor flow will often change between summer (inward) and winter (outward). Hence, the location of any low vapor permeance layer must be carefully considered, and it is often best to avoid low permeance layers in assemblies especially in climate zones with hot-humid summers. Because of the secondary importance of vapor diffusion control (relative to air and rain control), the vapor control layer is not always indicated on the drawings in this document.

Very low permeance (less than 0.1 US perm or 6 ng/Pa•s•m^2) vapor control layers (often called vapor barriers) are sometimes required for framed walls by building codes or manufacturer's specifications. This requirement is based on a limited amount of research conducted over 50 years ago, and has recently been re-evaluated in some detail. Based on a sizeable body of new research, the 2009 International Residential Code and International Building Code have recently been changed to remove the requirement for a low-permeance vapor control layer for all climate zones of North America[27] and specific types of wall designs.

27
Such low-permeance layers may still be required in special applications, and can be important for use below slabs on grade, as crawlspace ground covers, etc.

In Canada, Part 5 of the National Building Code of Canada does not require a vapor barrier, but does require vapor diffusion to be controlled to avoid damaging quantities of condensation in both summer and winter. The climate-specific rules in the IRC/IBC are also recommended for Canadian climate zones as the physics do not respect national boundaries.

To specify what level of vapor permeance is acceptable, it is useful to describe three levels, or classes, of vapor control depending on the ASTM E96 dry-cup vapor permeance of the vapor control layer:

- Class I - less than 0.1 perms
  - e.g., polyethylene sheet, sheet metal, aluminum or polyethylene facing on any product.
  - Virtually vapor impermeable, a vapor barrier. A Class I vapor control layer divides the wall assembly into interior and exterior zones with regard to vapor diffusion. Vapor diffusion wetting and drying can occur on both sides of the Class I vapor control layer, but not across the vapor control layer.
  - Required or desirable only in extreme climates and extreme occupancies. Useful for special applications such as ground covers.
- Class II - 0.1 to 1.0 perms
  - e.g., the kraft facing on fiberglass batts, some vapor control

Table 3.4: Vapor control layer classification; tested by ASTM E96 dry-cup (Method A)

Class	Permeance (US perms)	(Metric perms)	Description
I	Less than 0.1	Less than 6	Impermeable
II	0.1 to 1.0	6 – 60	Semi-impermeable
III	1 to 10	60 – 600	Semi-permeable
none	Over 10	Over 600	Permeable

paints, 2” of extruded polystyrene foam (XPS), over 2 to 3” of medium-density closed-cell spray polyurethane foam, and more than 3 to 5” of EPS

- Vapor semi-impermeable. A Class II vapor control layer throttles or limits diffusion wetting but allows a small amount of diffusion to help dry an accidentally-wetted assembly.
- Class II vapor control layer provides sufficient vapor diffusion control to avoid condensation for most climates and occupancies.

- Class III - 1.0 - 10 perms
  - e.g., latex or enamel paints on gypsum board/gypsum board, more than about 6” to 8” (150-200 mm) of open-cell spray polyurethane foam

Table 3.5 Climate zone descriptions

Zone	Description	City
1A, 2A, 3A south of warm-humid line	hot-humid	
1B, 2B	hot-dry	
3A north of warm-humid line, 4A	mixed-humid	
3B, 3C, 4B	mixed-dry	
4C	cool, marine	Vancouver, Seattle
5A, 6A	cold, humid	Toronto, Chicago, Halifax
5B, 6B	cold, dry	Lethbridge, Helena
7	very cold	Edmonton, Anchorage
8	subarctic	Whitehorse, Fairbanks

- o Vapor semi-permeable. A Class III vapor control layer restricts vapor diffusion, but allows significantly greater vapor diffusion wetting and drying than a Class II vapor control layer.
- o Although Class III vapor control layers allow significant drying assemblies will often require insulated sheathing in cold climates to control diffusion wetting.

The need for a vapor control layer, and which class, depends strongly on the enclosure design, the air permeance of the insulation layers,[28] the interior conditions, and the exterior climate.

28
Air permeable for an insulation layer is defined as having a permeance of less than 0.04 cfm/ft² @ 0.3 in wc (0.2 l/s/m² @75 Pa) when tested to ASTM E283 or similar.

The exterior climate is divided into zones from 1 to 8 based on heating degree days (HDD) (Figure 1.3). The zones are further sub-divided into different exterior humidity levels, indicated by appending the letters A to C (e.g., zone 6C).

The class of vapor control required can be prescribed for many common wall or roof assemblies for interior conditions of normal residential, school, retail, and office occupancy. This means indoor temperatures of around 72 °F (21°C) and indoor winter time relative humidity of less than 40% (less than 35% in zones 7 and 8).

3.4.1 Vapor Control Recommendations

Different types of assemblies have different vapor control requirements. Although the requirements can be developed through rational engineering analysis, a simplified summary of recommendations, many from the "I" codes, is presented below for the "normal" occupancies described above.

Buildings with indoor relative humidity levels of over 40%RH when the outdoor temperature is below 20°F (-7°C) for prolonged periods of time require special analysis and often unique enclosure details to prevent moisture problems.

Above-Grade Wall Assemblies

Four categories of above-grade wall assemblies include most of the possible enclosures:

a) framed assemblies with all or most of the insulation inside of the sheathing and between the framing members;

b) framed assemblies with some insulation outboard of the framing and some insulation between the framing members;

c) assemblies with all or most of the insulation outboard of the structure (framed or solid); and

d) assemblies with insulation comprised of only air-impermeable and Class II vapor control insulation between, within, or outside the structure.

a) Framed wall assemblies with all or most of the insulation value installed between the framing or structure (e.g., wood or steel stud)

as vapor permeable (more than 10 perm) insulation (e.g., fiberglass, rockwool, cellulose, or open-cell foam)

The goal of the vapor control design is to prevent vapor diffusing easily and condensing on either the cold sheathing in cold weather or the cold interior finish during warm weather.

- No vapor control layer is needed in climate zones 1, 2, 3, 4a or 4b.
- A Class I or Class II vapor control layer is required on the interior side of framed walls in zones 4c, 5, 6, 7, and 8, with the exceptions of basement walls, below-grade portion of any wall, and wall construction that is not sensitive to moisture or freezing (e.g., concrete).
- Class I vapor control layers, including non-perforated vinyl wallpaper, reflective foil, glass, epoxy paint, white boards, melamine, etc. are not recommended and *should be avoided* on the interior of air-conditioned building occupancies in climates with humid summers in zones 1-6. The dividing line between dry (B) and moist (A) can be found in ASHRAE 90.1. Enclosures clad with unvented water absorbent claddings (e.g., stucco, masonry, fiber cement, wood) are at especially high risk of summer condensation.
- A Class III vapor retarder can be used instead of a Class I or Class II when
    - in zones 4c, or 5, vented cladding is used over sheathing with a perm rating of more than 1 (wet-cup), i.e., OSB, plywood, or exterior gypsum sheathing, OR
    - in zone 6, vented cladding is used over high permeance (more than 10 perm) sheathings such as fiberboard and exterior gypsum

Vented claddings include vinyl siding, metal panels, terra cotta, wood or fiber cement siding on air gaps, and masonry veneers with clear airspaces and vent openings top and bottom. A clear gap of around $^3/_8$" (10 mm) will generally provide sufficient airflow to allow for ventilation, but at least 1" (25 mm) should be specified for masonry walls.

b) Framed assemblies with some insulation value outside of the framing or structure.

The recommendations in the previous assembly category may be used, but it is usually desirable to design for more drying, especially in warmer climate zones (4 and 5 especially). The use of insulation on the exterior of the sheathing increases its temperature in cold weather, thereby relaxing the need to control cold-weather vapor diffusion. Exterior insulation made of foamed plastic has the benefit that it also tends to reduce inward vapor drives during warm weather. MFI insulation work differently in that they allow more outward drying.[29]

A Class III vapor retarder can be used instead of a Class I or Class II in zones 4c, 5, 6, 7, or 8 where any of the criteria for the specific zone from the list below is met. These criteria may depend upon the climate zone and the ratio of the insulation value in the stud space to insulation value installed outboard of the sheathing.[30]

A Class III vapor control layer may be used on the interior of framed walls in zone 4c and higher, if the any of the following criteria are met:

Zone 4c (e.g., Vancouver, Seattle or Portland)

- Sheathing-to-cavity R-value ratio of >0.20
- Insulated sheathing with an R-value ≥ 2.5 on a 2x4 framed wall
- Insulated sheathing with an R-value ≥ 3.75 on a 2x6 framed wall.

Zone 5 (e.g. Chicago, Windsor, Boston)

- Sheathing-to-cavity R-value ratio of >0.35
- Insulated sheathing with an R-value ≥ 5 (e.g., 1" XPS) on a 2x4 framed wall
- Insulated sheathing with an R-value ≥ 7.5 (e.g., 1.5" XPS) on a 2x6 framed wall.

Zone 6 (e.g., Toronto, Ottawa, Helena, Montreal, Halifax, Minneapolis)

- Sheathing-to-cavity R-value ratio of >0.50
- Insulated sheathing with an R-value ≥ 7.5 (e.g., 1.5" XPS) on a 2x4 framed wall
- Insulated sheathing with an R-value ≥ 11.25 (e.g., 2" PIC) on a 2x6 framed wall.

Zones 7 and 8 (e.g. Calgary, Edmonton, Whitehorse, Anchorage, Fairbanks)

- Sheathing-to-cavity R-value ratio of >0.70
- Insulated sheathing with an R-value ≥ 10 (e.g., 2" XPS) on a 2x4 framed wall
- Insulated sheathing with an R-value ≥ 15 (e.g., 3" XPS) on a 2x6 framed wall.

29
Exterior insulation made of mineral fibers is highly vapor permeable and when combined with vapor permeable sheathing and membrane layers, enclosures will behave differently than less permeable sheathings/membranes: less R-value of such products are needed to perform well than the rules in this category.

30
The insulation value in the stud space is often a function of whether 3.5", 5.5", 6" or 8" framing is filled with insulation. Self-supporting spray-applied insulations such as glass-fiber and foam products, or batt products well supported on the open side DO NOT need to fill the whole stud space, and thereby reduce the required insulation value of the exterior continuous insulation layer. Cavity insulations need to be in tight contact on five surfaces: the sixth surface can be open to the interior.

Insulated sheathing can be installed in the form of expanded polystyrene (EPS) or semi-rigid glass fiber and rockwool with approximately R-4 per inch, extruded polystyrene (XPS) with R-5 per inch, polyisocyanurate (PIC) with approximately R-6 per inch, or closed-cell spray foam (usually around R-6/inch).

c) **Assemblies with all or most (more than 75% of the total) of the insulation value located outboard of the structure (framing or solid)**

This is the simplest and most robust wall to design with respect to vapor control. Such walls should ideally have all moisture sensitive components and materials located on the inside of the insulation. In this location, a Class I or II layer on the inside of all or most of the insulation value is acceptable and recommended if all outboard components are moisture tolerant.[31] A Class III layer on the interface of a high permeance (more than 10 perms) insulation layer outboard of a moisture-sensitive structure should only be used if warm weather and inward vapor drive condensation are not an issue or are controlled by other means.

31
Condensation during warm weather and condensation caused by vapor driven inward from wetted cladding heated by the sun will occur on the exterior of the vapor control layer: hence it is best detailed as a drainage plane water control layer.

d) **Assemblies comprised of only air-impermeable and Class II vapor control insulation between, within, or outside the structure.**

The EPS and urethane foam cores of wood or metal SIPS, board foams, and medium density closed-cell spray foam (between wood or steel studs), if installed as continuous layers, all provide their own vapor control layers and require no additional vapor-diffusion control in any climate. Wood studs themselves and small cracks between steel or wood studs do not allow significant vapor to flow by diffusion. Their use does not change the recommendations.

Cracks between framing members and insulation boards are often significant for airflow control and must be addressed in all of the systems described in the previous paragraph.

Below-Grade Wall Assemblies

Below-grade spaces, such as basements, are of particular concern with respect to improperly located Class I vapor control layers. Because the moisture drive in below-grade walls is always from the exterior to the interior in zones 6 and lower, installing a low-permeance layer on the interior side of insulation wall will cause moisture related durability issues by trapping moisture in the enclosure (see Appendix B–*BSD-103: Understanding Basements* or online at www.buildingscience.com). A Class I vapor control layer outside a concrete or masonry basement wall is recommended to control the flow of vapor from damp soil into the assembly. Installing a Class II vapor control layer on the inside of below-grade framed assemblies is recommended for zones 7 and higher.

In cold climates (zone 5 and higher), condensation may occur on the interior side of concrete/masonry structure of a below-grade wall assembly if insulated on the inside with an air and vapor permeable (Class III or more, e.g., fibrous insulation) layer. To control both inward and outward drives, it is recommended that a Class II or III insulation product (e.g. most foams) be used in contact with the interior face of the concrete/masonry, and any insulation installed between interior framing follow the rules of "Framed assemblies with some insulation value outside of the framing or structure."

Below-Grade Floor Assemblies

Locating all of the insulation below the structure (concrete slab, or framed) with a Class I vapor control layer between the structure and the insulation is the practical and economical manifestation of the perfect enclosure. Some insulations provide the same level of vapor control (foil-faced polyiso or EPS, plastic-faced XPS) but many products will require a special low-permeance layer (polyethylene sheet is commonly used, inexpensive, and effective).

If no impermeable floor finishes are likely to be used during the life of the structure, the vapor control class requirement can be relaxed to a Class II. See also *BSI-003: Concrete Floor Problems* and *BSI-009: New Light in Crawlspaces* online at www.buildingscience.com for more discussion of vapor control and flooring.

Roof Assemblies

Roofing behaves differently than walls from a vapor control point of view for a number of reasons: most roof membranes are located on the exterior and provide Class I vapor control. Roof membranes, shingles, metal roofing are all vapor impermeable. Only back-ventilated roofing such as concrete tiles perform in a similar manner to walls.

Assuming a vapor impermeable roofing membrane, four categories of roof assemblies will include most practical roofing systems:

a) unvented roof assemblies with all or most of the insulation outboard of the structure (framed or solid).

b) vented, framed roof assemblies with all or most of the insulation inside of the sheathing and between the framing members

c) unvented, framed roof assemblies with either some insulation outboard of the framing and some insulation between the framing members, or some air-impermeable insulation outboard of air-impermeable insulation, and

d) unvented roof assemblies with insulation comprised of air-impermeable and Class II vapor control insulation between, within, or outside the structure.

a) unvented roof assemblies with all or most (>75%) of the insulation outboard of the structure (framed or solid)

These types of roofs comprise most of the low-slope, rigid board insulated, commercial roof systems installed over metal and concrete decks. However, there is no technical difference between a low-slope roof built in this manner and a sloped roof. This type of roof has very little need for additional vapor control as all of the common board foam roofing insulations (PIC, EPS, XPS) provide sufficient vapor control for most climate zones.

- No additional vapor control layer is needed if a Class II vapor control insulation layer is used (>2” of XPS, PIC, >3” of EPS).
- To control diffusion at joints in board foam, all board insulation should be installed in two layers or more, with joints offset vertically and horizontally. This is especially important in zones 5 and higher.

Such roofs should ideally have all moisture sensitive components and materials located to the inside of the insulation. In this location a Class I or II layer on the inside of all or most of the insulation value (e.g., 75%) is acceptable but will restrict desirable and effective inward drying. For zones 5 and lower, a vapor control layer of Class III or higher is recommended to allow for inward drying of incidental moisture due to solar heating of roof membranes (e.g., gypsum board on metal deck).

b) Vented, framed roof assemblies with all or most of the insulation value installed between the framing (e.g., wood/steel stud, metal buildings) as vapor permeable (more than 10 perm) insulation (e.g., fiberglass, rockwool, cellulose, or open-cell foam)

If a compact roof is well-vented above the insulation (i.e., with soffit and ridge vents connected by an open air-space of 1.5”/38 mm or more),

- o no vapor control layer is required in zones 1 through 3.
- o Class III vapor control in zones 4 through 6.
- o Class II vapor control in zones 7 and higher.

If a roof is sloped over a well-vented attic (i.e., with soffit and ridge vents connected by an open air volume of at least 12” (300 mm) average height, and no less than 1.5”/38 mm anywhere),

- o no vapor control layer is required in zones 1 through 4,
- o Class III vapor control layer is recommended for zones 5 through 6,
- o Class II vapor control layer is recommended in zones 7 and higher.

Unvented, framed assemblies with all the insulation value comprised of air permeable insulation (fiberglass, rockwool, cellulose) between the framing are not recommended due to the potential for air leakage condensation.

c) Unvented framed roof assemblies with some insulation provided by air-impermeable insulation.

Roofs need not be vented if diffusion and air leakage related wetting can be strictly limited. To accomplish this, some air-impermeable, Class II insulation can be located outside of air and vapor permeable (more than 10 perm) insulation (e.g., fiberglass, rockwool, cellulose).

The ratio of the insulation value of the exterior air- and vapor-control insulation to the insulation value of the interior air- and vapor permeable insulation increases as the climate becomes colder, and the interior more humid. For normal commercial and residential occupancies the following recommendations can be made (Note: painted gypsum board or more than about 150 mm/6" of open-cell foam are Class III retarder).

For zones 1 through 3A, 3B

- Outer air-impermeable insulation value >15% of total
- No vapor control layer required.

For zones 3C/4C (e.g. Vancouver, Seattle, Portland, San Francisco)

- Outer air-impermeable insulation value >25% of total
- Class III vapor control

Zone 4A, 4B, 5 (e.g. Windsor, Kansas City, Boston)

- Outer air-impermeable insulation value >35% of total
- Class III vapor control in zone 5

Zone 6 (e.g. Toronto, Ottawa, Montreal, Halifax, St. Johns, Minneapolis)

- Outer, air-impermeable insulation value >50% of total
- Class III vapor control

Zones 7 and 8 (e.g. Edmonton, Calgary, Winnipeg, Whitehorse, Fairbanks)

- Outer, air-impermeable insulation value >65%
- Class III vapor control

Air-impermeable insulation can be in the form of expanded polystyrene (EPS) with approximately R-4 per inch, extruded polystyrene (XPS) with R-5 per inch, polyisocyanurate (PIC) with approximately R-6 per inch, 2 pcf (32 kg/m^3) closed-cell spray foam (usually around R-6 per inch), or 0.5 pcf (8 kg/m^3) open-cell spray foam (usually around R-4 per inch). All of these products, other than open-cell spray foam and EPS, provide a Class II vapor control layer in thicknesses of 1.5" or more. EPS of 1 pcf density (16 kg/m^3) which requires between 3" and 5" (depending on density) to reach Class II performance, and open-cell foam requires a special vapor control coating.

d) unvented roof assemblies with insulation comprised of only air-impermeable and Class II vapor control insulation between, within, or outside the structure.

The EPS and urethane foam cores of wood or metal SIPS, board foams, and medium-density closed-cell spray foam (between wood or steel studs), if installed as continuous layers, all provide their own vapor control layers and require no additional vapor diffusion control in any climate with normal residential and commercial interior humidity conditions. Wood studs and cracks between steel or wood studs do not allow significant vapor to flow by diffusion and hence do not change the recommendations.

Cracks between framing members and panels are significant for airflow control and must be addressed in all of the systems described above. A very effective airflow control layer is required on the inside of the insulation and/or framing for all unvented roof assemblies.

3.4.2 Vapor Control Layers: Materials and Systems

All roof membranes, peel-and-stick membranes, flashings, and bituminous coatings act as vapor barriers (Class I vapor control) and should be located with respect to the thermal control layers with this in mind. Most foamed plastics provide some vapor resistance, and often act as, and/or obviate the need for, a separate vapor diffusion control layer. For example, layers of 2" (50 mm) or more of extruded polystyrene, polyisocyanurate, and closed-cell medium density polyurethane foam all provide sufficient vapor resistance for all but the coldest climates (climate zone 7 and lower) and most interior occupancies. Expanded polystyrene (EPS) may require 3" to 5" (75–125 mm) thickness to provide Class II vapor control, and open-cell low-density spray polyurethane foam cannot provide sufficient vapor resistance for many purposes without special analysis in climate zones 5 and higher.

Medium to high-density (2 lb/ft^3) closed-cell spray foam can be used on the interior of non-insulating sheathing as a replacement for insulating sheathing in a "flash and batt" or "flash and blow" application. For example, in zone 6, medium-density closed-cell spray foam with an R-value of 11.25 or more can be used on the interior cavity side of exterior sheathing installed on a 2x6 framed wall with the remainder of the cavity insulated with a vapor permeable fibrous insulation (such as sprayed or blown glass fiber, rockwool or cellulose) and a Class III vapor control layer. As new spray foam products proliferate, the designer must ensure that the vapor permeance of the foam products is understood.

All fibrous insulations, such as cellulose, fiberglass, rockwool, cotton, polyester, etc. are highly vapor permeable, and can be assumed to have a permeance of well over 10 perm/600 metric without further investigation or testing.

3.5 RELATIONSHIP OF AIR, VAPOR, AND THERMAL CONTROL

Installing the incorrect vapor control layer in the incorrect location can lead to premature enclosure failure, however, it is very important to remember that air leakage is usually a far more powerful mechanism of water vapor movement. Figure 3.40 plots the results of simple steady-state calculations for a 8 ft (2.4 m) high and 16" wide (0.4 m) stud bay in a very cold climate (e.g., zone 7) with a month of outdoor temperatures averaging 0 °F/-18 °C.

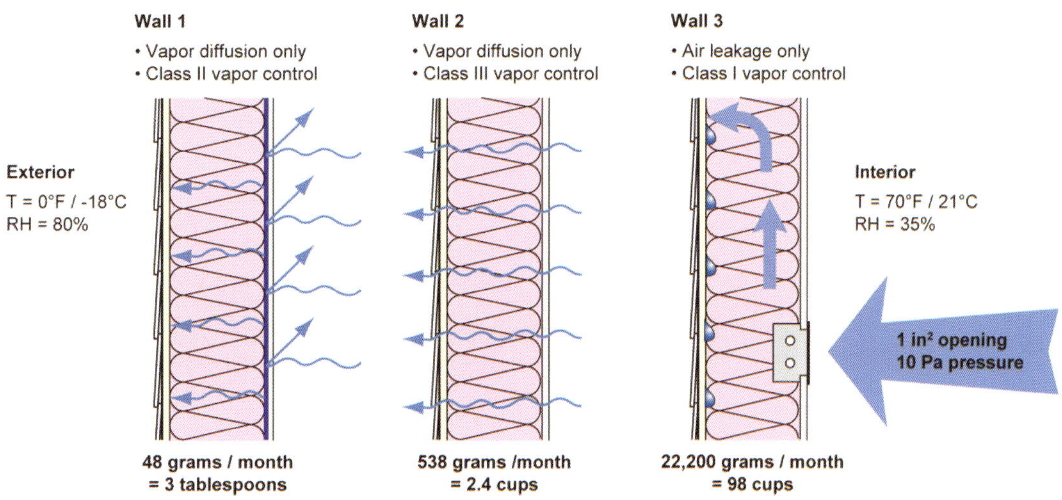

Figure 3.40: Comparison of cold-weather moisture accumulation in a single stud bay due to diffusion and air leakage.

It should be clear from the figure that even a modest air barrier flaw (i.e., a one square inch opening is formed by a crack 8” long x $^1/_8$” wide) can deliver truly significant quantities of moisture: in this extreme example, air leakage transports from 40 to 400 times more moisture than diffusion through painted gypsum board. Diffusion through a 1 perm (57 ng/Pa s m^2) Class II vapor control layer transports only 3 Tbsp. (48 g) in a sunless and extremely cold month! On the other hand, a thin coat of latex paint on primer (Class III) would allow a damaging 2.4 cups of water to diffuse through. In such a cold climate, some vapor diffusion control (Class II) is sufficient, but an extremely high level of airtightness is required. This simple theory has been supported by both experience in the field and more sophisticated computer analysis and field measurements.

Thermal control interacts with both vapor control and airflow control as they relate to condensation. If insulation warms a surface, condensation is less likely to occur (or impossible) and subsequent drying will be faster. Although insulation installed between framing does not reach its rated R-value because of thermal bridging, it can act at its rated R-value over a significant area between the framing.

Insulation within framing or inside of the structure, places the structure and sheathing close to exterior conditions, and violates most theory of durable enclosures. The reason this bad practice remains common is primarily historical: as solid masonry walls developed into framed systems, insulation was easiest and cheapest to add within the hollow spaces that formed. The consequence of adding this insulation was a cooling of the exterior portions of the wall during cold weather. Condensation could occur more easily, and drying occurred more slowly. Both changes resulted in reduced durability and required changes to building practice: air barriers and vapor barriers were invented and deployed, and superior rain control and enhanced drying became required (see also Appendix B–*BSI-028: Energy Flow Across Enclosures* or online at www.buildingscience.com). Adding insulation to the exterior of moisture-sensitive sheathing and structure reverses these trends, and placing all the insulation on the exterior results in very durable enclosures, often more durable than in the past.

That exterior insulation had these beneficial effects has long been recognized (e.g., Timusk 1983, CBD 1964, CBD 1967). However, there is often confusion between condensation fed by air leakage (the critical issue) and the air barriers needed to control this and condensation fed by vapor diffusion (rarely much of a problem) and the vapor barriers used to control diffusion.

The impact of adding exterior insulation on both air leakage and diffusion condensation can easily be assessed with approximate methods, usually with conservative results. One such approach is described below.

3.5.1 Controlling Cold Weather Condensation

Condensation in cold weather can be supplied by water vapor that is transported by air leaking outward, or by water vapor diffusing outward. Cold-weather outward vapor diffusion can be managed by the appropriate specification and location of vapor control layers with respect to insulation layers described in Section 3.4. Air leakage condensation is usually more significant, and its control is described in this section.

Air leaking outward through a framed enclosure in cold weather will contact the back of the sheathing. To avoid condensation in a framed enclosure with air- and vapor permeable insulation inside of the sheathing, the temperature of the inside of the sheathing should be maintained above the temperature at which indoor air could condense, the "dew point temperature" of the interior air.

In walls with sufficient exterior insulation, the dew point temperature of the interior air will be below the temperature of the back of sheathing and thus condensation due to air leakage cannot occur within the stud space. If condensation is not calculated to occur because of air leakage (using the method described below), then *diffusion* condensation *cannot* occur, even if no vapor resistance (no vapor barrier or other control layer) is provided inside of the sheathing, or the sheathing is a vapor barrier (such as foil-faced insulations).

Condensation itself is not usually a failure: if air leakage condensation occurs only under extreme conditions (e.g., the 2.5% design condition), air leakage during the following hours will quickly *dry* the wall when sheathing temperatures rise above the interior dew point. Hence, the choice of conditions for the analysis is important. Although data for air temperatures is readily available, even north-facing walls will have some exposure to diffuse solar radiation and warm above the air temperature for many hours of cold winter months. For materials with some tolerance to moisture exposure (e.g., glass-matt faced exterior grade gypsum sheathing is quite resistant to moisture), and/or some capacity to safely store moisture (e.g., plywood sheathing), it is recommended that the average monthly outdoor temperature be used as the design criteria.[32]

32
The average monthly temperature is recommended for most applications based on exeprience. The 99% design temperature, or lowest expected temperature is too low, as it presumes even short duration condensation events are damaging (they are not).

The interior conditions within a building during cold weather are critical variables in understanding the risk of condensation, and necessary if predictions and calculations are to be made. Interior temperatures are often in the 70 °F/21 °C range, but relative humidity levels, and thus air moisture content, can vary significantly. In most office, school, and retail occupancies, ventilation rates are high enough that the RH during winter months is in the range of 25 to 35% and significantly lower during very cold weather. In some residential occupancies, the interior moisture generation is higher and exterior air ventilation rate lower than commercial occupancies, and hence the RH will often be higher. In special occupancies, such as swimming pools, both the interior temperature and relative humidity levels will be higher (78°F/25 °C and 60%RH), resulting in very high absolute humidity levels.

The moisture content of the exterior air always drops at very cold conditions as the maximum air moisture content drops. As the weather, and the climate, becomes colder the interior RH drops in non-humidified buildings because moisture generated inside is diluted by increasingly dry exterior air brought in for ventilation or by air leakage. This effect provides some protection against condensation as the coldest week of the year is likely to coincide with some of the lowest interior humidity levels.[33]

33
The correlation of indoor humidity levels and outdoor temperature would be much more direct if not for the moisture storage capacity of the building fabric, and the changing moisture production rates within a building. Sharp sudden drops in outdoor temperature are much more likely to result in condensation because the building stores a higher moisture level. If the temperature outdoor slowly drops over a period of days, the building can be slowly dried with outdoor air.

The interior moisture content is usually defined by a combination of temperature and relative humidity. More direct measures are absolute humidity or humidity ratio, usually expressed in grams of water per kg of dry air (grains of water per pound of dry air). However, practically speaking the most useful measure is the dew point temperature of the interior air.

Given an agreed upon set of interior and exterior design conditions, it is easy to calculate the level of insulation required on the exterior of the framing space/sheathing to control air leakage condensation by ensuring the temperature at the back of the sheathing is more than the interior air dew point temperature. If one assumes that interior finishes and exterior cladding have little thermal resistance (almost always a reasonable assumption) then the back of the sheathing temperature can be found from (Figure 3.41):

$$T_{back\ of\ sheathing} = T_{interior} - (T_{interior} - T_{exterior})\frac{R_{batt}}{R_{total}}$$

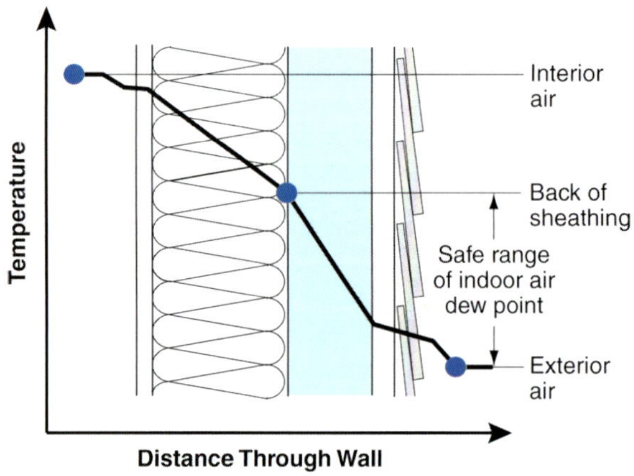

Figure 3.41: Insulating sheathing reducing air leakage condensation

It should be clear that any insulated sheathing provided on the exterior of framed structures will provide better protection against air leakage condensation in cold weather than no exterior insulation. For a fixed R-value of exterior insulation, the risk of condensation within the enclosure will also drop as the interior insulation R-value drops. Thus, if there is no insulation within the framing (i.e., reducing the inboard R-value to that of the interior finish and empty stud space only, approximately R-2) almost any reasonable level of exterior insulation R-value provides complete protection from air leakage condensation and diffusion during cold weather.

Table 3.6 provides an estimate of the level of insulation (sheathing plus airspace and cladding) that should be provided outside of a stud space filled with air permeable insulation (i.e., batt or blown fibrous insulation) to prevent cold-weather exfiltration condensation. It can be seen that mild temperatures and dry interior air require little exterior insulation to control condensation, whereas very cold temperatures or higher interior humidity levels demand significant proportions of insulation on the exterior of the framing. As experience has shown, a museum maintained at 50%RH in very cold climates (e.g., Fairbanks, Alaska or Yellowknife, Northwest Territories) should have essentially all of the insulation on the exterior.

Indoor		RH	20	25	30	35	40	50	60
Dew point		°C	-3.0	0.0	2.5	4.7	6.6	9.9	12.7
		°F	26.6	32.0	36.6	40.5	44.0	49.9	54.8
$T_{outdoor}$	°C	°F							
	0	32	0.00	0.00	0.12	0.23	0.32	0.47	0.60
	-5	23	0.08	0.19	0.29	0.37	0.45	0.57	0.68
	-10	14	0.23	0.32	0.40	0.48	0.54	0.64	0.73
	-15	5	0.33	0.42	0.49	0.55	0.60	0.69	0.77
	-20	-4	0.41	0.49	0.55	0.60	0.65	0.73	0.80
	-25	-13	0.48	0.54	0.60	0.65	0.69	0.76	0.82
	-30	-22	0.53	0.59	0.64	0.68	0.72	0.78	0.84
	-35	-31	0.57	0.63	0.67	0.71	0.74	0.80	0.85
	-40	-40	0.61	0.66	0.70	0.73	0.76	0.82	0.86

Table 3.6: Ratio of exterior-to-interior insulation R-values required to control air leakage condensation

Consider a house in Toronto (January average temperature = -4 °C/25 °F) with an interior RH of 35%. From Table 3.6, the dew point can be seen as being approximately 40°F/5°C, and that about 37% of the total insulation value of the wall should be on the exterior in the form of insulating sheathing, air gaps, and cladding.

Targeting a nominal enclosure R-value of 20, would require 0.37 * 20 of the total, or R-7.5 on the exterior to avoid condensation if air leakage occurred.[34] This leaves R-12.5 on the inside, which could be comprised of an R-12 batt and interior finish. The exterior cladding and airspace add some

34 This approach does not eliminate air leakage condensation under all hours. However, the approach of winter-time average conditions and January average conditions has given good results in field practice as only a small amount of condensation occurs on cold nights and re-evaporates during warm days.

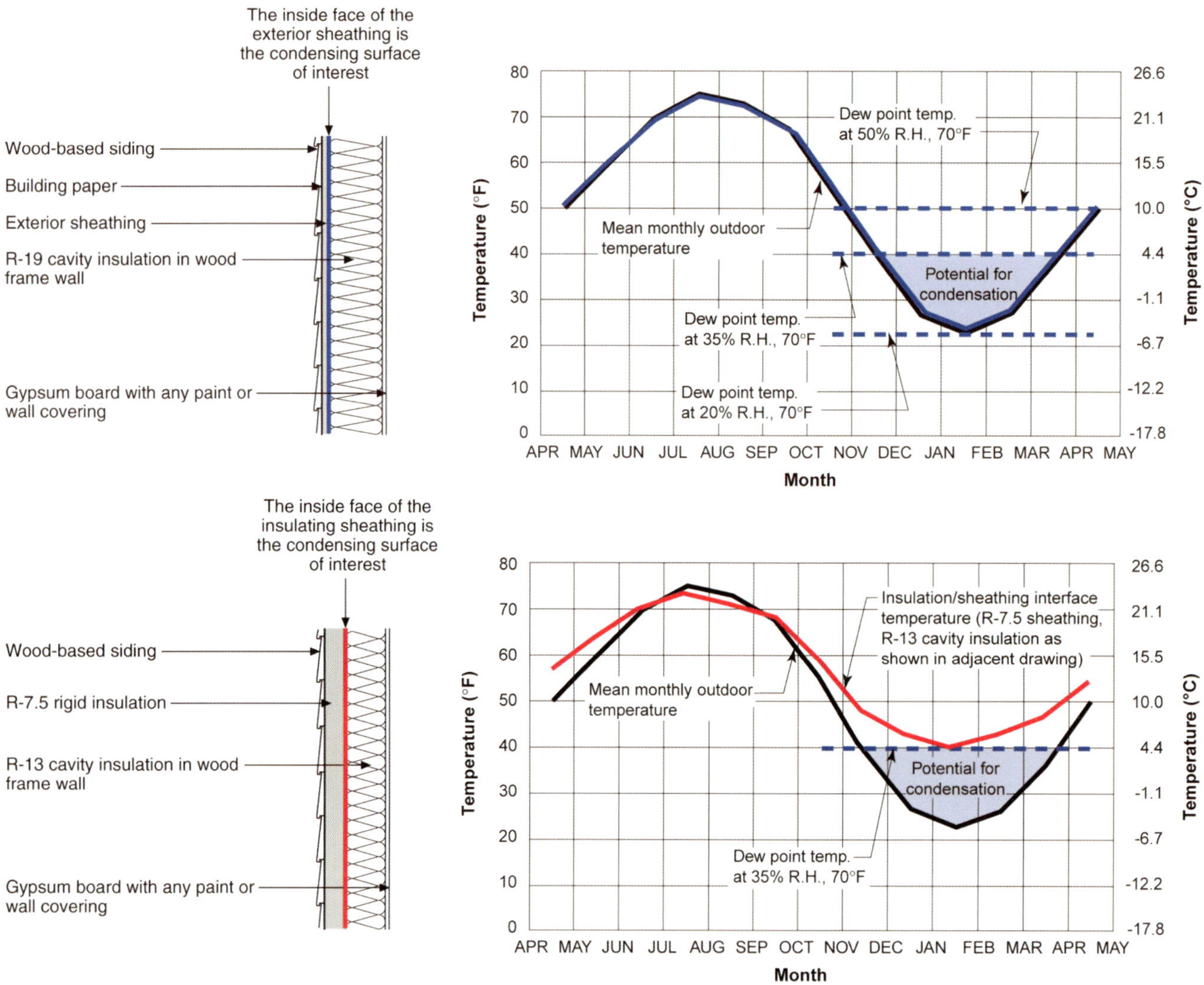

Figure 3.42: A monthly analysis of the condensation potential of two walls in the Chicago climate.

R-value on the exterior, but these can be conservatively ignored. This solution, R-12 batts between 2x4 studs with R-7.5 exterior insulation sheathing, is very safe against air leakage condensation. If R-30 were the target, 0.37*30= R-11 of exterior sheathing and R-19 stud space insulation would be one solution. The exterior cladding and airspace add some R-value on the exterior, but these values are now a basis for design.

Adding more air-permeable insulation to the stud bay (e.g., if a 6" stud is required structurally, a well-intentioned contractor may fill the stud cavity with an R-20 batt) of the assembly with R-7.5 exterior sheathing will of course lower the safety against condensation, dangerously in this case. Adding significantly more insulation to the exterior, raising the R-7.5 to R-15 for example, will significantly reduce risk. Regardless of the wall assembly, exterior climate and indoor humidity the same trends will always hold: adding insulation value to the exterior decreases risk, adding air-permeable insulation value to the stud space increases risk.

It is important to note that the R-values in question are the center of stud bay R-value, as condensation will occur at the coldest part of the sheathing, and this will be between the studs. Hence, although the actual R-value of a R-12 batt installed between 3.5" steel studs at 16" o.c. (90 mm studs at 400 mm spacing) will be around R-5, the batts will be effective over the middle of each stud bay, and it is here that condensation fed by air leakage or diffusion would begin to occur.

This type of simple analysis can be conducted on a monthly basis and plotted to help visualize the condensation risk. An example of a wood-framed wall in the Chicago climate is shown in Figure 3.42.

Given the results of this analysis and the knowledge that steel studs with insulated stud bays inserted between concrete and steel frames have R-values of only around R-5 to R-7, leads to the general recommendation of placing all the insulation on the exterior of light-gauge steel enclosures.

Consider the light-gauge steel framed wall designs in Figure 3.43 during a period of cold weather. Applying R-10 (RSI 1.76) of insulating sheathing (continuous insulation of any type) on the exterior of the framing will result in a sheathing temperature of above 60 °F (15 °C) everywhere within the stud space, including at the sheathing during a night when the outdoor temperature is 4 °F (-15 °C). Hence condensation is practically impossible within the stud space or on the sheathing (typically one of the moisture sensitive components in the assembly) even if air leakage occurs as all the surfaces are above the dew point temperature of the interior air.[35] If R-19 (RSI3.5) insulation is placed between the framing, the sheathing temperature will be approximately 10 °F (-12 °C), well *below* the temperature at which condensation can occur. The latter design relies on perfect air barriers (air-impermeable spray foams are one solution) and, if the cavity fill is highly vapor permeable (e.g., fiberglass, rockwool, or open-cell, half pound density foam), a vapor control layer (Class II) is necessary to reliably manage vapor diffusion.

The design with all of its heat flow control as a continuous layer of insulation on the exterior can perform very well even if some air leakage occurs,

35
This conclusion is valid even for high humidity interior environments, such as museums, as 70°F/50% RH air has a dew point of about 50°F/10°C. Only throughfasteners such as screws, brick ties and roof screws would be at risk in such high RH environments. Swimming pools may have dew points exceeding 60°F/15°C and hence would require more R-value on the exterior to avoid interstitial condensation in cold climates.

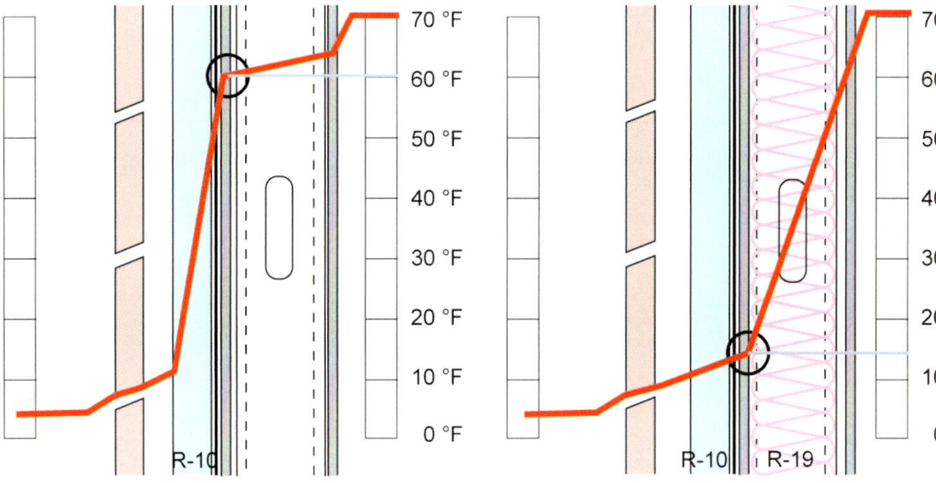

Figure 3.43: Insulating sheathing as condensation control measure. Continuous exterior insulation on the left, framing cavity insulation on the right. The red line plots the temperature through the two assemblies on a 4°F (-15°C) night.

and requires no special care in the selection of interior layers for vapor control. It should also be recalled that the wall with only exterior insulation would have an overall R-value of around R-12 (RSI2.1), whereas the wall with only framing cavity insulation would have an overall R-value of between R-6 to R-8 (RSI1.1 to 1.4) (depending on floor and wall framing intersection details and the type of cladding).

In many situations, a hybrid of exterior insulating sheathing and stud cavity insulation may be considered. Figure 3.44 plots the temperature through two hybrid solutions under the same conditions as the previous two walls. Applying R-12 (RSI2.1) within the stud space will improve the thermal performance of the wall by about R-6 (increasing the assembly to an R-value of over 16/RSI2.8), but will reduce the sheathing temperature to 35 °F (2 °C) on this cold night. In many commercial occupancies, the interior dew point temperature will fall below 35 °F (2 °C) during cold weather and so condensation is unlikely, but far from impossible (i.e., the risk is higher). If the R-12 were added in the form of air-impermeable spray insulation (e.g., SPF), air could not practically reach the sheathing and there would be no risk of air-leakage condensation.

If R-19 (RSI3.5) of air permeable insulation was added to the stud space, the R-value of the assembly would rise by about R-7 over the empty stud space scenario: that is, almost $^2/_3$ of the insulation value of the R-19 batt would still be lost. However the sheathing temperature would drop below 30 °F (-1 °C) and the risk of condensation would be higher. The small increase in heat flow control is achieved at the cost of much increased condensation risk.

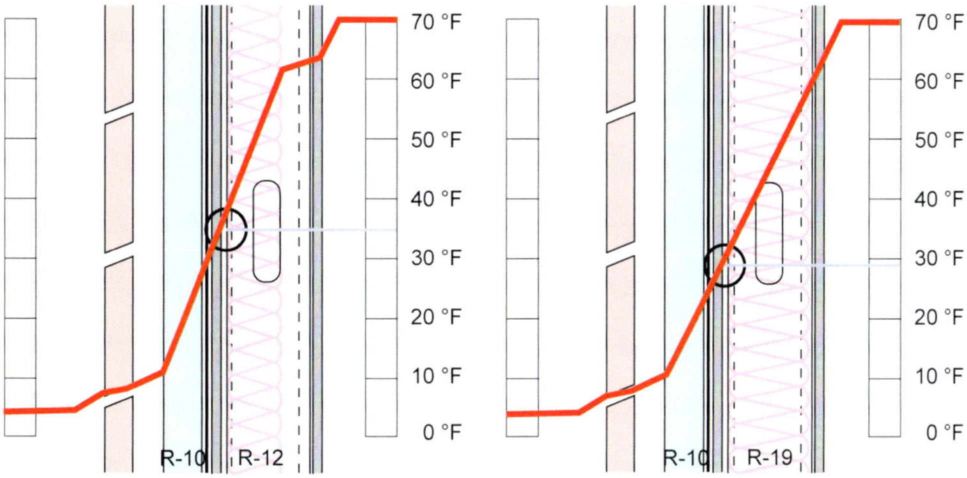

Figure 3.44: Hybrid insulation approach. Although riskier, particularly in cold climates and higher interior humidities, hybrid walls offer slightly higher R-value and can be moisture-safe in many applications. Note that the ratio of exterior insulation value to the framing cavity R-value defines the cold-weather condensation risk.

The same solutions that avoid air-leakage condensation will also completely solve cold-weather condensation due to vapor diffusion even if the exterior sheathing is a perfect vapor barrier (e.g., foil-faced or plastic faced insulation boards). If the sheathing layers (including structural sheathing, water control, and insulation) chosen are somewhat vapor permeable (such as expanded polystyrene over building paper and plywood), less R-value can be used and diffusion condensation will still be controlled (because much of the vapor that diffuses into the stud bay will pass harmlessly through to the outside). If the sheathing layers are very vapor permeable (e.g., rockwool, over fiberboard or gypsum sheathing and housewrap) then very little insulation value is required outboard of the stud bay to control diffusion. However, while these permeable layers can essentially eliminate vapor diffusion condensation risks with lower exterior sheathing R-values, the risk of air leakage condensation does not drop as much: air leakage may still deliver more water vapor to the back of the sheathing than can be removed by diffusion through the sheathing and hence condensation could still occur and accumulate

For important projects or situations in which the design team has little historical experience, investigation using widely available computer models such as WUFI-Pro would be prudent if the necessary time and skills are available.

3.5.2 Controlling Warm Weather Condensation

Many cold climates also have warm humid weather during the summer. It can be argued that summer condensation is perhaps just as significant a performance problem as cold-weather condensation in zone 5. Summer condensation problems also occur in zone 6 but are increasingly unlikely in zone 7 or 8. Summer condensation can result if humid airflows inward and/or if water vapor diffuses inward from rain-wetted cladding.

As the source of moisture in warm weather is the exterior air or wet exterior layers, condensation will begin to occur on the exterior surface of the first cold solid surface below its dew point that inward flowing air (or inward diffusing vapor) encounters while moving from the exterior to the interior. This surface might be an interior vapor barrier such as polyethylene film or vinyl wall paper, or even a large plastic whiteboard attached to the interior of an exterior wall. In these examples, condensation will occur within moisture sensitive spaces (the stud space) or on materials adjacent to paper-faced gypsum wallboard. These materials can quickly be damaged and generate objectionable mold growth.

The temperature analysis described in the previous section can also be used to assess this scenario. If the interior layers of an enclosure assembly are sufficiently vapor permeable (the best example is gypsum wallboard painted with latex paint) short-term condensation can be tolerated as drying to the inside can occur very quickly before harm occurs. The "self-drying roof" approach advocated by Desjarlais (1995) relies on summer heating of the roof membrane to dry small leaks to the inside before damage occurs.

Alternatively, placing moisture tolerant materials (including insulation) outboard of a low vapor permeance air and water barrier can ensure that all summer condensation occurs outside of the moisture sensitive components of an assembly. This is the approach taken to control warm-weather condensation in Section 4.1.

3.6 OTHER CONTROL FUNCTIONS

Enclosures must control much more than heat, air, vapor, and water of course. The two other control functions that tend to dominate the design of enclosures are fire control and sound control.

Many assembly choices cannot be used in non-combustible or high-rise construction. For example, wood studs are often a superior choice, even for high-rise construction, but codes require non-combustible framing.[36] This is an example of fire control dominating enclosure choices. Similarly, enclosures must have 1 or even 2-hour fire resistance ratings if used close to lot lines, or containing certain high-risk occupancies.

Sound control, usually expressed as Sound Transmission Class, can be a requirement of some enclosure designs. A newer measure, the Outdoor-

36
It is important to recognize that a wood framed wall has equal and often better performance in a fire than light-gauge steel stud walls. The codes simply exclude wood for historic reasons, most of which are no longer valid. The interiors and contents of modern commercial buildings have a tremendous amount of combustible material that provide many times more fuel load to a fire than does wood framing. True high-performance assemblies are likely to contain wood or wood products in the future (wood infill walls are already being used in Europe, Scandinavia and Oceania for large and tall buildings).

Indoor Transmission Class, is often more appropriate for enclosures intended to exclude outdoor noise from the indoor environment.

The control of impact, solar radiation, corrosion, etc. may all become sufficiently important in an enclosure design to dominate choices. However, the four main control layers described earlier will remain critical to the performance of the enclosure, regardless of the additional restrictions other control functions may impose.

Chapter 4 High Performance Enclosures

4.1 THE PERFECT ENCLOSURE

To summarize the concepts of control layers and continuity, an abstract example of the "perfect" building enclosure can be useful. Figure 4.1 presents a conceptual "perfect" enclosure showing the drained and ventilated exterior finish (the "cladding"), the thermal, air, vapor, and water control layers, the structure, as well as optional interior finishes (often also fire control layers) and service distribution spaces (see also Appendix B–*BSI-001: The Perfect Wall* or online at www.buildingscience.com).

By locating the heat flow control layer (insulation) on the exterior of the structure and by locating the combined air, water, and vapor control layers between the structure and the insulation, the risk of interstitial condensation due to air leakage of diffusion is essentially eliminated. The structure and air, water and vapor control layers are well protected from UV exposure, impact, and temperature extremes, thereby ensuring their durability. Such a strategy works well in all climate zones, from the bitter cold of an Alaskan winter to hot humid Houston summers. Only for special-use buildings, such as a freezer-warehouse in Miami, would other arrangements be considered. In typical commercial applications, this should be the high-performance enclosure design of choice.

The same physics apply for roofs (Figure 4.2), basement walls (Figure 4.3), and slabs on grade (Figure 4.4).

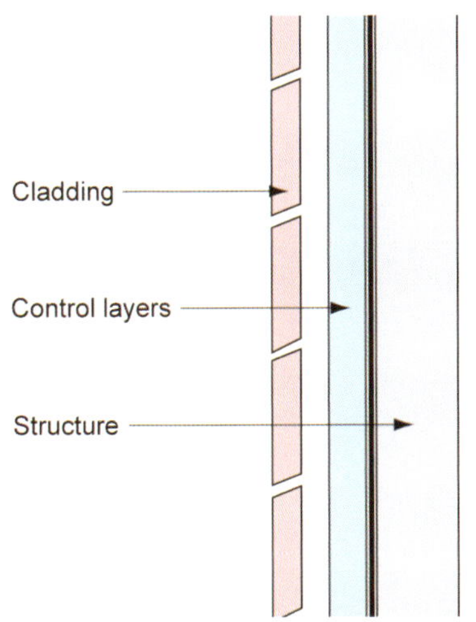

Figure 4.1: Perfect Wall: Note that the drainage gap and tolerance are accommodated behind cladding.

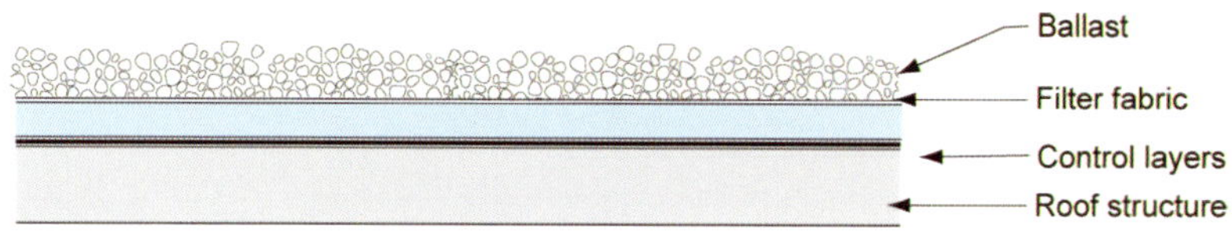

Figure 4.2: Perfect roof. Note drainage gap provided above roof membrane.

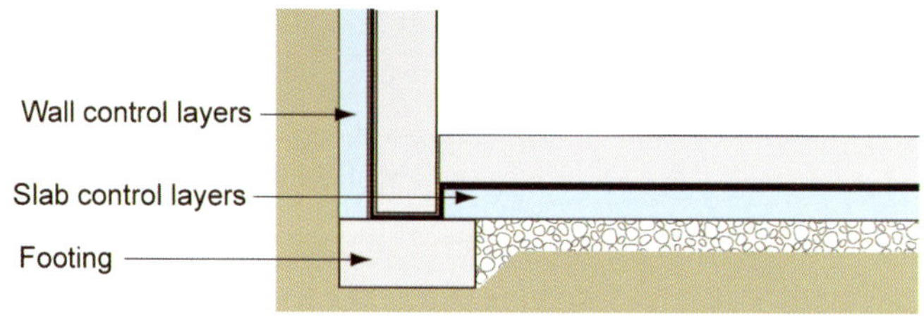

Figure 4.3: Perfect basement wall enclosure. Note drainage gap provided at outside of wall membrane, no exterior finish required.

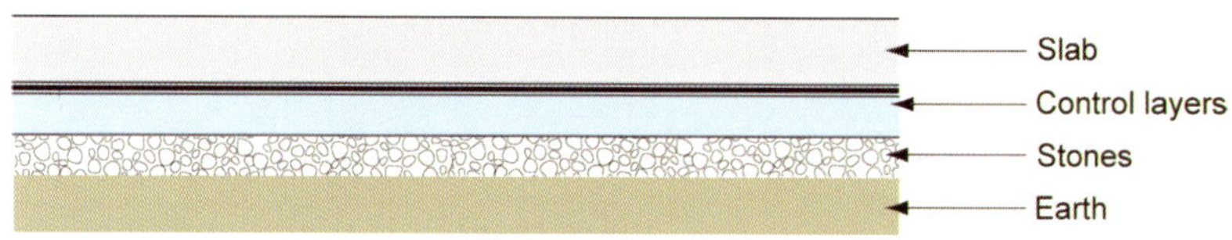

Figure 4.4: Perfect enclosure slab-on-grade. Note drainage gap provided by clear stone, no exterior finish.

The idea of the perfect enclosure is intended to guide designers on the proper principles during concept design, and to aid the comparison of competing enclosure designs. More specific choices depend on the specifics of each project, and the goals of the design.

4.2 ABOVE-GRADE WALLS

Numerous examples of "perfect walls" have been built for some time. The primary difference between some of the commonplace designs shown below and high-performance enclosures for low-energy buildings, is the amount of insulation and the care and attention to achieving continuity of the water, air, and thermal control layers.

In wood-framed assemblies, a common compromise involves filling some or all of the stud space with fibrous insulation (Figure 4.8). Adding an R-20 batt to a wood-framed wall will provide around R-12 to R-13, substantially better than if the same batt were installed between steel framing. Thus, it is practical to add some exterior continuous insulation, but the vapor permeance of the water-air control layers must now be carefully chosen for the climate and interior use.[37]

This wall assembly deviates from the perfect wall concept and hence requires more careful consideration in the selection of the vapor control layer and the relative thermal resistances of the cavity fill and the exterior continuous insulation layer. Higher R-values can be achieved in less wall

37
In low-rise residential buildings, the enclosure surface to floor area tends to be high, and hence higher overall R-values are desirable. This makes the desired R-value for the exterior insulation layers high, which also reduces the risk of moisture problems.

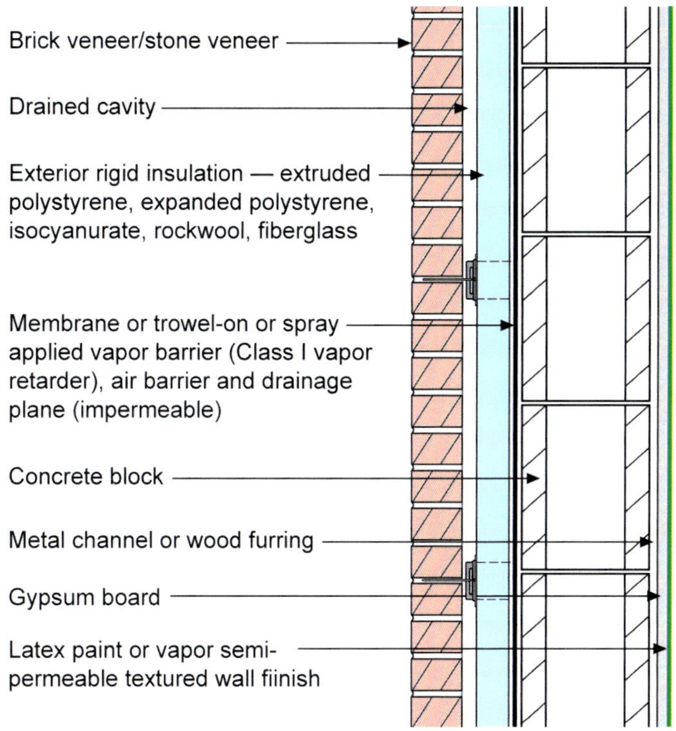

Figure 4.5: This "perfect wall" assembly can perform well in all climate zones. Although the cladding is shown as brick, a designer has a very wide choice. The class of the vapor control layer, which can be provided by the insulation product, is not critical but Class I or II provides the widest range of application. Highly vapor permeable insulation (e.g.,rockwool) and Class III vapor control requires consideration of the control of inward vapor drives.

thickness and less cost using wood stud walls which have fewer thermal bridges than steel and concrete structural systems.

A commercial hybrid system comprised of rigid polyisocyanurate board with an integrated drainage plane provides continuous insulation and rainwater control (Figure 4.10). An interior closed cell spray foam layer adds some R-value, provides air control continuity, and limits the possibility of condensation on any through fasteners.

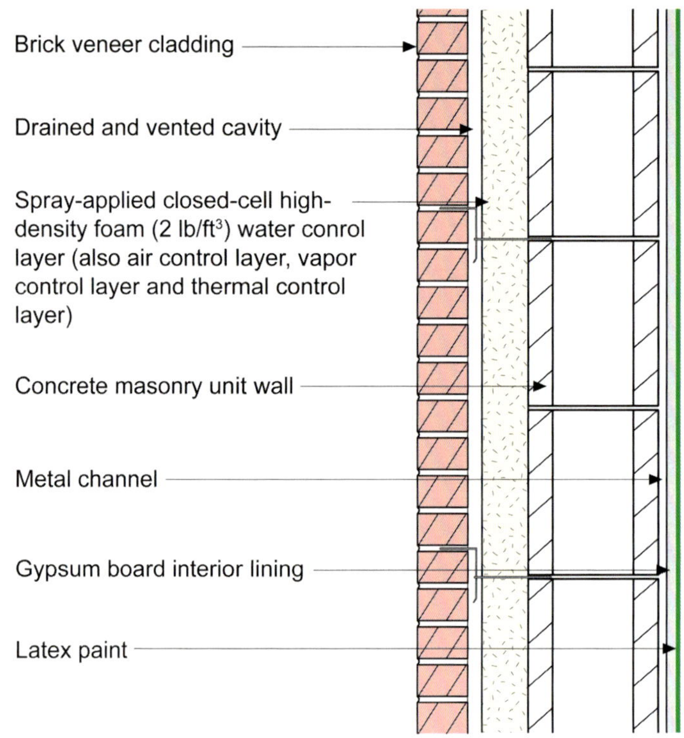

Figure 4.6: This assembly employs the medium-density closed-cell spray polyurethane foam as the water, vapor, and air control layers. It can be used in all climate zones and a low-permeance vapor control layer (e.g., polyethylene) is neither required nor desirable for occupancies with low to moderate winter-time relative humidities.

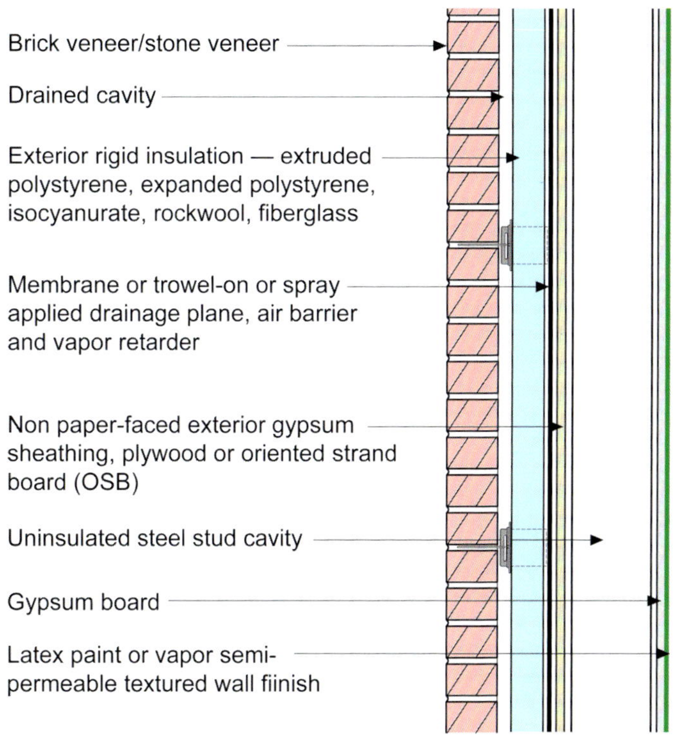

Figure 4.7: Replacing the CMU support structure with steel stud is an increasingly common lightweight solution. To protect the backup wall, the vapor control layer should have a permeance of Class I or II. The vapor control layer could be combined with the thermal layer or with the water and air control layer.

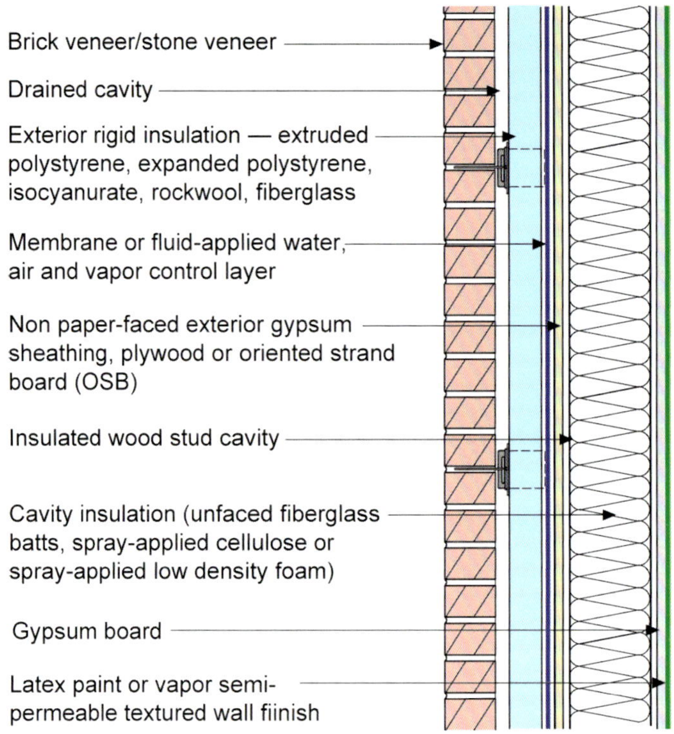

Figure 4.8: This wood-framed wall assembly deviates from the perfect wall concept and hence requires more careful consideration in the selection of the vapor control layer and the relative thermal resistances of the cavity fill and the exterior continuous insulation layer.

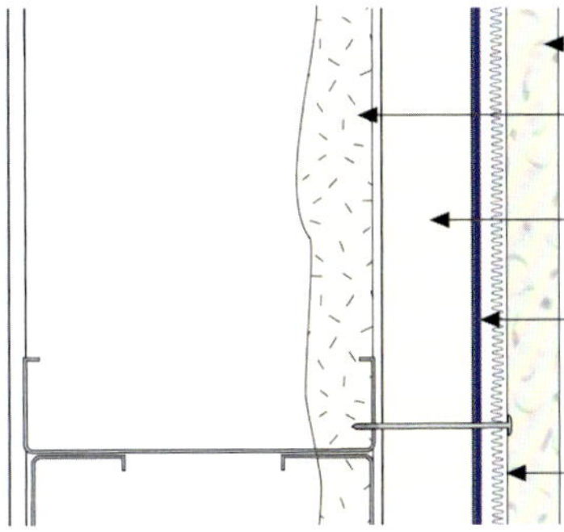

Figure 4.9: This hybrid system is comprised of spray foam and rigid board foam on the exterior of steel studs. Any cladding can be used.

Figure 4.10: The proprietary Thermax™ Wall System uses insulation facers and special tapes to locate the primary water and air control layers on the exterior. (Photo: Trinity Christian School courtesy Dow Building Solutions).

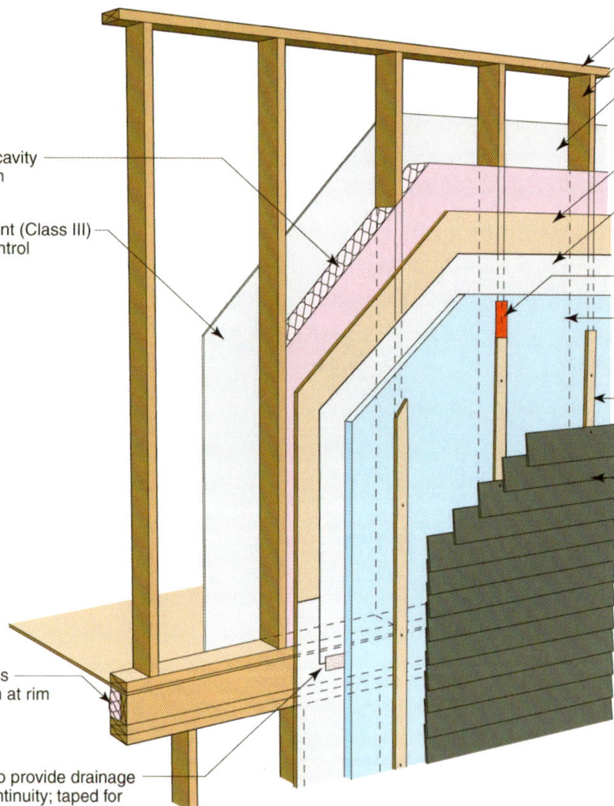

Figure 4.11: For combustible buildings of 1 to 4 storeys, a wood frame assembly with sufficient insulation outside the framing can provide an economical and high performance enclosure.

Cladding attachment is a common question when assemblies with exterior continuous insulation are proposed. Although solutions are not currently widely used in all parts of North America, they are simple, straightforward, and time tested in either specific regions or Europe or both.

In general there are two approaches to attaching cladding through continuous exterior insulation: the first takes advantage of the strength of the exterior insulation to support the cladding and the second provides a redundant system independent of the insulation.

The former approach is preferred as it reduces cost and the number and size of thermal bridges passing through the insulation. The strength of the insulation is taken advantage of to form a compressive strut and resist the tension generated by the screw (Figure 4.12). To spread the loads from the cladding over a wide area of the exterior insulation, a furring strip is used. Wood furring strips of 2.5" width are commonly used in low-rise applications as furring, and metal hat channels can be used for taller buildings.

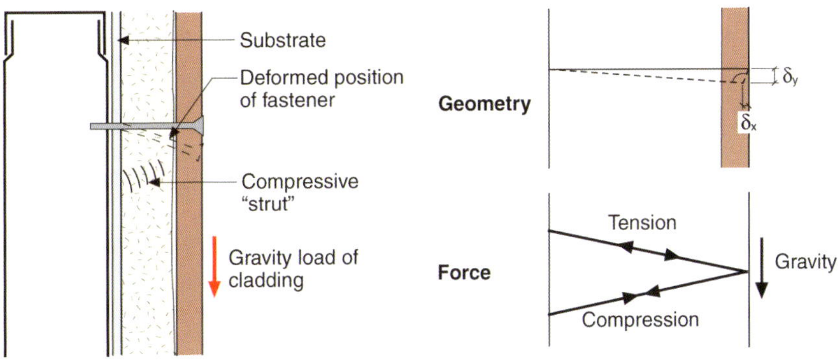

Figure 4.12: A "strut-and-tie" model can be used to explain the remarkable strength and stiffness that is achieved through the use of standard furring strips screw fastened through rigid sheathing. This low-cost and high-performance approach can be used to support gravity loads for most lightweight claddings.

As the weight of the cladding does not increase with height (it being transferred back to the structure at every screw),[38] the same solutions can be used for very large buildings as used in small buildings. Tall buildings will have much higher wind loads however, and so more and/or larger screws (which transfer the peak wind loads collected by the cladding to the structure) may be needed to resist the tension loads which result from the peak wind-induced suction loads.

38
The screws need to connect back to the structure, either concrete, CMU, steel studs or wood studs. Attachment to the sheathing is rarely acceptable, except if the sheathing is thick enough wood (e.g. ¾" true, 1" nominal wood products) or metal sheet sheathing designed to hold a screw and transfer tension to the studs.

The second approach, a separate redundant structure, is more traditional, and is often necessary for heavy claddings such as natural stone and masonry veneers. In the case of masonry veneers, relieving or shelf angles are used to collect the gravity load of the cladding at every one to three floors, and transferred back to the primary structure via "knife-edge" engineered connections (see also Chapter 5). Wire ties, usually at 16" centers

(400 mm) penetrate the insulation in the body of the wall and have a small impact on the thermal control. Heavier ties made of 18 or 16 gauge sheet metal have a larger impact (approximately 10%), especially if attached directly to the side of steel studs.

Proprietary clip systems are available from numerous metal panel, terra cotta and fiber-cement panel cladding manufacturers (Figure 4.13), although standard cold-formed galvanized angles as clips and rails can be designed by a competent structural engineer to transfer loads.

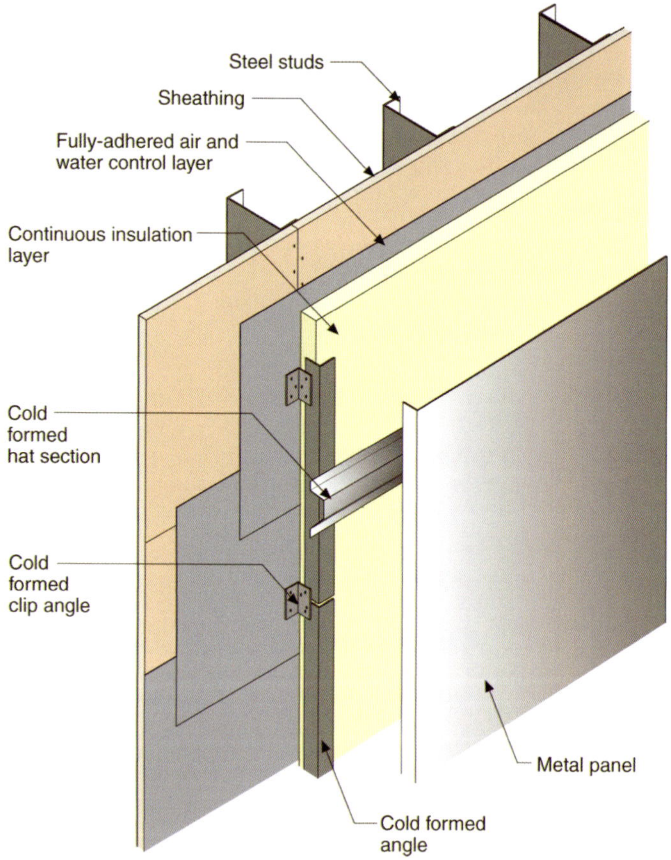

Figure 4.13: Thermal bridging can be managed by using a "clip and rail" approach behind a wide range of light-weight panel cladding products (such as metal, composite, and fiber cement).

Exterior Insulation and Finish Systems (EIFS) are a popular product that can be easily modified to provide high performance (Figure 4.14). EIFS are comprised of a glass-mesh reinforced lamina adhered to insulation (usually EPS) applied over sheathing on framing or solid masonry/concrete (Figure 4.15).

EIFS have the fundamental benefit of providing a layer of continuous insulation and a finish at a modest cost. Too many EIFS are constructed with only 1 to 2" of EPS insulation (R-4 to R-8) and no drainage or secondary rain control, seriously limiting performance. To deliver a high performance enclosure, thicker layers of insulation, a drainage layer, and a drainage gap should be included in the design.

The drainage plane is also often the air control layer. The drainage gap is formed by grooves in the insulation or adhesive strips run vertically. The drainage gap must be kept small to limit or eliminate the potential for wind-washing (e.g., under about 3 mm or $^1/_8$").

Almost all manufacturers have systems with drainage planes, gaps and weeps. Systems with up to 12" (300 mm) of insulation are available from several manufacturers. As described earlier, when a high proportion of R-value is placed outside the sheathing (and the applied drainage plane), an interior vapor control layer is not needed and should not be used, even in cold climates.

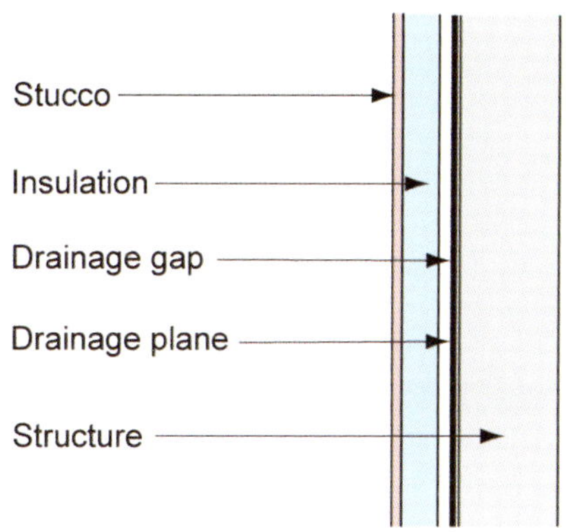

Figure 4.14: Drained EIFS are a modified version of the perfect enclosure. A gap of even $^1/_{16}$" is all that is needed for drainage.

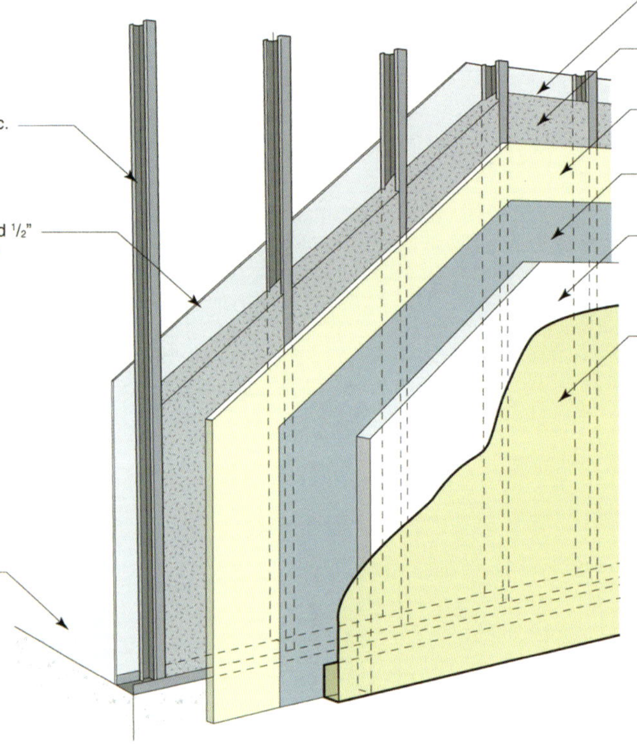

Figure 4.15: Exterior Insulation Finish Systems, when using a drainage gap and air-water control membrane, are a practical implementation of the perfect wall.

Figure 4.16: Drained EIFS can be easily integrated with other drained enclosures with exterior insulation, as the air, water, and thermal control layers are all located outside of the structure in a continuous plane.

Insulated Concrete Forms (ICFs) have become popular and can be a high-performance enclosure. Most have a true R-value of nearly 20 (RSI3.5), which is not always high enough for very cold climates or smaller buildings (like homes), but some products are available at up to R-60 (!) and additional insulation can be added to most systems on the exterior. No additional vapor control layer is required or desirable for most occupancies in all climate zones.

Although not following the perfect wall approach exactly, the addition of a drained cladding system (such as metal panels, or drained EIFS) can result in the same performance. Even with direct-applied EIFS lamina on the EPS (i.e., a drained-joint perfect barrier element approach to rain control), the moisture tolerance of the materials (EPS) and the continuity of the control layers has resulted in very good performance in practice. Rain control problems have been observed at penetrations such as undrained window

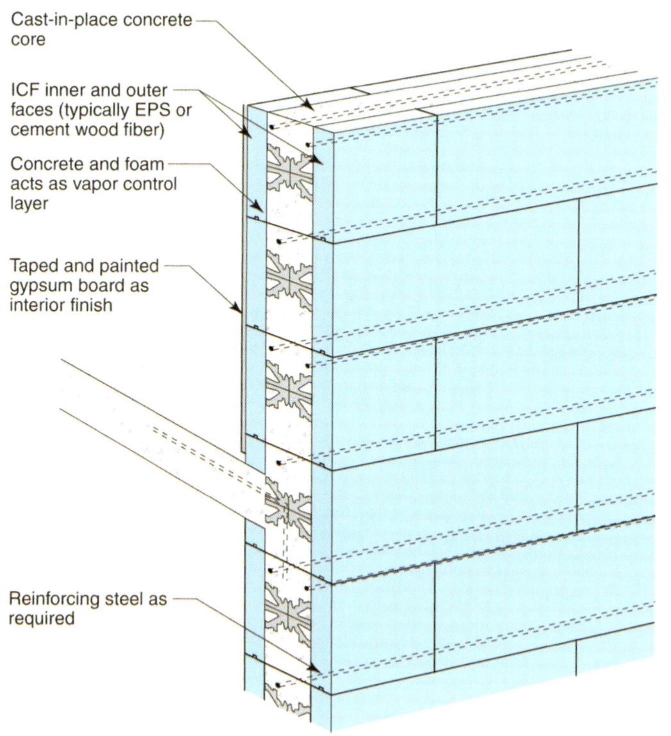

Figure 4.17: Insulated Concrete Forms are a new breed of products that use the EPS foam as the thermal control layer and reinforced concrete as the air and vapor control layer.

openings, and wall to lower roof connections (at which point the water control layer should be taken from the roof back to the concrete core).

The high continuous R-value of an ICF is compromised a small amount at the floor slab intersection in multi-storey buildings (Figure 4.19). Although the resulting enclosure is still high performance, floor slabs can reduce the overall thermal performance of a typical ICF enclosure by around 7-10%.

Figure 4.18: The high strength of ICF construction lends itself well to multi-storey buildings. The modest window area, good airtightness and continuous insulation of the apartment building shown means this is a high-performance enclosure.

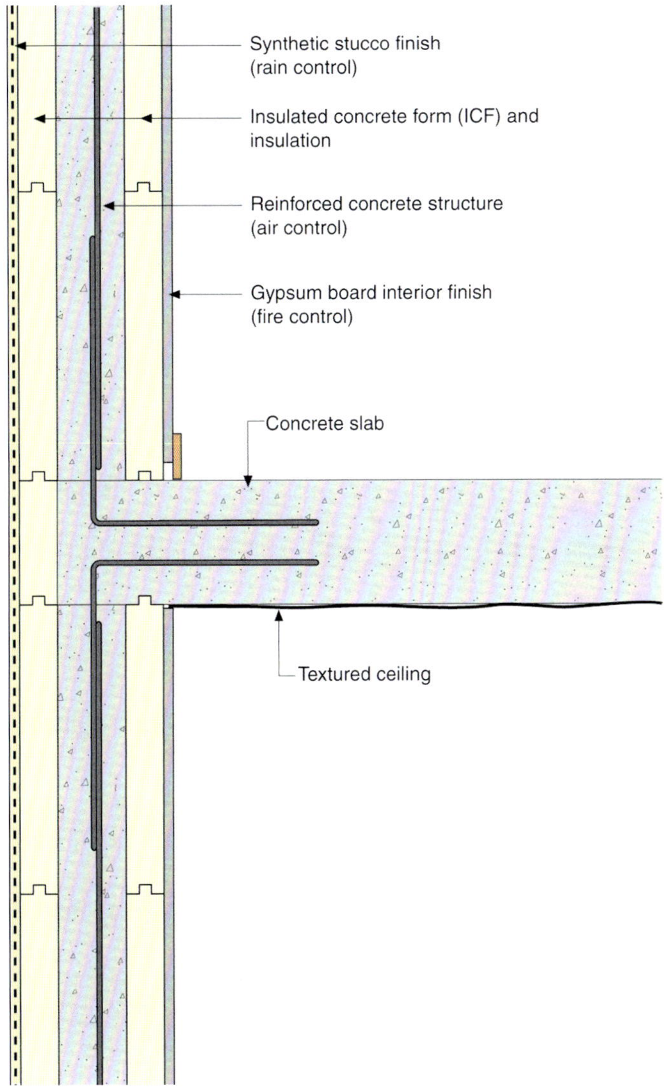

Figure 4.19: The floor structure penetrates the interior thermal control in multi-storey buildings.

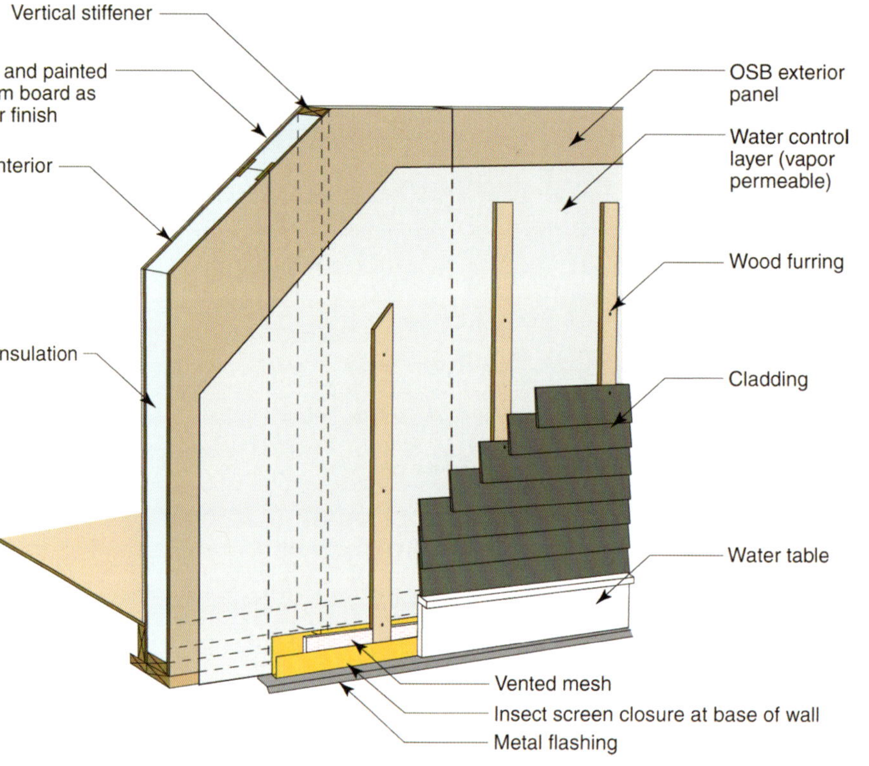

Figure 4.20: Structural Insulated Panel Systems (SIPS) are another new enclosure system. The foam core can provide the air and vapor control layer as well as the thermal control. A vapor permeable drainage plane is required (fluid applied or sheet goods), as is a ventilated space to encourage outward drying of the vulnerable outer layer of OSB. Joints must be air sealed both inside and outside.

Interior insulation is often installed on the interior of CMU, precast and reinforced concrete structures (Figure 4.21). These enclosures are highly problematic in cold climates because of air leakage condensation problems,[39] rain leakage at sealed joints and window penetrations and thermal bridging. Rain penetration can be managed by using two-stage drained joints. Air leakage condensation can be solved by applying board foam with sealed joints or spray foam direct to the concrete. Thermal bridging at large structural connections is more difficult to solve.

39
The foil-faced semi-rigid mineral fiber insulation commonly used in precast systems does not perform as an effective air barrier, or even a convective loop suppression in tall buildings.

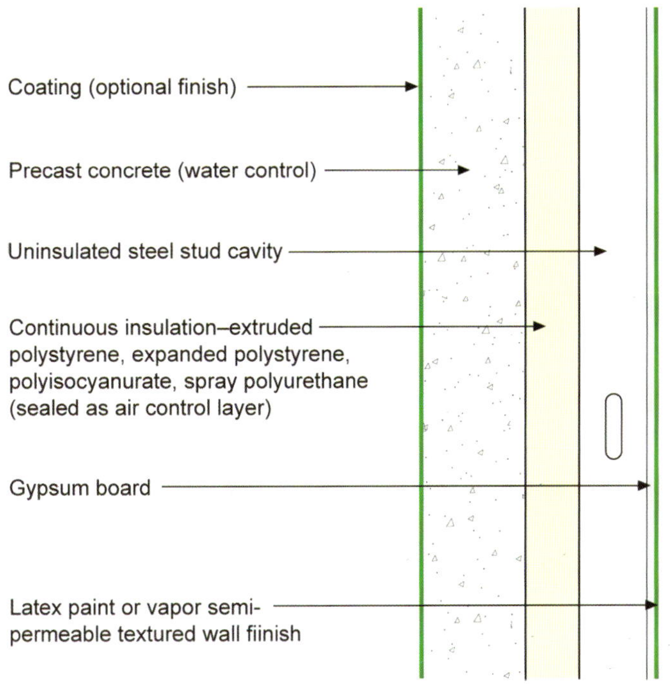

Figure 4.21: Insulation is often installed on the inside of tilt-up, precast and reinforced concrete structures. As a minimum, air- and vapor-permeable insulation should not be used.

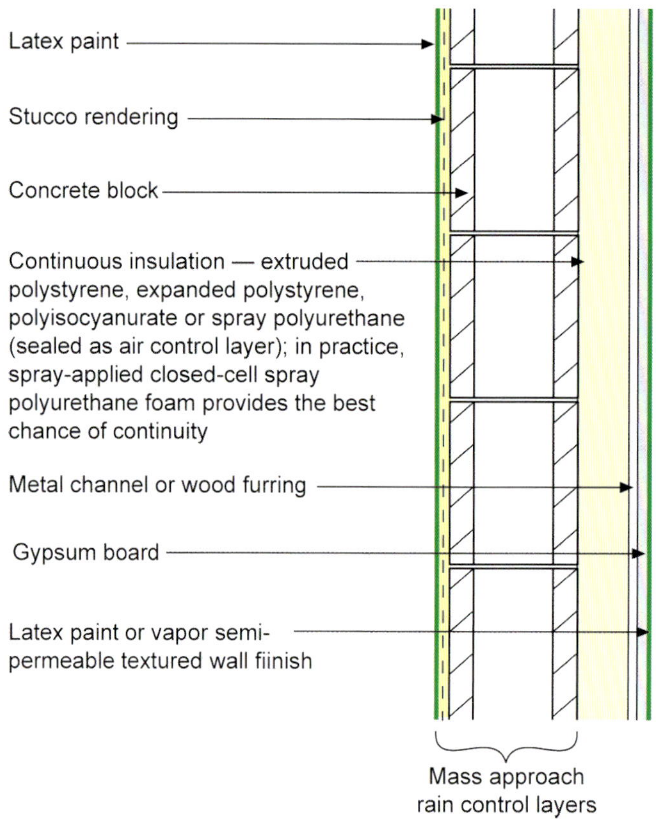

Figure 4.22: Interior insulation, if air-impermeable and vapor semi-permeable, can be successfully used to create a high-performance enclosure of CMU walls. Ensuring air, thermal and rain control layer continuity is the challenge, especially at structural penetrations and, hence, this assembly is limited to simple, usually one-story structures, such as strip malls.

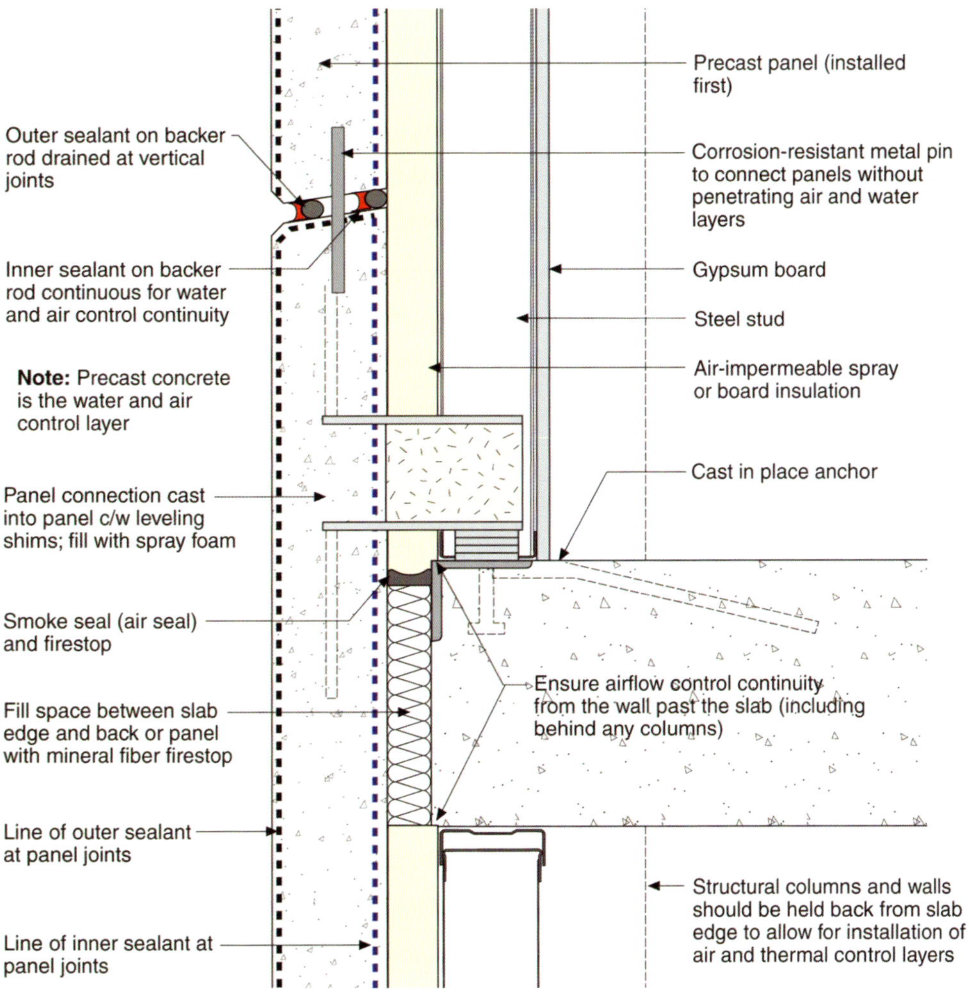

Figure 4.23: Architectural precast enclosure walls can be designed as high performance assemblies. The precast is a perfect barrier rain control layer with drainage used at the vulnerable sealant joints and window penetrations. Key to high performance is the use of continuous air-impermeable insulation as both thermal control and airflow control (foil-faced fibrous insulation has not been successful at controlling convective loops in practice).

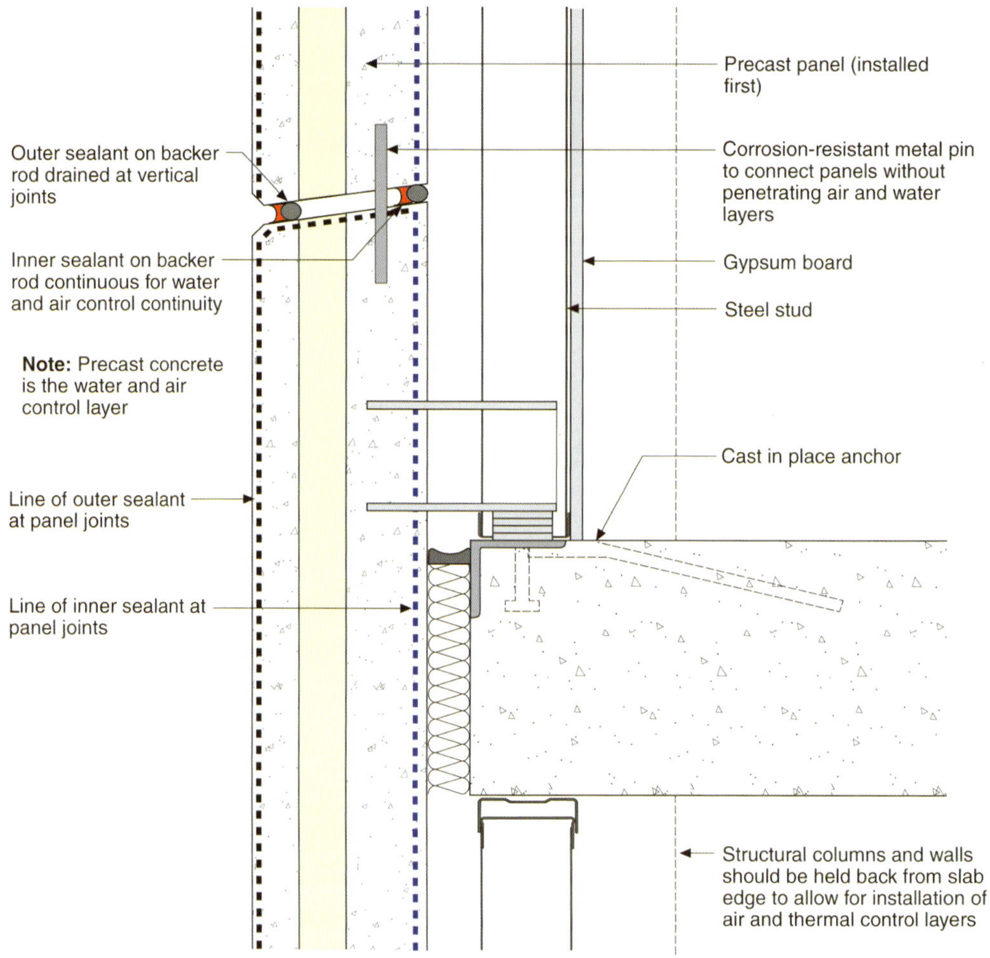

Figure 4.24: Precast concrete sandwich panels can easily provide a high-performance enclosure wall. The outer precast layer is a perfect barrier rain control layer with drainage used at all panel joints and through-wall penetrations at windows. The interior concrete layer is an excellent air barrier and vapor control layer. Continuous insulation is provided so long as connections between wythes are made of fiberglass ties or small stainless steel wire connections.

Curtainwalls are inherently low-performance enclosure systems. The thermal performance of most is a true R-value of less than 4, often only 3. High performance triple glazing, large thermal breaks (over $^1/_2$" or 13 mm) and limited amounts of framing can produce systems with overall R-values of 5 to 7 (U=0.8 to 1.2 W/m^{2}K). While solar heat gain can provide free heat in cold weather, the same solar gain can be problematic in many weather conditions if large areas are used.

The rain control strategy that should be used for all curtainwalls is the drained joint, perfect barrier system. Although many systems are available

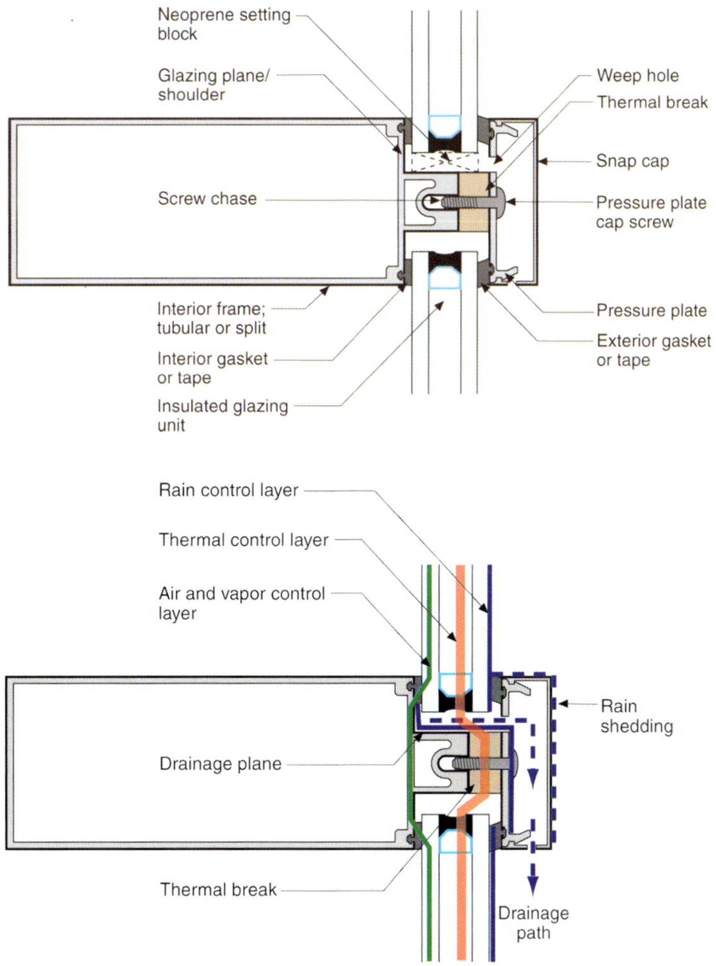

Figure 4.25: Curtainwalls can only approach high-performance if systems with large thermal breaks ($^1/_2$" or larger), exceptional glazing units, and drained rain control are used. Understanding the control layers is important to guide the design of connections.

as drained (even so-called "pressure equalized") the complexities of the systems are such that achieving a continuous drainage plane and drainage gap, especially at framing intersections, requires constant diligence.

The airtightness of curtainwalls, provided by the interior glass layer and aluminum, can be good, especially if a wet-applied interior sealant is provided. However, transitions and intersections of vertical and horizontal aluminum elements provide plenty of opportunity to allow uncontrolled air leakage. The air-vapor control layer should be provided by a strong, stiff metal backpan at spandrels: foil-faced insulation has been found to be quite unreliable in practice, and is not recommended.

Spandrel panels control solar heat gain very well, but the thermal resistance is rarely more than twice as high as a vision wall section even with R-12 to R-20 of installed insulation (Figure 4.26).[40]

40
Typical-sized thermally-broken aluminium framed spandrel panels typically achieve R-5 with R-16 of installed insulation. Although the R-value can be increased with interior-to-backpan insulation, the benefits are small (R-1 to R-2), the effort significant, and risks of problems increase. Choosing a higher performance product is a better approach.

The connections of the thermal, water, and air control layers within a curtainwall to those in adjoining enclosure assemblies is challenging, and should not be left to the curtainwall manufacturer during the shop drawing phase. Details should be developed for the construction drawings, and the control layers and the intended means of achieving continuity shown.

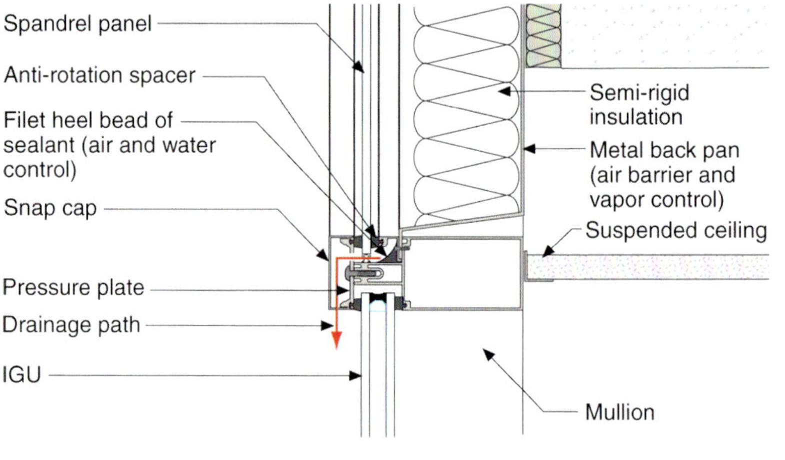

Figure 4.26: Although the insulation value installed behind spandrel glass is often rated at R-12 to R-20, thermal bridging through framing and via the metal backpan reduces the true R-value to between R-5 and R-7 depending on spandrel size, framing thermal breaks, and installed insulation value.

4.3 BELOW-GRADE ENCLOSURES

The thermal performance of below-grade enclosures is important: it makes little sense to have highly insulated enclosure walls and not insulate the basement and slab well. An additional challenge is the need to provide durable and healthy assemblies that control liquid water below and at grade, do not allow condensation (of humid air on cold below-grade surfaces) in winter or summer, and allow inward drying of built-in moisture, condensate, and capillary wetting.

Below grade assemblies have the advantage that the thermal mass of the soil significantly reduces the temperature differences between the interior and exterior. Hence, insulation levels for basement walls are typically about half of above-grade walls. Given that basements often have some proportion projecting above grade, higher R-values for these types of basements, at least over their upper parts, is recommended.

Slabs are currently the only part of the building enclosure that do not require insulating in most residential codes. In cold climates, this is usually a mistake for high performance enclosures. Although the R-value might only be a quarter of the above-grade values, even a small amount (R-5 to R-10) around the perimeter (4 to 10 ft or 1.2 to 3.0 m) can be justified. In climates with humid summers, continuous insulation under the slab is recommended to control the formation of condensation and/or high humidity at the slab, especially if carpets or furniture are placed on top of the slab.

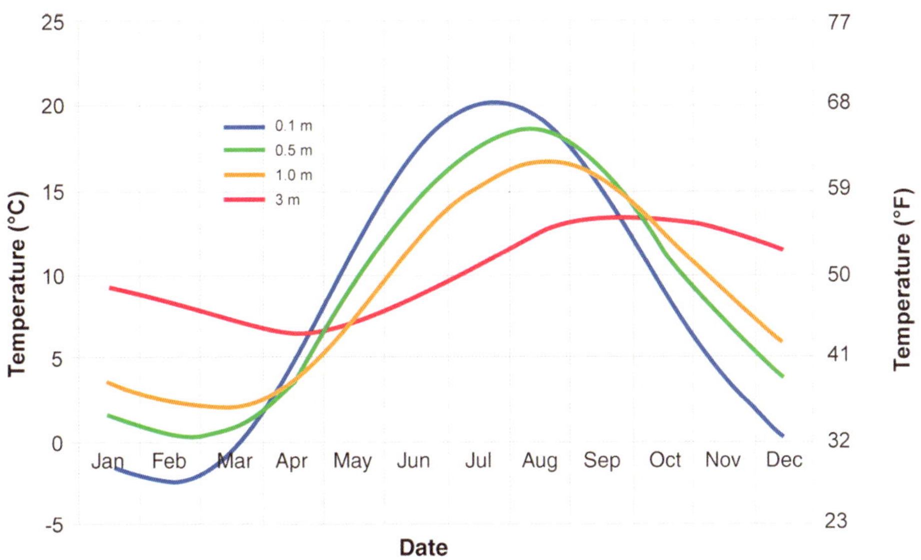

Figure 4.27: Typical ground temperature distribution for zone 5/6 climate.

Figure 4.28: Insulating below a slab-on-grade is relatively inexpensive and effective. Note the vertical stem wall insulation and vapor barrier in this photo.

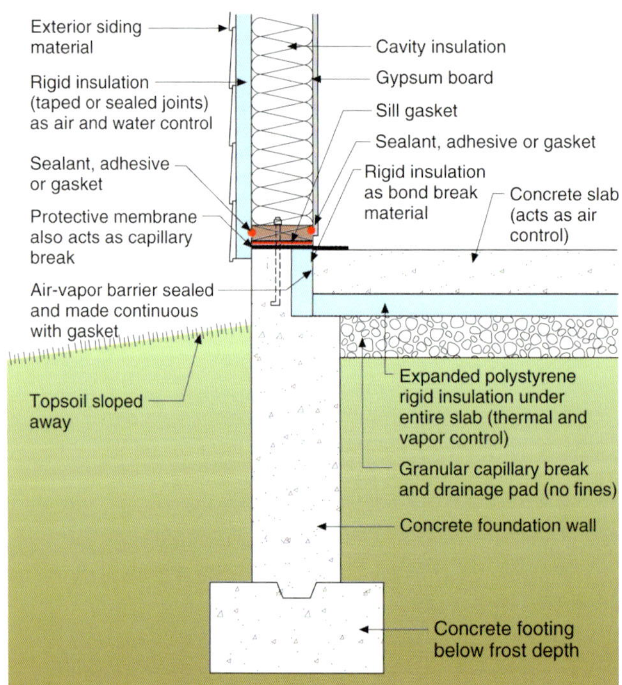

Figure 4.29: An insulated slab with a frost wall supporting a framed wall.

To control inward vapor drives and outward vapor drives, insulation with some vapor diffusion resistance and airtightness is required. (See *BSD-103: Understanding Basements* at www.buildingscience.com). In practice this means that at least some rigid foam insulation as part of the insulation approach is used as a vapor control layer. In many cases, spray foam or taped and sealed board foam is used to produce an airtight layer.[41]

41
Although concrete is an air barrier, it is critical that convection loops do not allow air from inside the basement to reach the cold basement wall where condensation can then occur.

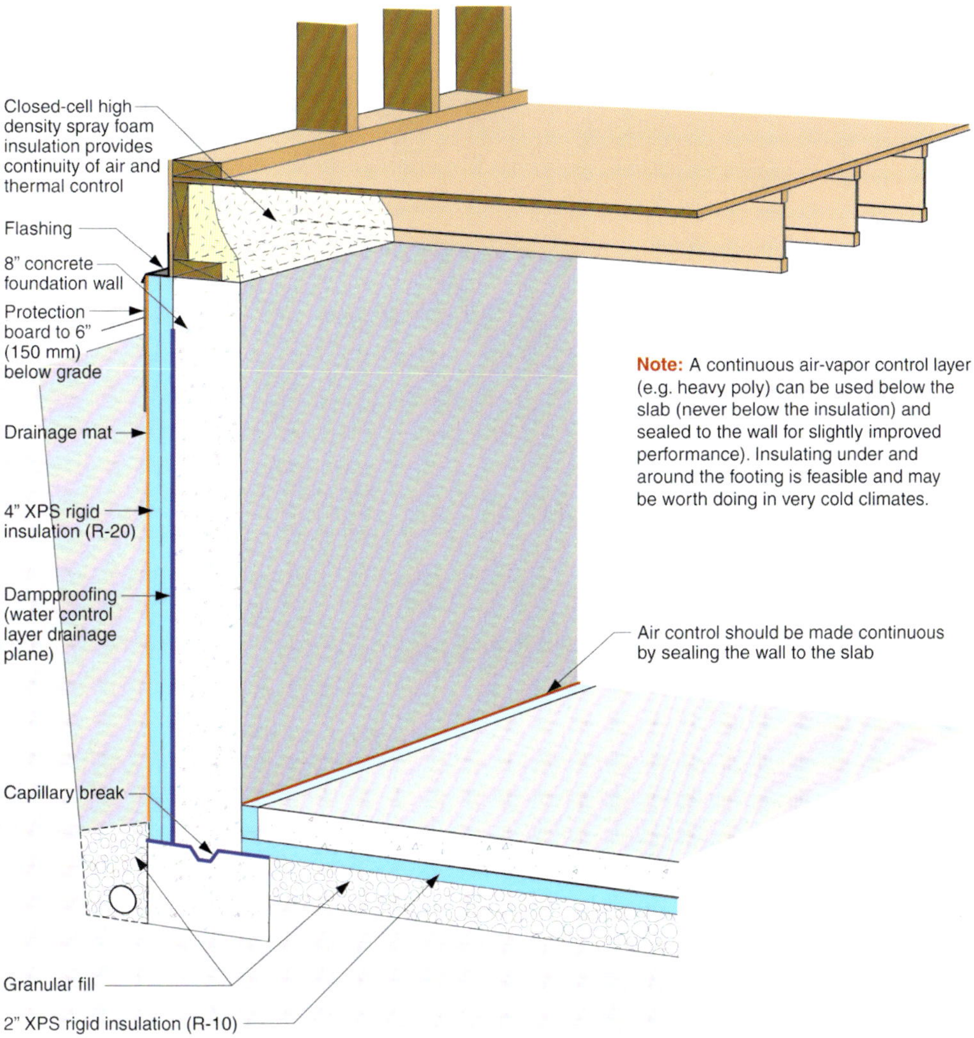

Figure 4.30: Basements that provide continuity of the above-grade perfect wall are usually preferred.

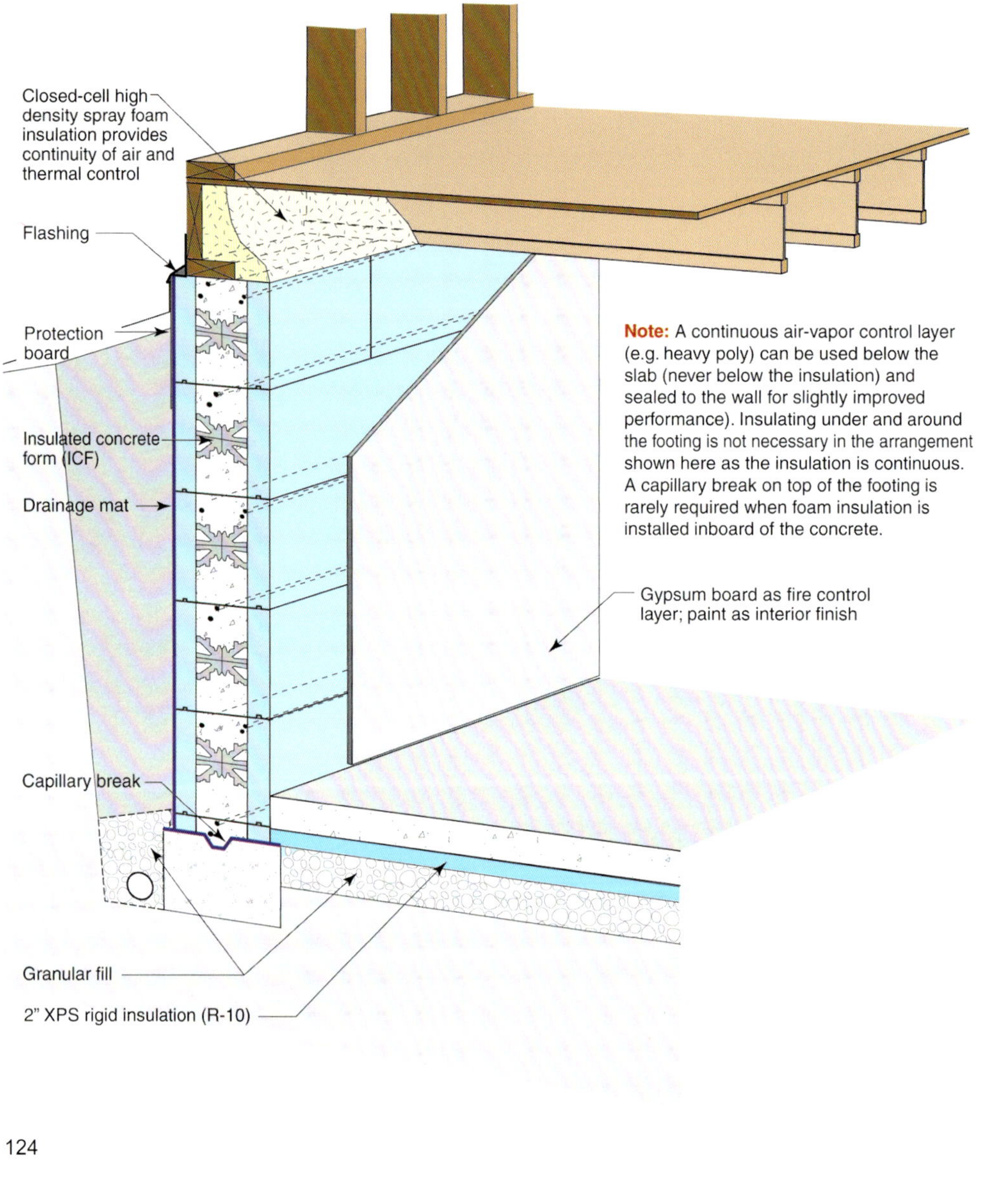

Figure 4.31: ICF Basement enclosure systems can be ideal high-performance basement enclosures.

Control of below grade liquid water is clearly a critical function. Waterproofing (perfect barrier approach) is not preferred as it requires excellent workmanship to succeed, but may be needed in areas with persistent high water tables. Drained approaches employ a drainage plane (often a spray applied "dampproofing" and a drainage gap (often "free-draining" soil, but preferably a drainage gap membrane, e.g. dimple sheet or drainage mat). The drainage plane and gap must lead to a drain tile and an exit. Drained approaches reduce hydrostatic pressures and keep the soil around the building drier.

Figure 4.32: An ICF basement with sub-slab insulation installed. The interior of the walls will be finished with gypsum board for fire control, and the floor slab will be poured on top, resulting in very good continuity of control layers from floor to walls.

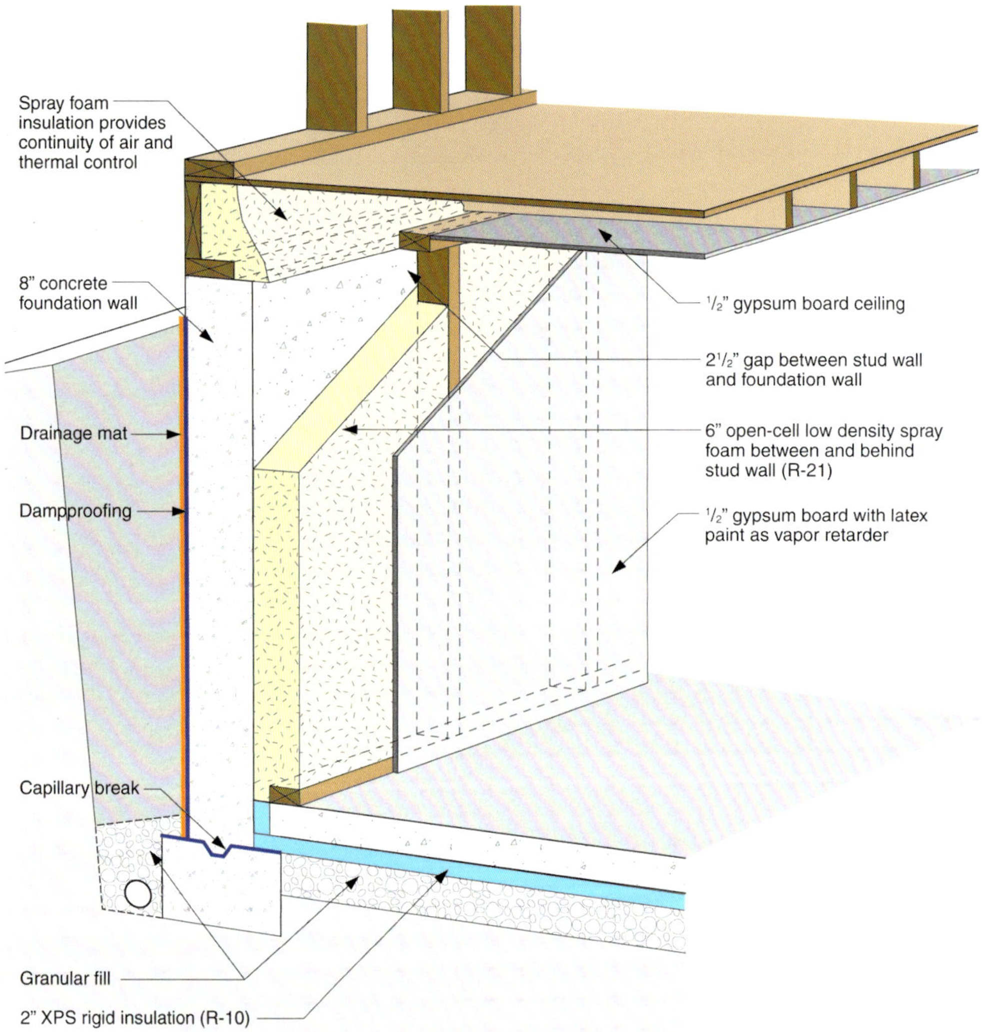

Figure 4.33: A basement assembly using open-cell foam to control air movement, vapor control and thermal control. This assembly is limited to climate zone 6 or lower and dry commercial interior spaces.

Figure 4.34: A basement assembly using closed-cell spray foam applied to the inside of an existing foundation, with thin steel studs installed to the interior to allow for service distribution and to apply fire-control layers and finishes.

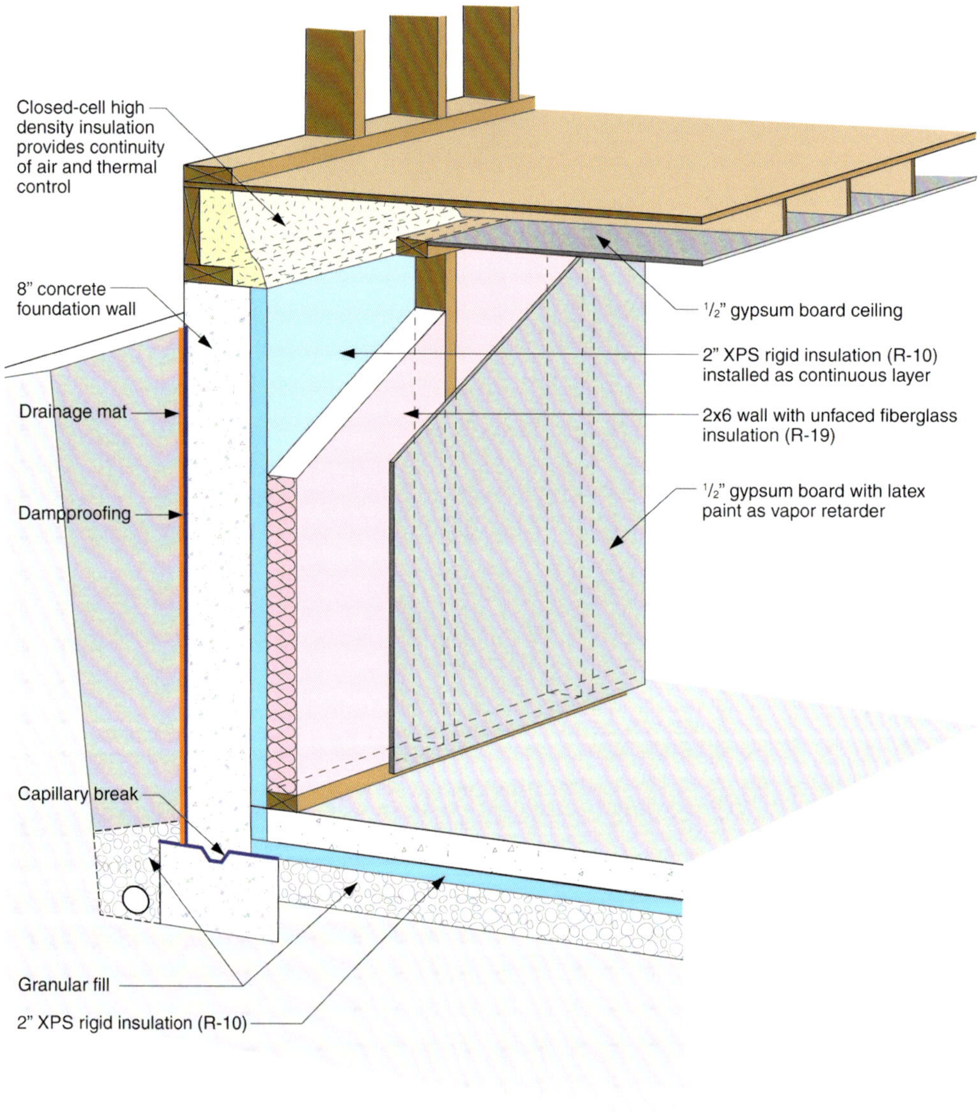

Figure 4.35: A hybrid insulation approach that can be successful in all climate zones (below the permafrost line). No interior low-permeance vapor control layer or finish should be used.

Photo courtesy of Stephanie Finegan

Figure 4.36: A basement assembly showing blown fiberglass behind netting over extruded foam board insulation.

4.4 ROOFS

Two broad categories of roof enclosure can be defined: pitched (more than about 1:12, and often 3:12 and higher) and low-slope. Low slope roofs are widely used over larger building areas and longer spans. Pitched roofs have many benefits but tend to be limited to shorter spans and smaller buildings.

4.4.1 Low-Slope Roofs

Low-slope roof systems control rain by using the perfect barrier approach: as such they are prone to failure. Exposed membrane roofs (face-sealed perfect barriers) have especially short lives, usually under 20 years before complete replacement is necessary. The other challenge with exposed membrane roofs is the location of the vapor control and air control layers (the roof membrane) on the cold and exposed side of the enclosure. Installing an interior vapor barrier then results in a sandwich in which any incidental moisture from construction, condensation, and small leaks is trapped.

Despite these major limitations, exposed membrane roofs are the most common low-slope roof. To achieve the best performance possible, an interior air barrier system, multiple layers of insulation, a cover board, and a fully-adhered or ballasted membrane are recommended (Figure 4.39).

Inverted (IRMA) or protected membrane roof (PMR) systems locate the primary water, air and vapor control layers below the insulation. They have many advantages: the primary one being that the moisture, air, and vapor control layers are protected from UV radiation, temperature swings, and mechanical damage, thus ensuring excellent durability. Another advantage is that the air and vapor barriers are on the inside of the insulation, and no vapor trap is created. The two primary disadvantages of such roofs is that they require ballast (to prevent the insulation from floating) and limit the insulation choice to extruded polystyrene. To achieve the full performance potential of PMR, several aspects are important: a drainage gap over the membrane and under the insulation,[42] it must be sloped, and a high quality geotextile should be installed over the insulation and under the ballast (this should be made continuous to avoid the entry of dust that can form mud in the drainage gap over decades).

The slope of all low-slope roofs should be indicated on the design drawings and 2% is recommended as good practice. Higher slopes are better, although some roof membranes will creep and slide at slopes of 4% or more. The drains should be located in the middle of the roof span, since this is where the roof will sag structurally under load.

42
This drainage gap is often resisted by insulation manufacturers, but without it, experience has shown that a film of water will always be in contact with the foam insulation on its underside, and then slowly work its way into the extruded polystyrene insulation over 15 to 25 years driven by cyclic temperature gradients. To extend the life of the insulation to that of the membrane, a drainage gap of around ⅛" (3 mm) is recommended. The gap should be sufficiently sealed at all penetrations to avoid the potential for wind washing (pretty limited, but real) to erode the thermal performance.

Air barriers are important for all enclosure systems and roofs are no exception. Although roof membranes are watertight air leakage can still occur at parapets, HVAC curbs, expansion joints, and other penetrations. Hence, all high-performance low-slope roof assemblies should use a fully-adhered air barrier membrane at the deck level (or at least below most of the insulation value). Roof membranes can be excellent air barrier systems as they are strong continuous and air impermeable. However, loose-laid or mechanically-fastened membranes can billow in the wind–the resulting pumping action can draw significant amounts of air into the roof, causing condensation and energy loss (see also Figure 3.17).

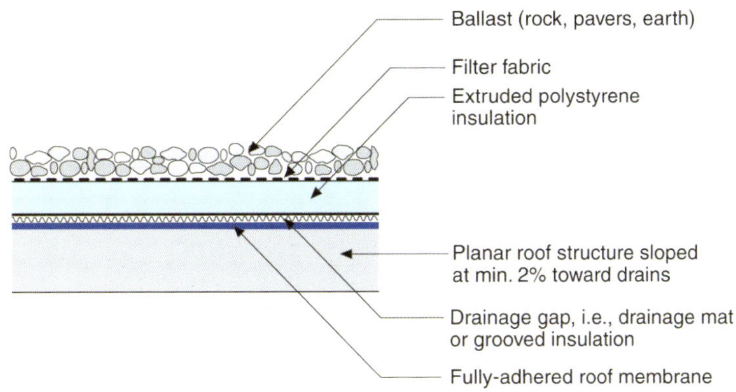

Figure 4.37: The protected membrane roof can be used with ballast of light-colored or pavers (both reflective thermal mass) or soil and plantings.

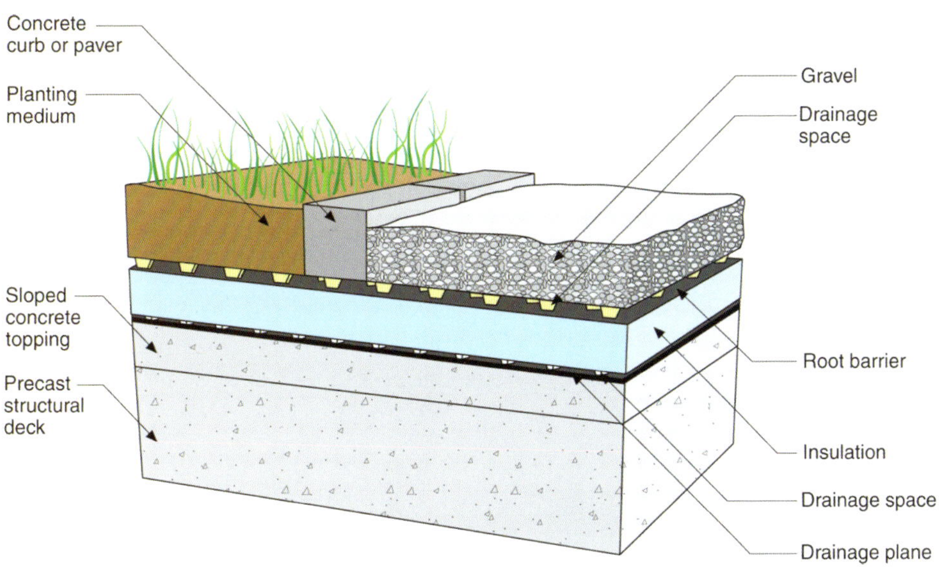

Figure 4.38: The protected membrane approach accommodates different finishes (plantings, pavers, river rock) with ease and is preferred for green roof systems.

In most commercial occupancies (that is, those in which the interior dew-point temperature is kept below 40°F/5°C during winter) in zone 5 or lower, a deck level vapor control layer is not needed, as foam insulation provides the required vapor control. To ensure that vapor control is provided at insulation joints, it is important that a least two layers of insulation with the

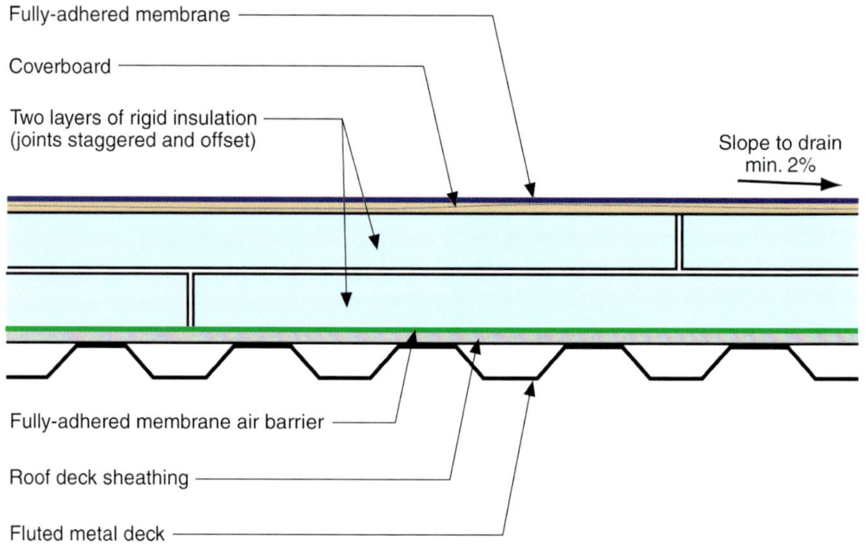

Figure 4.39: A common low-cost, high-performance compromise low-slope roofing system. The structure should be sloped to drain, an air barrier provided at deck level, and installed insulation in at least two layers with joints offset.

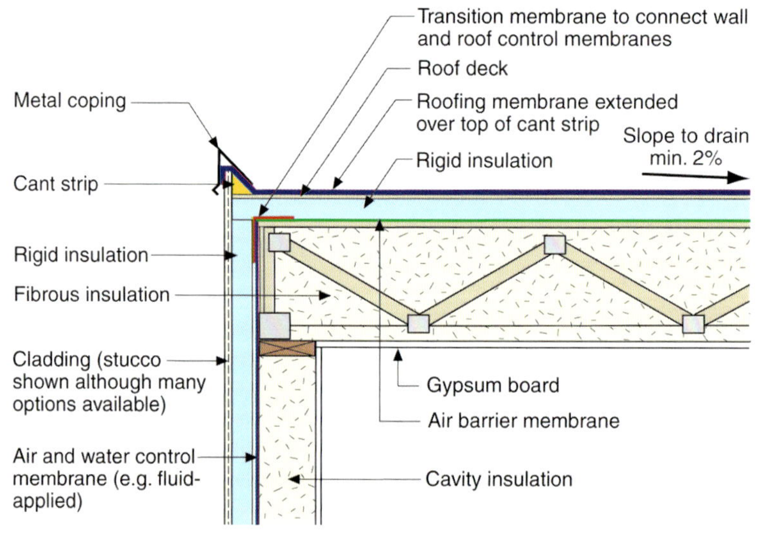

Figure 4.40: Although it is always desirable to have continuous insulation outside the structure, in the case of wood trusses it is sometimes acceptable to tolerate the small thermal bridges that result. The ratio of the exterior insulation R-value to fibrous cavity fill must be chosen carefully to respect the rules of vapor control.

joints offset and staggered be used. In colder climates and higher interior relative humidity environments, a vapor control layer is usually recommended. Concrete decks (especially those poured on metal decks) can be a major source of construction moisture and may require a vapor control layer even in warmer climates.

4.4.2 Pitched Roofs

Ventilated, pitched roofs with the insulation at the deck level are commonly used in housing but are applicable to many commercial buildings as well (Figure 4.41). These roofs have the advantage that they are very easy to insulate to very high levels because low-cost blown fibrous insulation can be used. The ventilation of the underside of the roof does little to reduce the roof deck or membrane temperature: its purpose is to remove small amounts of moisture that may enter the roof during construction, because of ceiling air leaks, or even because of small roofing leaks. Ventilation can remove the small amount of heat that may conduct through a good insulation layer, and thereby be effective for preventing ice-dams. Ventilation can rarely solve ice-dam problems caused by air leaks, ductwork in the attic space, or insufficient insulation.

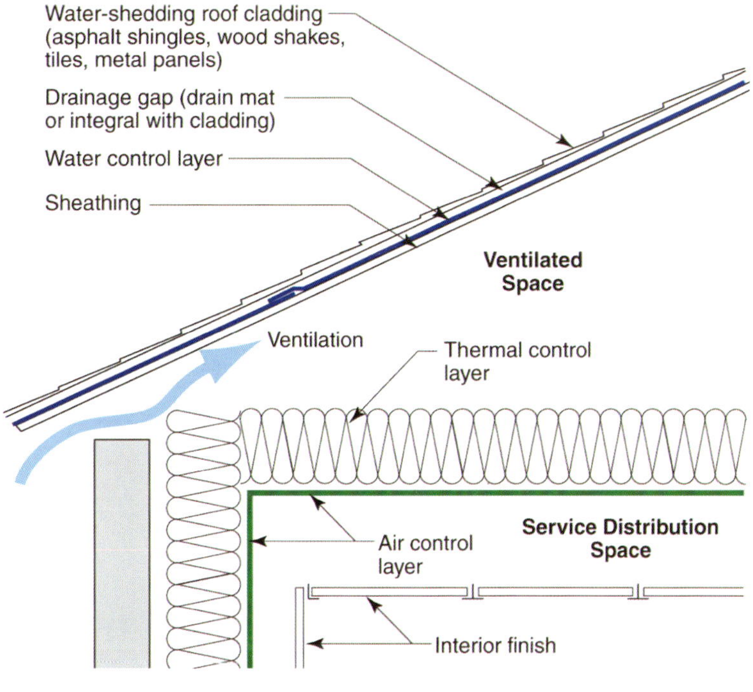

Figure 4.41: Conceptual drawing of ventilated and pitched roof system.

The design and construction challenge of pitched roofs is to ensure that the ceiling plane (the air control layer or the plane of airtightness) is built in a very airtight manner. Practical experience with less than perfect airtightness has driven some designers to abandon this type of roof. However, if airtightness can be achieved, performance will be excellent. Penetrations at partitions, wiring, lighting, etc. must be sealed as airtightly as possible. In no case should ductwork or airhandling units be allowed in the attic space.[43] To achieve exceptional airtightness, one solution is to install a thin (e.g., 2"/50 mm) layer of spray foam[44] from above onto the gypsum board and then apply R-50 to R-100 of blown insulation.

43
The exception to this rule are airtight ducts that are exhaust only (and hence there can be no energy implication to poor duct insulation or small air leaks). Large leaks in exhaust ducts can deliver significant amounts of moisture into the attic space and overwhelm ventilation.

44
Open-cell or closed-cell foam can be used as both are air impermeable. If vapor control is required, either closed-cell foam of sufficient thickness or a paint coating can be used. A sheet of polyethylene could be used, but this eliminates the potential for bonding the spray foam to the gypsum board, which slightly increases the risk of air leakage through flaws in the foam layer.

The thickness of loose fill insulation (often 24"/ 600 mm) required to achieve high R-values will generally require a "raised heel" truss design that maintains nearly the full insulation thickness where the roof rests on the wall. It may not always be economical to provide a very high heel, e.g., over 12"/300 mm, and in these cases, spray foam insulation can be used at the perimeter to take advantage of its higher R-value per inch. For example, a 12" heel height with 10" of insulation will provide "only" about R-38 with loose fill, but around R-60 with ccSPF (see Figure 4.43).

Pitched roofs can also be used in applications where the attic space is part of the conditioned volume of the building: this can be occupied space or space for mechanical equipment and ducts. Although more expensive to build, unvented cathedral ceilings can be constructed which follow the same building science as low slope roofs (Figure 4.45). It is again critical

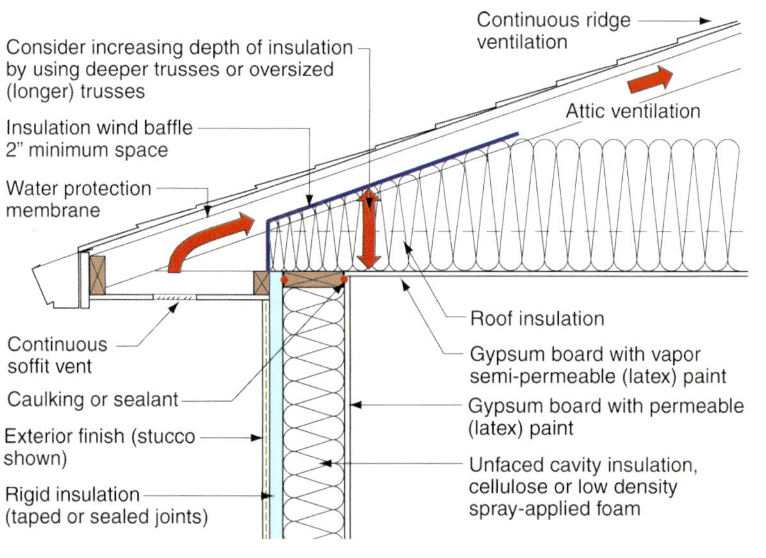

Figure 4.42: A pitched ventilated roof. These economical roofs require exceptional airtightness at the ceiling air barrier plane to perform well. Recessed lights, wiring, partitions, plenums, etc are all sources of leaks that need to be sealed.

that airtightness be provided at the interior: practical experience has shown that very high levels airtightness cannot be easily or reliably provided by sealed polyethylene or even using the airtight gypsum board approach (both of these can deliver sufficient airtightness for ventilated roofs), but can be achieved with spray foam and peel-and-stick membranes.

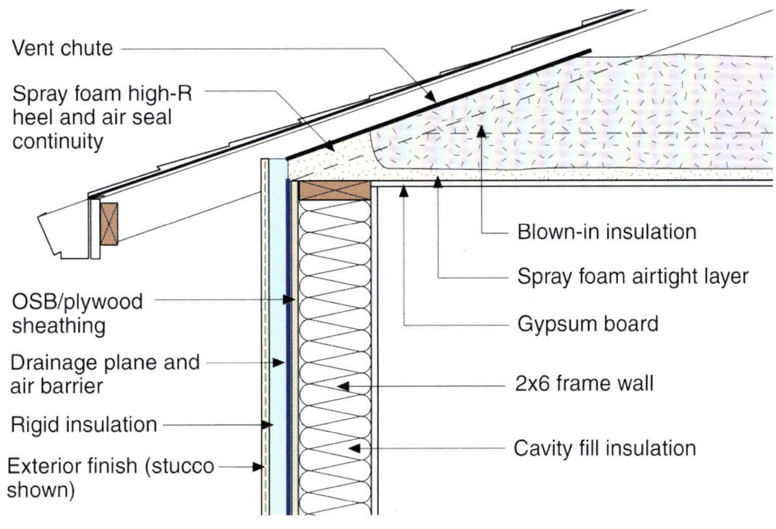

Figure 4.43: To achieve air control continuity and increase R-value at lower-than-desired heel height systems, closed-cell spray foam can often be used.

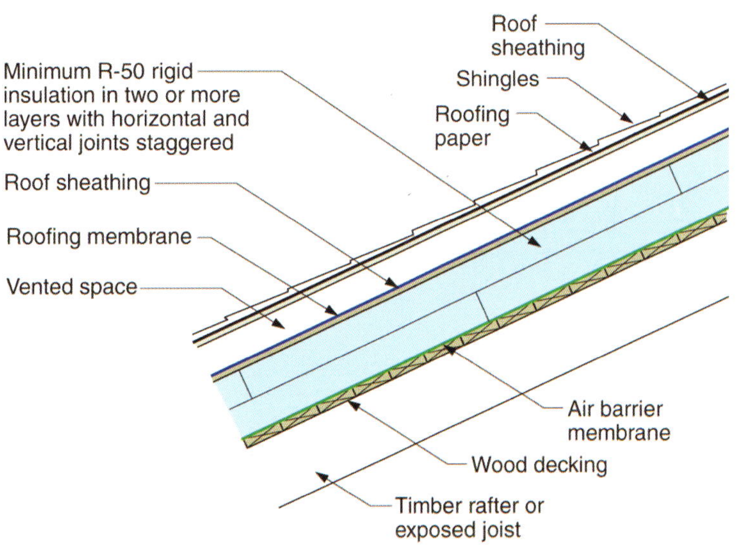

Figure 4.44: A vented-unvented hybrid compact pitched roof assembly. Suitable for all climate zones, but necessary only in high snow load areas which experience serious ice dam problems.

By using a combination of air- and vapor-permeable insulation, hybrid unvented roof solutions are possible whether for pitched or low-slope applications. The ratio of the R-value of the air-impermeable and vapor impermeable insulation to the R-value of the fibrous insulation (see the earlier vapor control section) is critical to success.

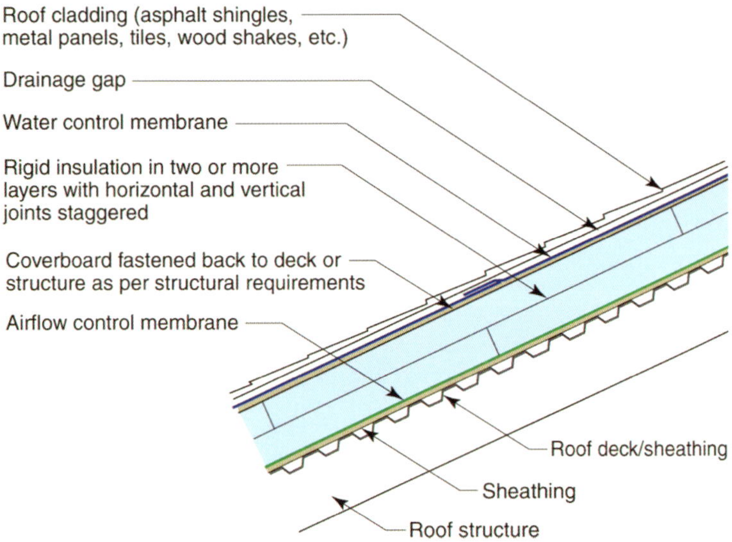

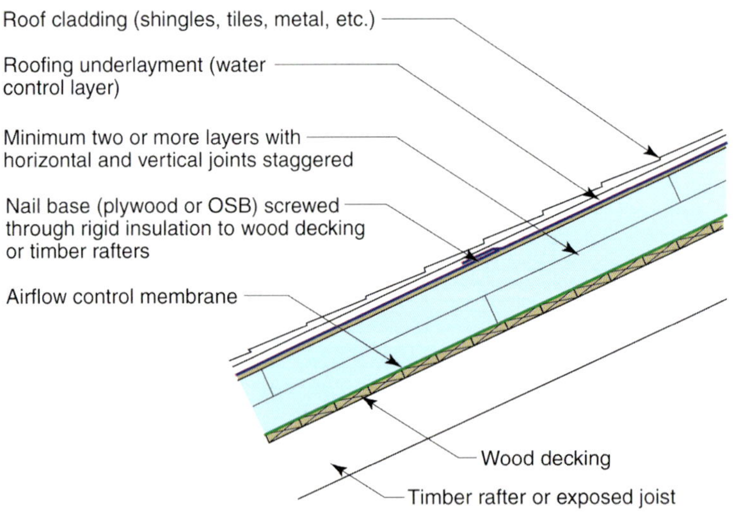

Figure 4.45: A pitched roof using perfect enclosure principles. No ventilation or special measures are necessary if similar approaches and materials to low-slope roofs are used, provided ventilation of the roof deck is not needed for ice-dam prevention in extreme climates. The air control layer on the deck is critical to successful performance.

Figure 4.46: A construction photo showing some of the layers of an unvented externally insulated roof. The lower level peel and stick is the air control layer, polyiso provides the thermal control layer, and the top OSB layer provides the nail base as in standard roofs.

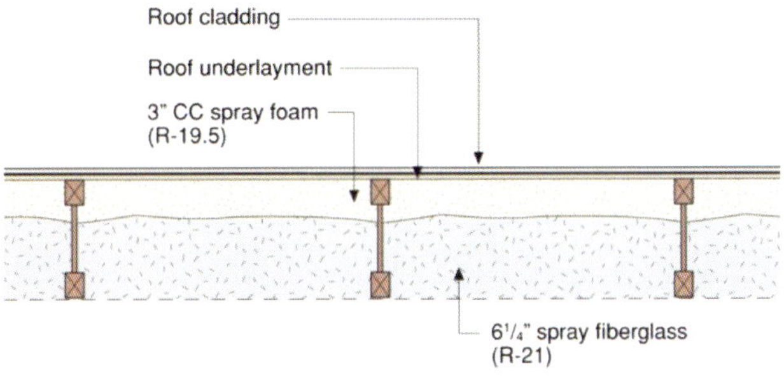

Figure 4.47: An example of a hybrid compact (unventilated) sloped roof. This roof could be a low-slope roof with the proper membrane as the physics are no different. Thermal bridging limits the performance, but thicker layers of insulation (of course, the proper ratio of air and vapor impermeable insulation to air and vapor permeable insulation must be maintained to control condensation) can be used to reach levels of R-60 or higher.

Figure 4.48: A spray foam application to an unvented pitched roof. The SPF provides the air, thermal and vapor control layers, but is interrupted at each of the rafters. Careful sealing of all wood-to-wood joints can reduce the risk of air leakage, but only layers of exterior insulation will reduce the thermal bridging.

Chapter 5 Enclosure Details

The transition of walls to roofs (parapets), walls to grade, and penetrations such as windows, balconies, and canopies consume most of the detailing time in design offices. This effort is justified as it is at these locations in the building enclosure where most performance problems occur. The problems are usually due to a lack of control layer continuity: i.e., thermal bridging, water and air leaks.

The following pages provide a few examples of how to provide continuity at some of the more common details.

5.1 ROOF TO WALL TRANSITIONS

The parapet is a common location of water and air leaks, and is often a thermal bridge as well. Ensuring continuity of the control layers from the wall to the roof requires more than just good detailing, but due consideration of construction sequencing and the division of trade responsibilities as well (Figure 5.1).

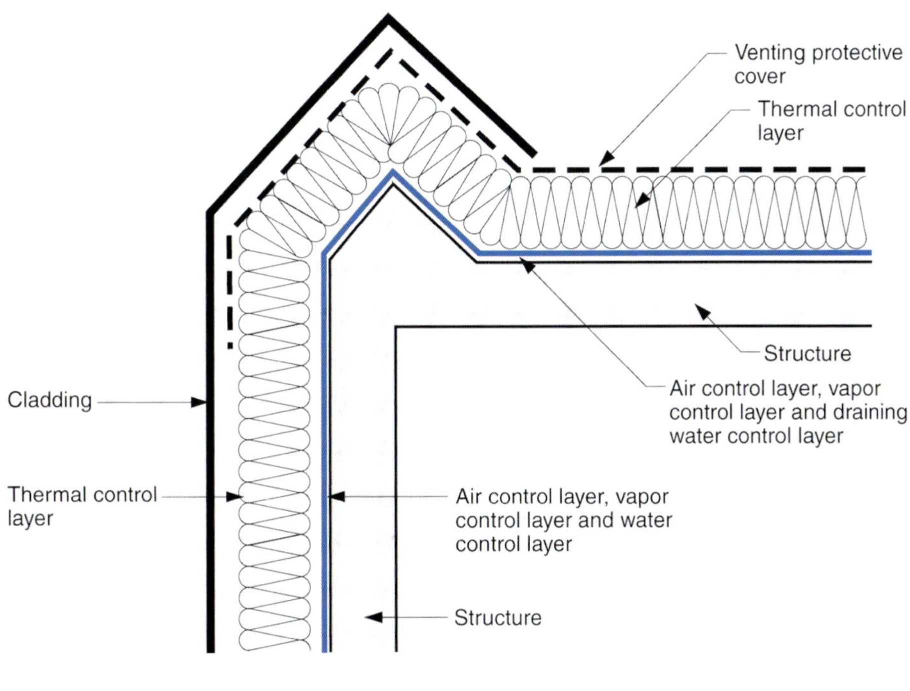

Figure 5.1: An idealized parapet (wall-to-roof) detail demonstrating the continuity of control layers (inspired by Max Baker).

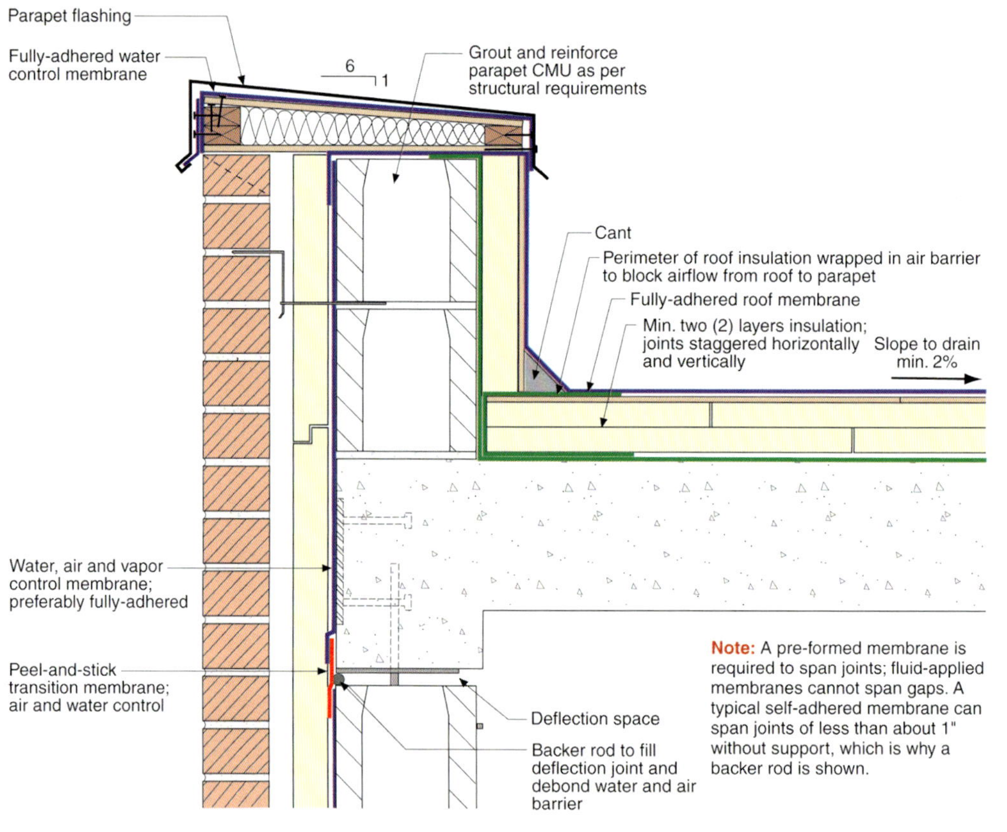

Figure 5.2: A typical parapet detail for an institutional project. Note the care in detailing the deflection space and parapet cap. Provided the parapet is tall enough, less heat will flow vertically than horizontally and so less insulation can be used on top of the parapet than the walls.

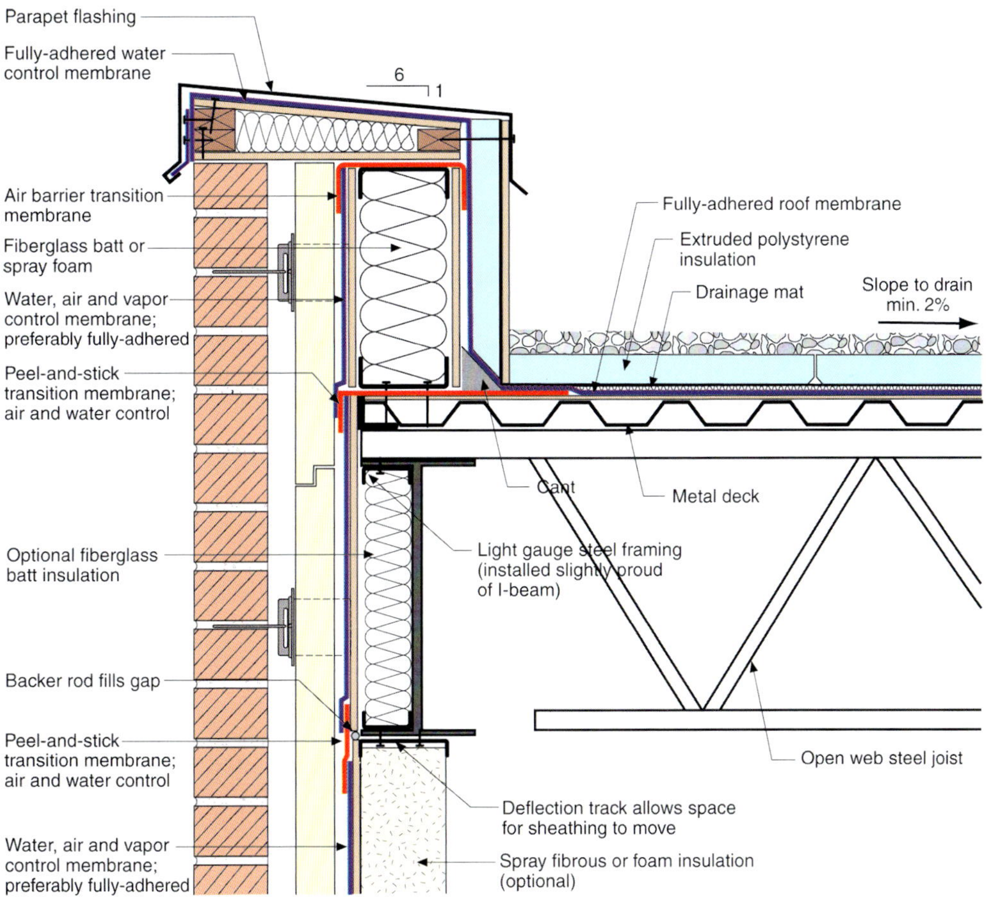

Figure 5.3: An inverted roof design allows both the designer and builder to easily ensure continuity of the water, air, and vapor control layers (the fully-adhered roof membrane) while also protecting these layers from extreme environmental conditions.

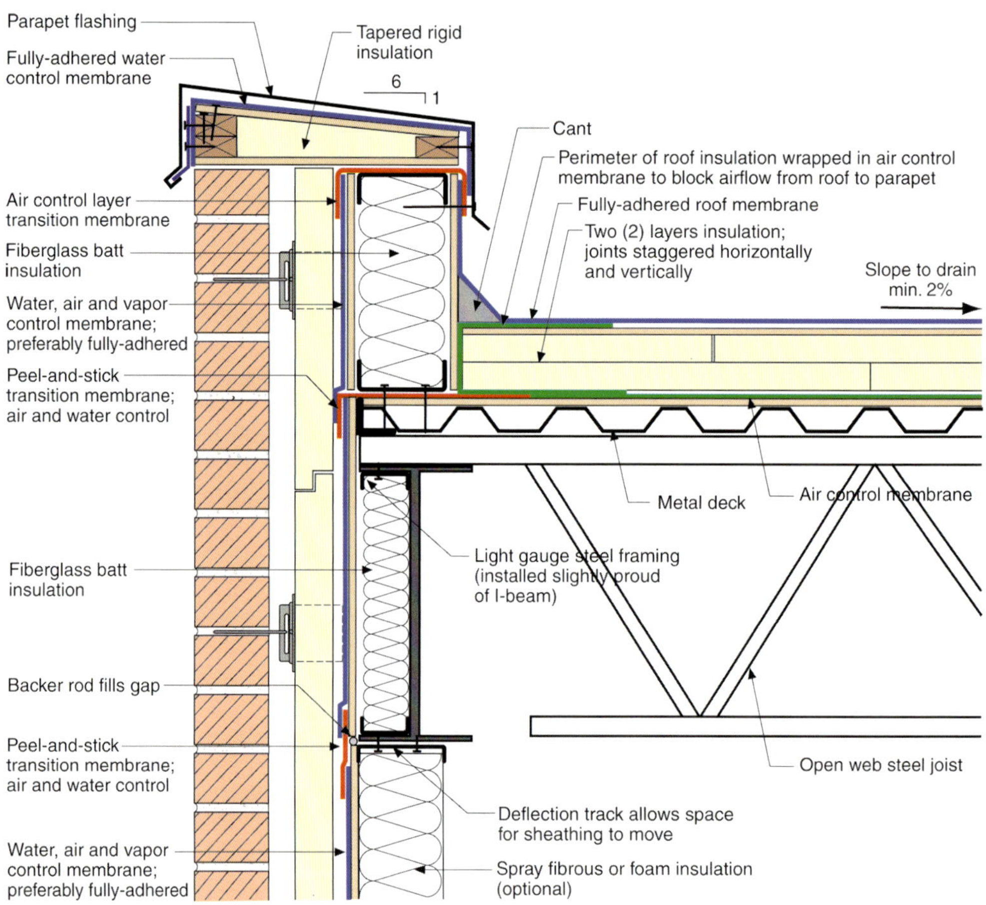

Figure 5.4: Exposed membrane roof detail at brick-veneer clad steel-stud wall in a steel-framed building. Note the continuity of the four control layers and the laps of membranes.

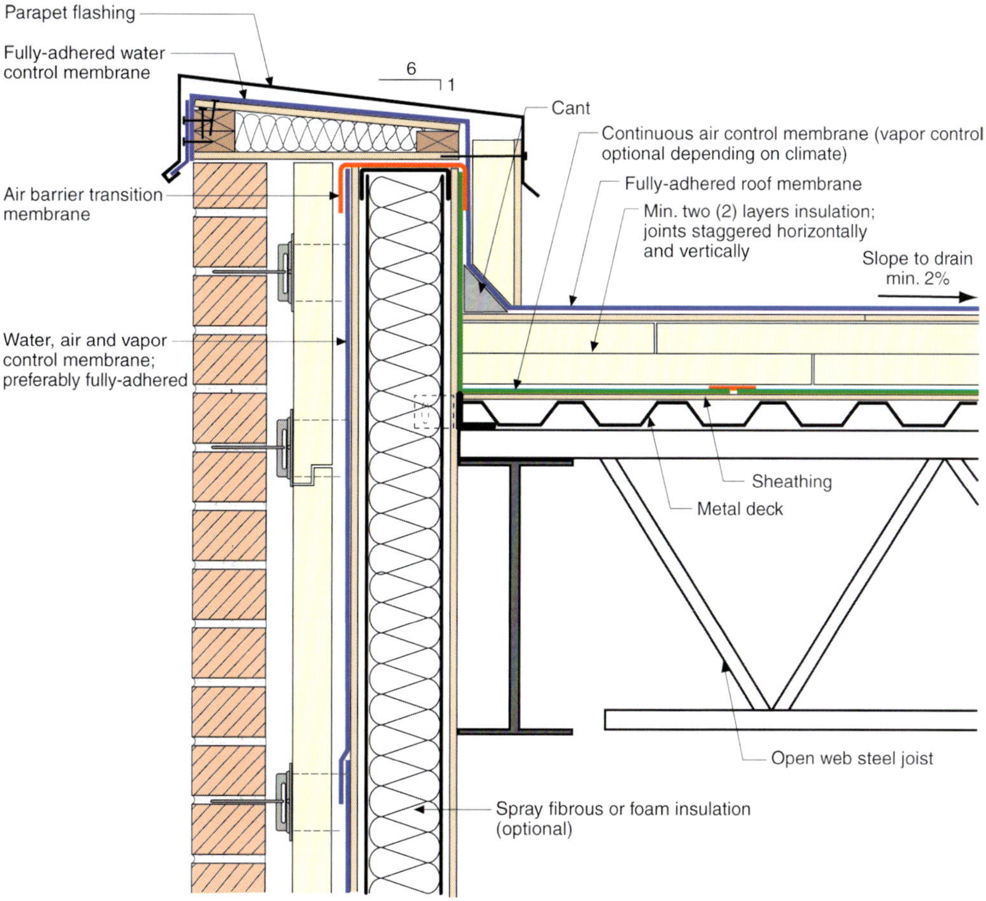

Figure 5.5: Extending the structure upward as a cantilevered parapet is possible, but great care must be taken to ensure continuity of the air and thermal control layers.

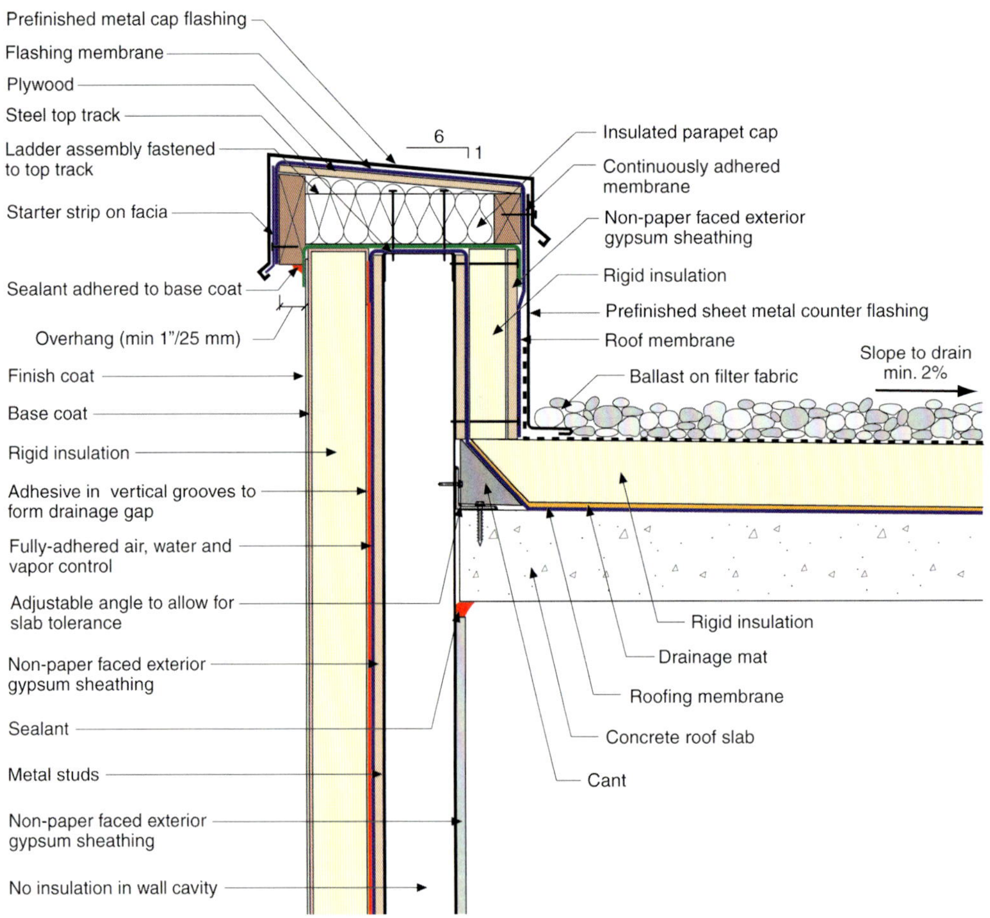

Figure 5.6: EIFS on a cantilevered parapet joining a protected membrane roof.

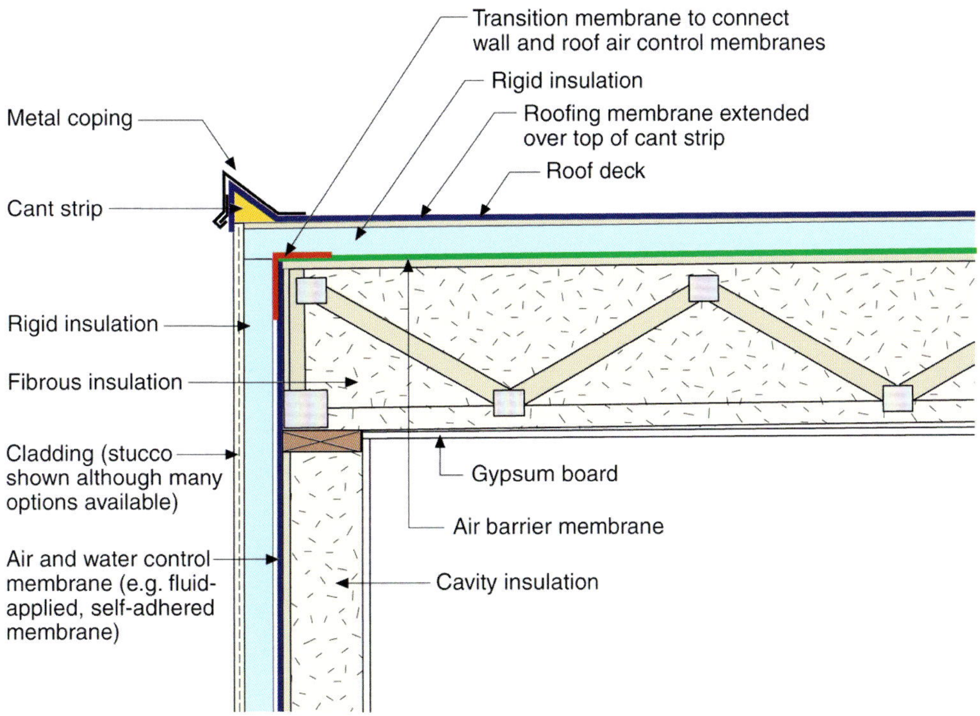

Figure 5.7: Low slope roofs can be built on wood structures to achieve high thermal resistance levels. Provided air and water barrier continuity is provided, the result is a high performance enclosure system.

5.2 AT GRADE TRANSITIONS

The transition from above-grade to below-grade walls is challenging for a number of reasons. Usually two different enclosure assemblies join here, and the design must take account of liquid surface water, melting snow, capillary wicking from soil, impact damage from use and landscape maintenance, etc.

If the exterior insulation "perfect wall" approach is used, it becomes relatively easy to continue the other control layers from the above-grade system to that below grade (Figure 5.8).

In those enclosures that use interior insulation, it too should be continued downward to the footing and connect with the sub-slab insulation.

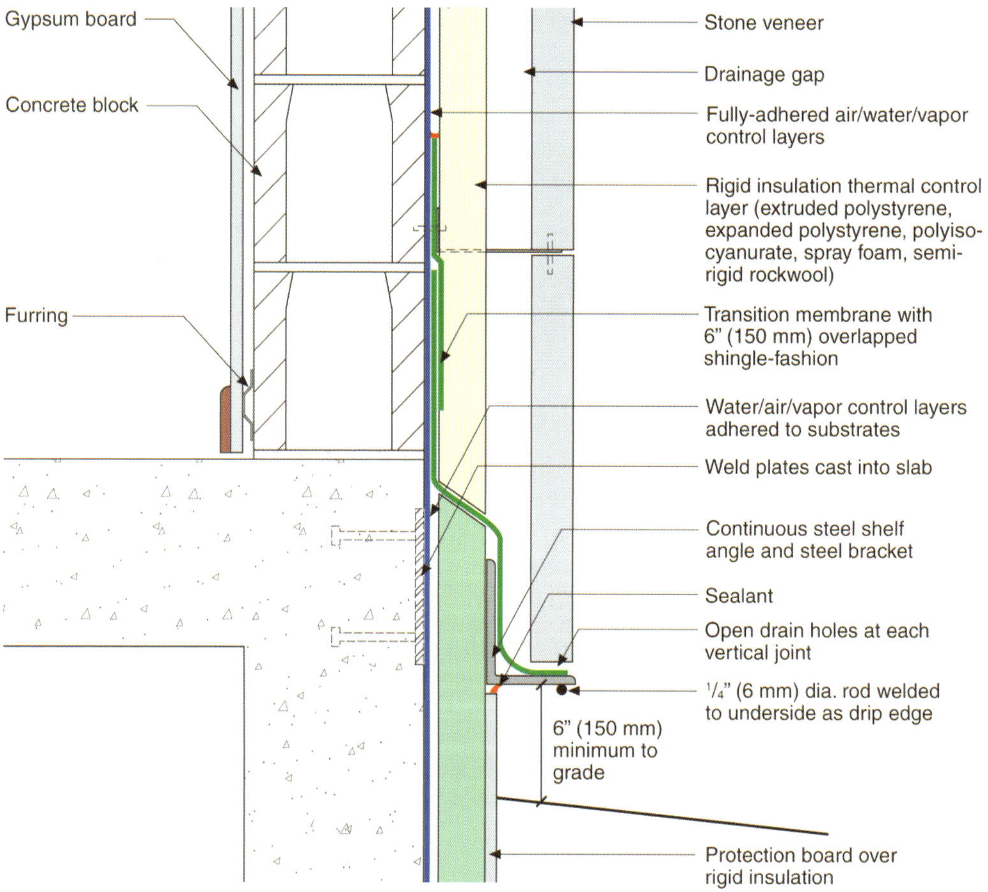

Figure 5.8: This base detail, as suitable for wood or steel construction as for CMU or concrete, allows for continuity of the control layers at this critical junction.

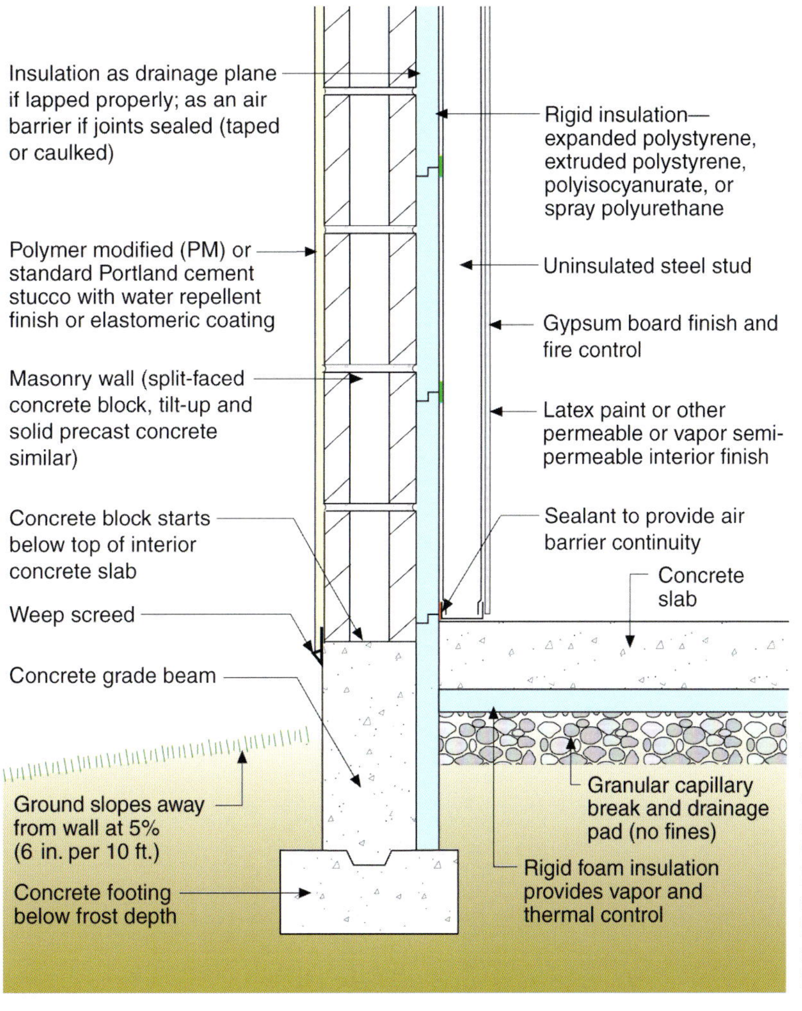

Figure 5.9: Load bearing assemblies on the exterior such as CMU and tilt-up can be insulated on the interior. The base detail shown provides for continuity of the control layers, but thermal bridges will exist at the roof to wall structural connection. Note: there is no need to support the slab by anything more than the soil and insulation below it.

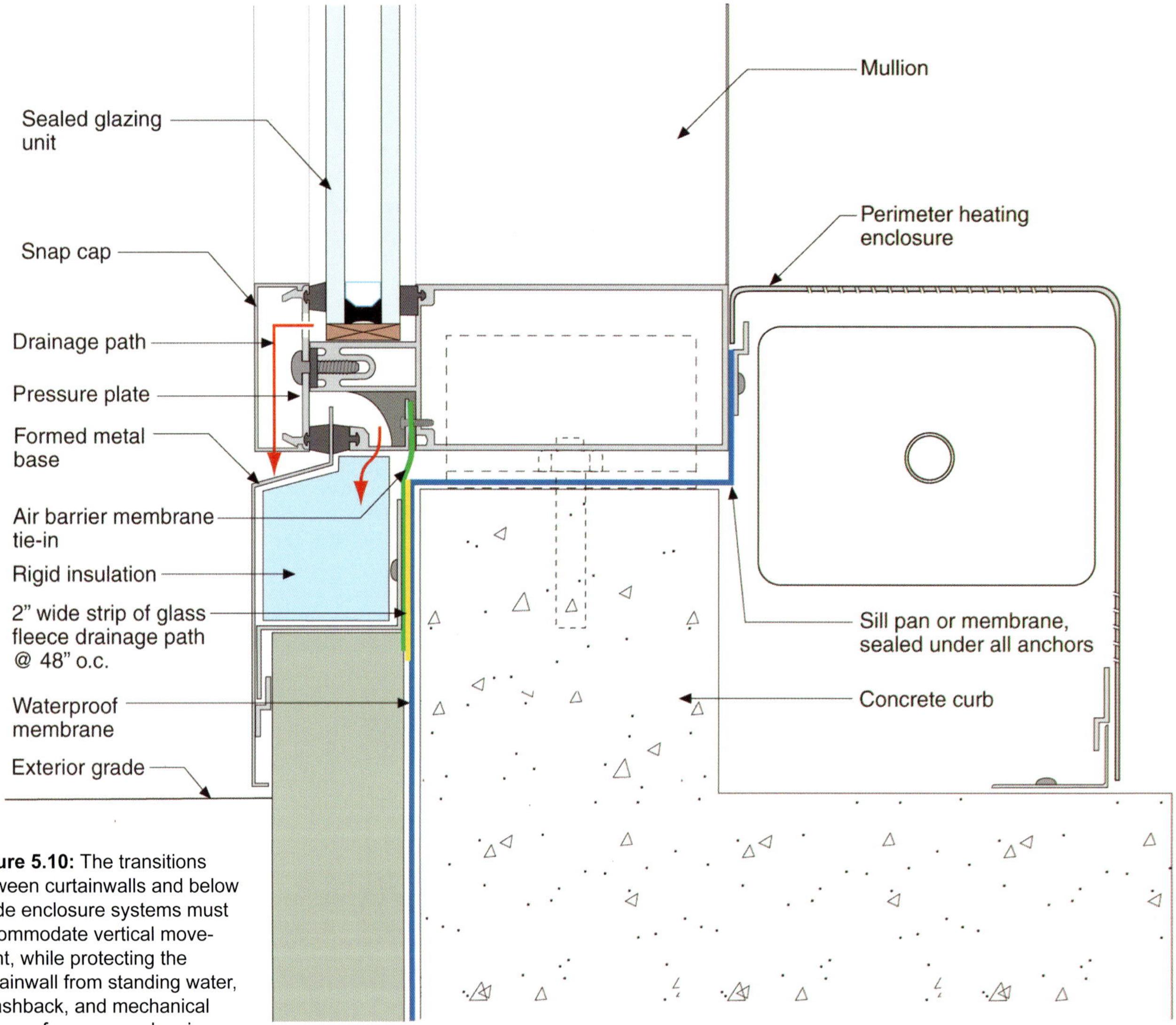

Figure 5.10: The transitions between curtainwalls and below grade enclosure systems must accommodate vertical movement, while protecting the curtainwall from standing water, splashback, and mechanical damage from use and maintenance typical of at-grade locations.

5.3 WINDOW AND DOOR PENETRATIONS

One of the most challenging penetrations to detail and construct is a punched window or door. The thermal, water, and air control layers in most products are not well defined, and hence maintaining continuity is especially challenging.

The water control layer of window and door products should generally be assumed to be at their innermost surface, as leakage of the unit, hardware or installation is a common occurrence. The rough opening should be drained to remove water that might penetrate. Drained flashing at the sill and two-stage sealant at the head and jambs are the common and recommended approach to ensuring control of rainwater penetration. It is

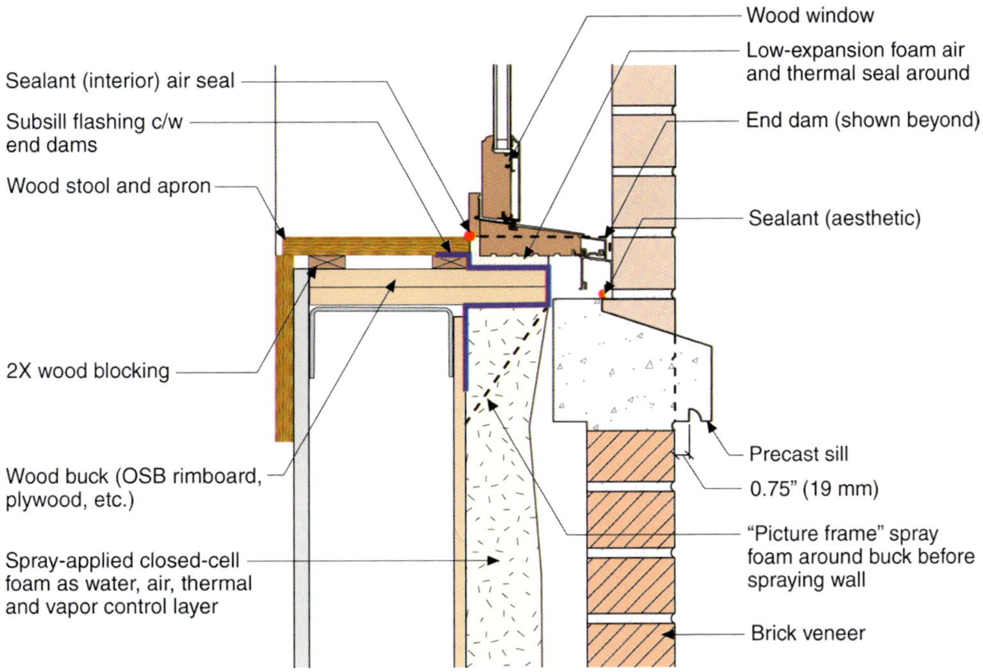

Figure 5.11: Ensuring continuity of the control functions across window rough openings is the most common detail at which high performance can be compromised.

convenient to also connect the air control layer to the innermost surface as well, but not necessary depending on the product.

Low-expansion foam is widely used to fill the rough opening with an air, vapor, and thermal control product.[45] However, it is difficult to achieve air flow control continuity using spray foam if the rough opening is narrow and deep. Voids and gaps regularly form during installation in rough openings of $^1/_2$" (13 mm) width and 4" (100 mm) depth. Hence, many high-performance details use an inner sealant bead and/or an adhered or clamped membrane to form the air control continuity. An alternate method is to specify larger rough openings so as to ensure about 1" (25 mm) or more space. This allows for easier installation and inspection, and generally requires multiple passes with the foam gun, all of which results in a much higher quality air seal.

45
Filling this void with fiberglass insulation provides poor performance. The insulation value provided is modest, the air flow control almost useless, and the vapor control non-existent. Combined with exterior and interior air-water-vapor sealant beads, stuffed fiberglass works in theory, but poorly in practice.

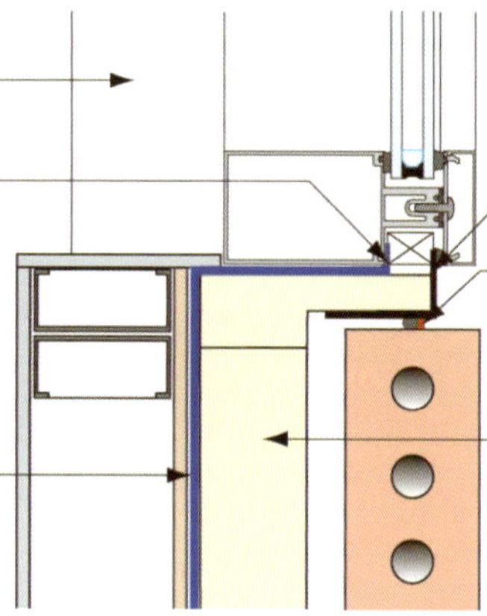

Figure 5.12: Curtainwalls often have better defined water and air control layers than windows making the design of control continuity easier.

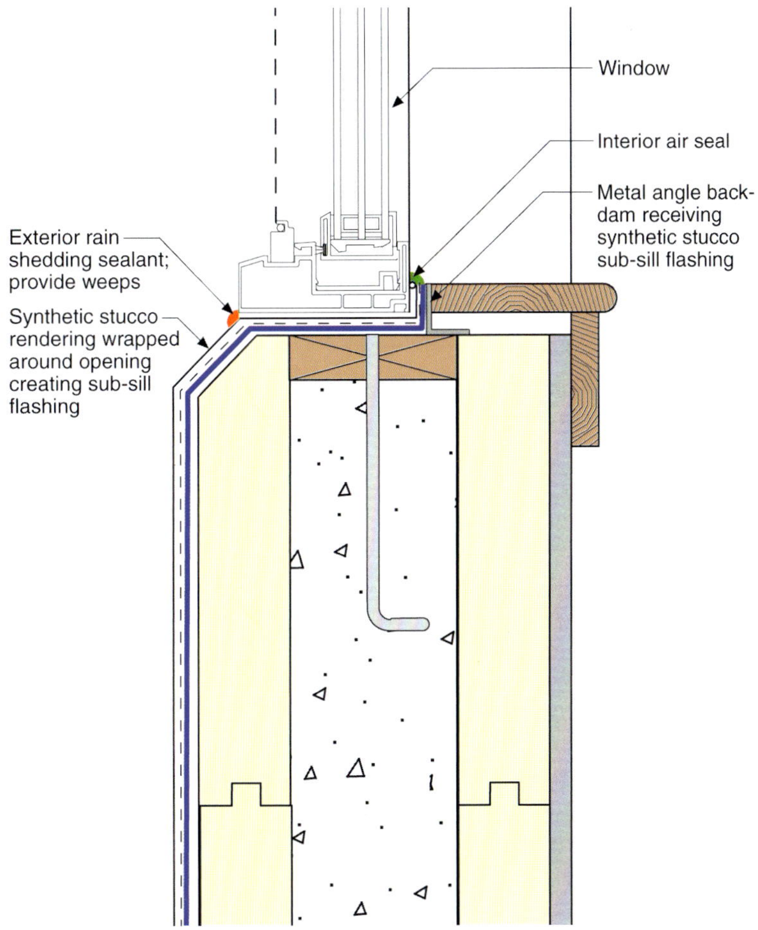

Figure 5.13: ICF systems allow for relatively easy integration with windows.

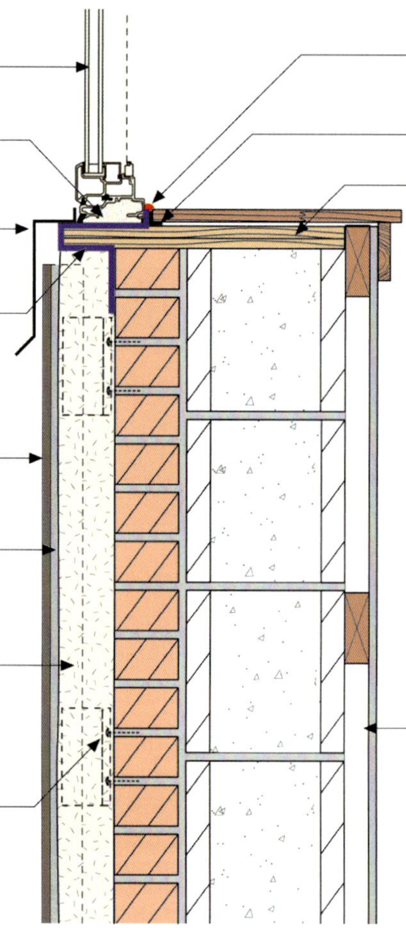

Figure 5.14: Overcladding of existing buildings, with a range of panel-based claddings, can produce a high-performance "perfect wall" retrofit of an existing enclosure. Rain control at windows is critical as for all window details.

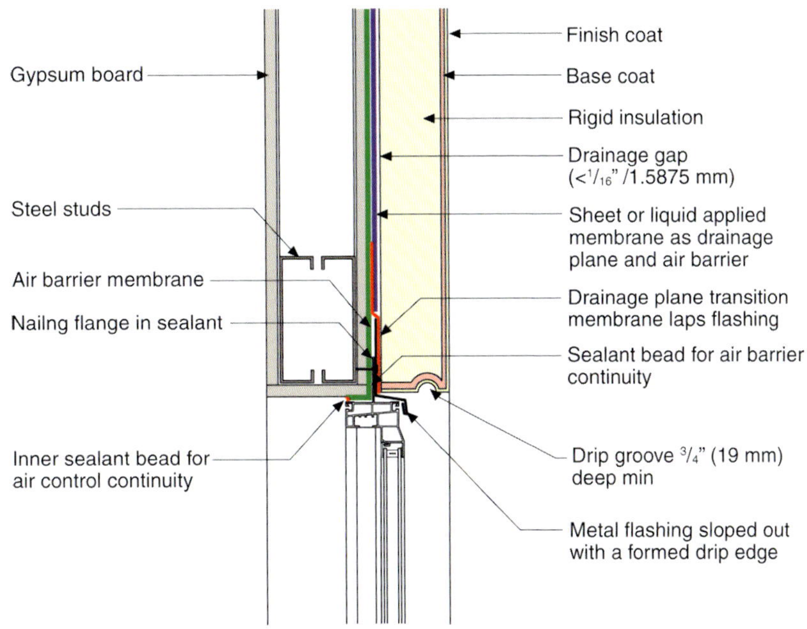

Figure 5.15: Window penetrations in EIFS wall systems are critical details.

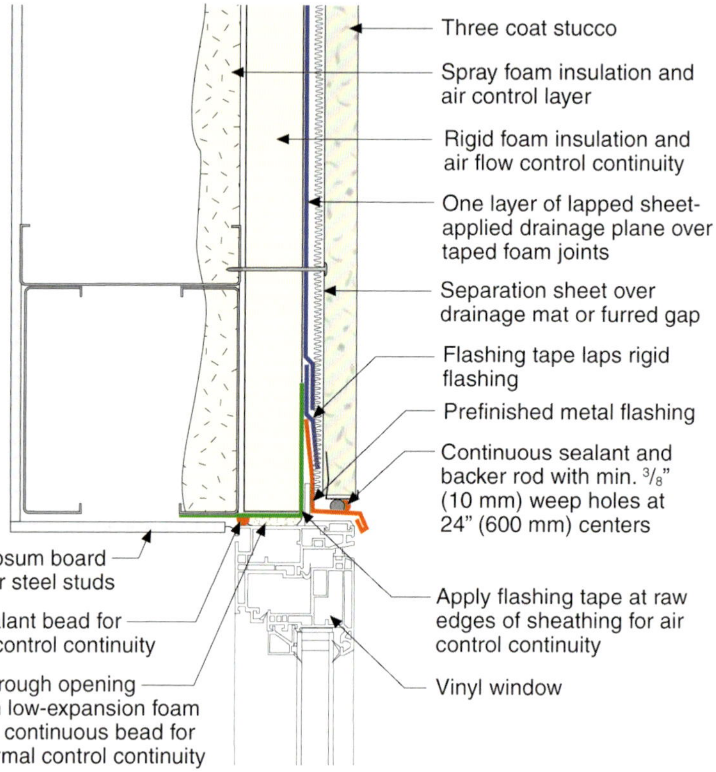

Figure 5.16: This drained stucco system splits a water-air control layer so that the airflow control continues into the jamb and the water control layer moves out onto flashing at a window penetration.

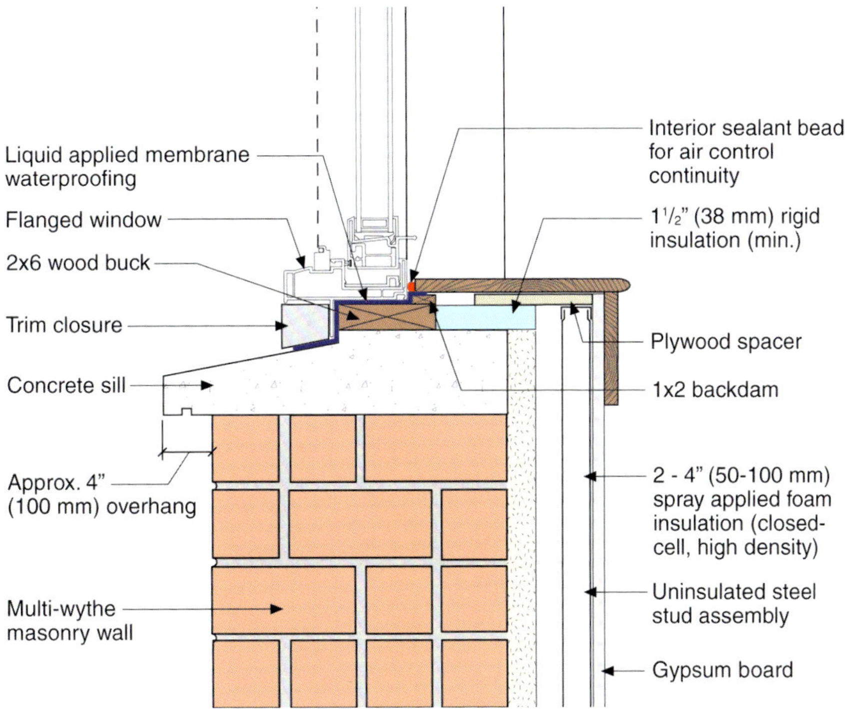

Figure 5.17: Control continuity at a window penetration in an interior insulated retrofit masonry wall can be achieved if care is taken.

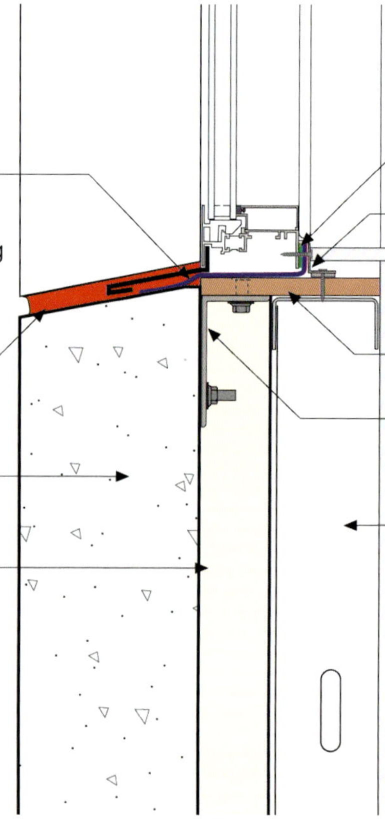

Figure 5.18: Failure to align the thermal and air control layers between windows and architectural precast is a common performance problem. The alignment of the window, the use of wood bucks (low-thermal conductivity) and sub-sill flashing reliably solves these problems and allows for high performance.

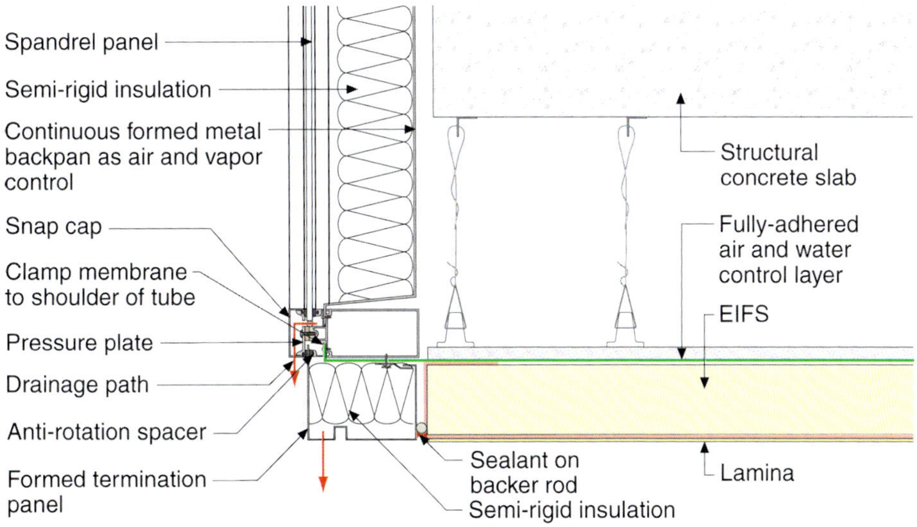

Figure 5.19: The continuity of the thermal and air control layers can be difficult to achieve at the transfer from a curtainwall to soffit. Note the accommodation of differential movement, and the insulation alongside the bottom of the curtainwall tube.

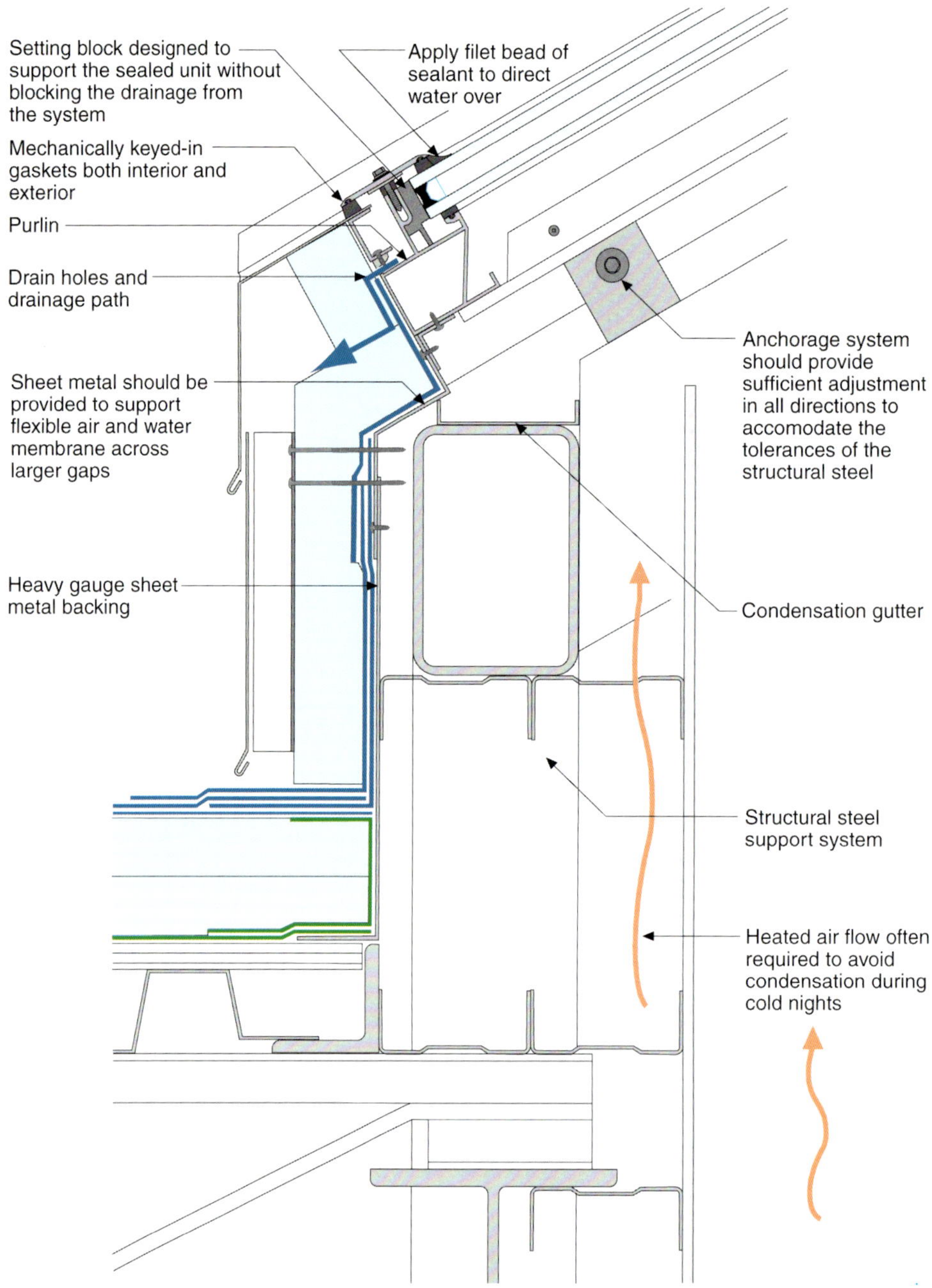

Figure 5.20: A skylight detail at a roof. Because of the poor thermal control of the curtainwall, heat must be applied locally and a condensation gutter provided to deal with intermittent condensation.

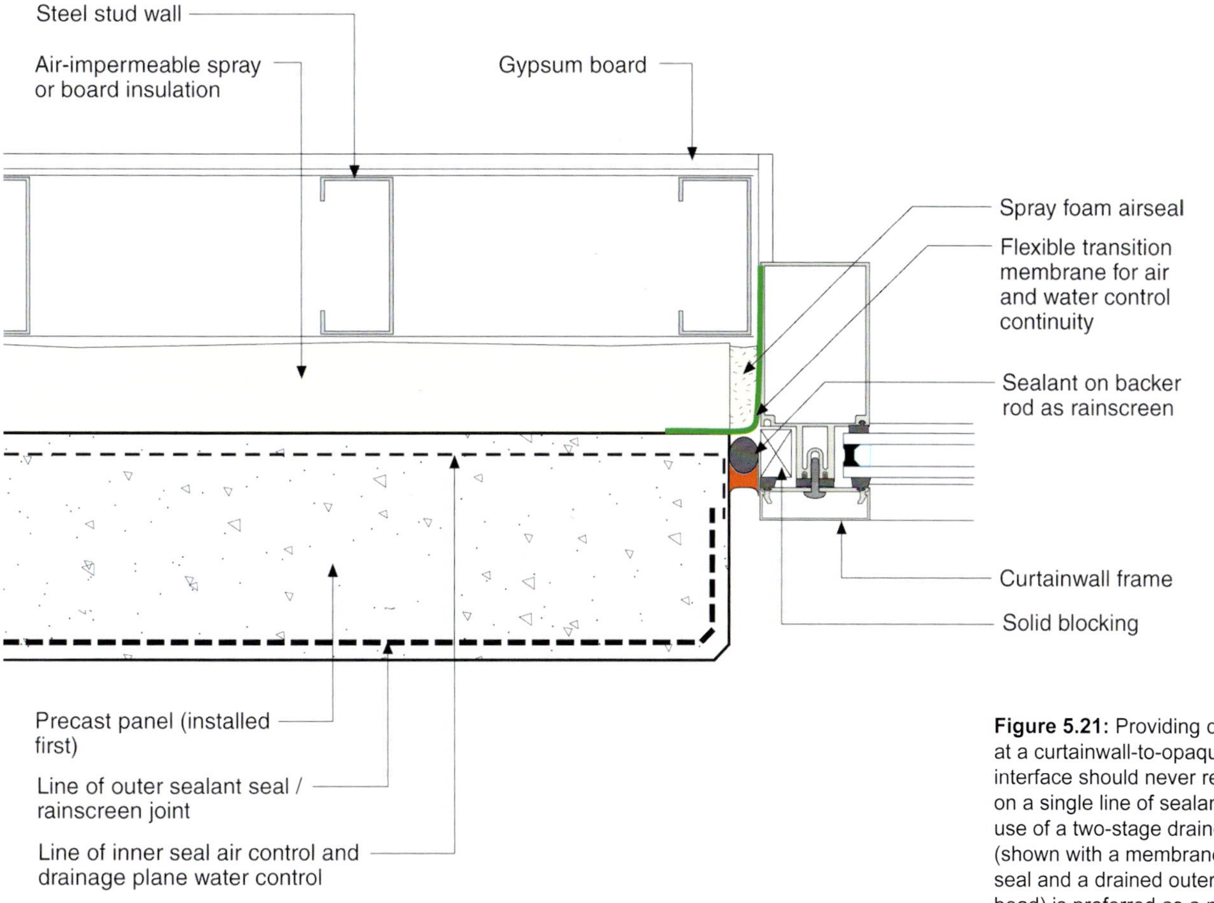

Figure 5.21: Providing continuity at a curtainwall-to-opaque-wall interface should never rely solely on a single line of sealant. The use of a two-stage drained joint (shown with a membrane inner seal and a drained outer sealant bead) is preferred as a more durable and reliable transition.

5.4 STRUCTURAL PENETRATIONS

Thermal continuity is always compromised at structural penetrations. However, the air and water control continuity need not be. The design of the structural connection can greatly reduce the impact of the thermal bridge.

Where low loads make it possible, wood should be used for window and door bucks, blocking, and similar conditions. The difference in thermal performance between steel and wood is usually a factor of ten. Of course for higher loads metals will be necessary for higher stiffness, strength and durability.

As recommended earlier in this text, only small amounts of steel (under 0.1% of enclosure area for high-performance) or concrete (under 2% of enclosure area) should be allowed to penetrate the insulation. If the proposed design exceeds these approximate limits, more detailed heatflow analysis should be conducted.[46]

46
Common analysis tools include the free Therm 2D steady-state computer program from Lawrence Berkeley National Labs, and the 2D and 3D dynamic HEAT2 and HEAT3 programs from Blocon software.

To reduce the impact on thermal control, the best strategy for penetrations is to use a few point penetrations. For example, short clips are always preferred over continuous Z-bars when supporting cladding or windows. The goal is to use as few highly stressed metal connections as possible.

As stainless steel has about one-third the thermal conductivity of carbon steel consideration should be given to using this material when two- or three-dimensional thermal modeling shows it to be beneficial. Aluminum is about 5 times more thermally conductive than carbon steel, and has well less than half the strength and stiffness. Hence, its use should be avoided when penetrating the thermal control layer.

Pultruded fiberglass is emerging as a durable material that can transfer significant structural loads with limited thermal bridging. Used for decades to produce window frames and as structural components in specialty corrosion-resistant structures such as cooling towers, the material is finding more and more use in high-performance commercial enclosures.

Proprietary cladding stand-off clips (Figure 5.22) are available, and many standard shapes (such as angles, channels, I-beams, plates) with defined structural material properties are available to create custom solutions for a wide range of loads.

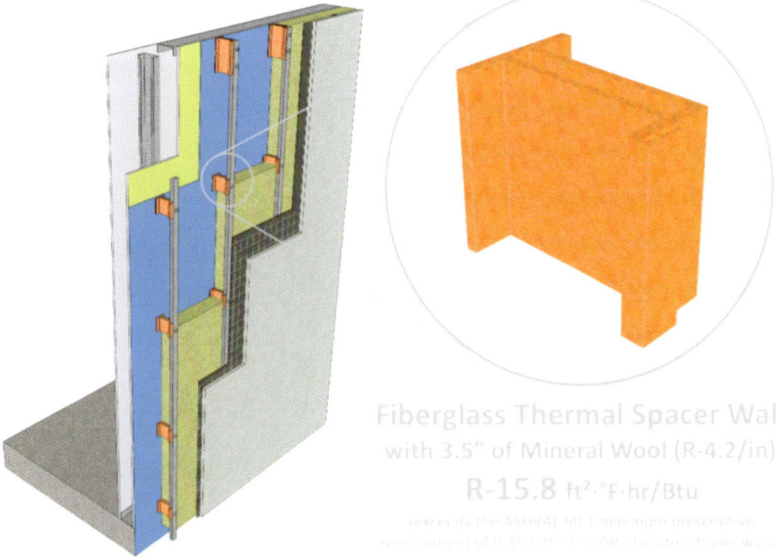

Figure 5.22: Proprietary fiberglass clips are available that can transfer significant cladding loads through insulation with very little thermal bridging. *(Courtesy of Cascadia Windows Ltd.)*

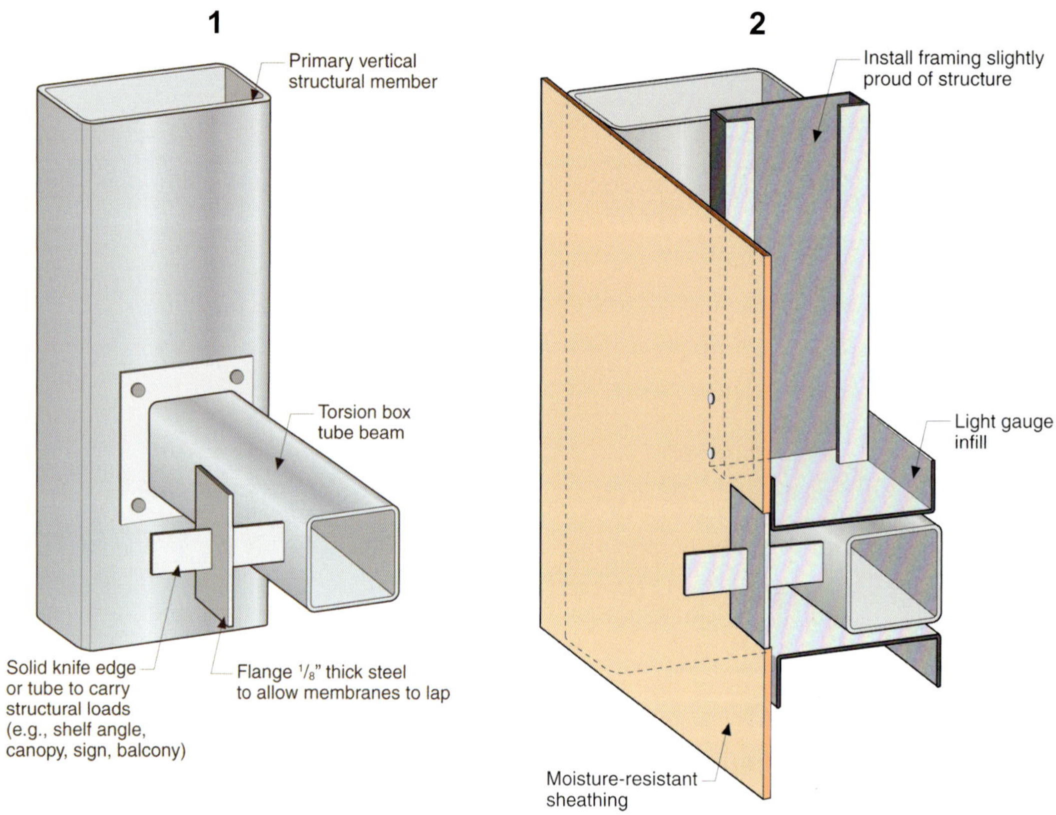

Figure 5.23: The thermal control layer can be penetrated to support a range of structural loads: signs, balconies, shelf angles, or shading devices.

3

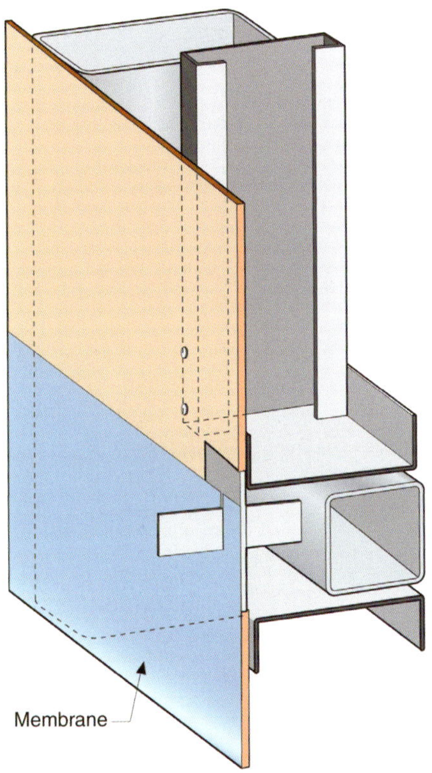

4

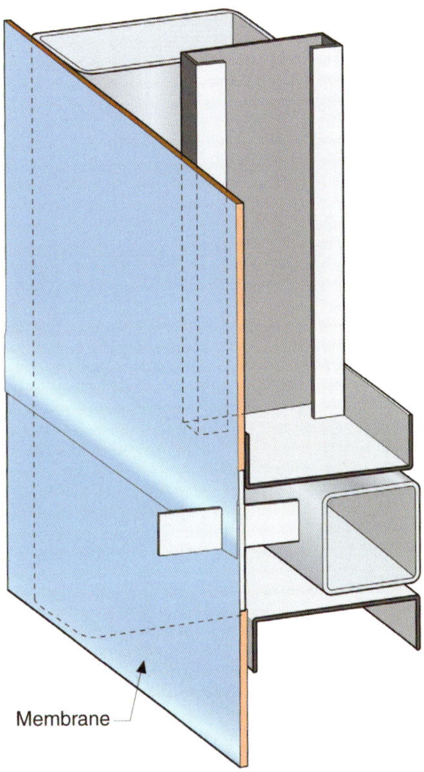

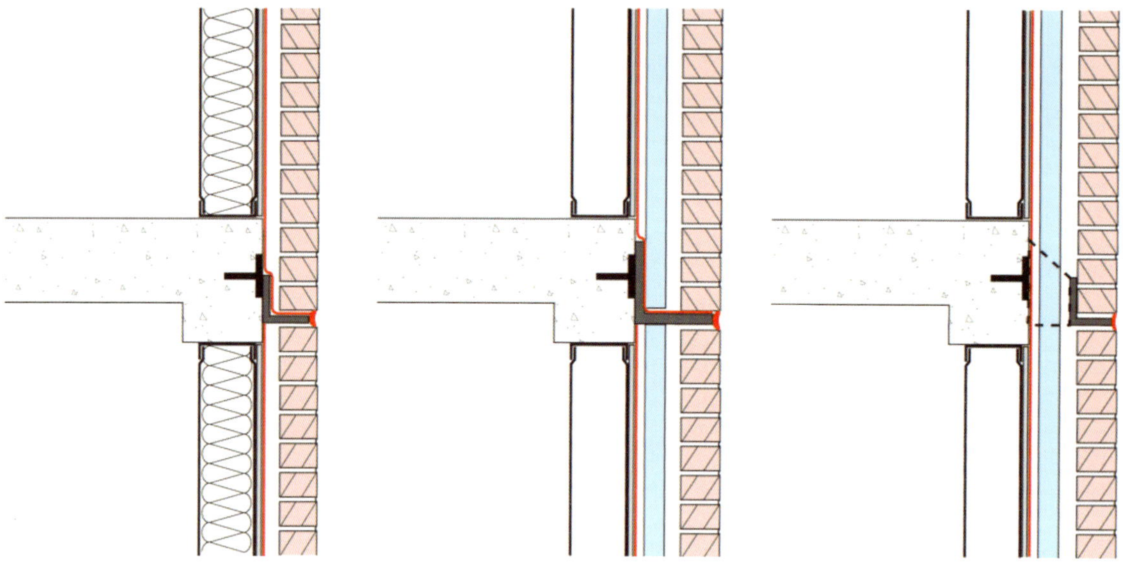

Traditional: no thermal control | Expensive: poor performance | Modest cost: good performance

Figure 5.24: Shelf angles can be significant sources of heat loss. Use of offset shelf angle ensures thermal, air and, water control continuity.

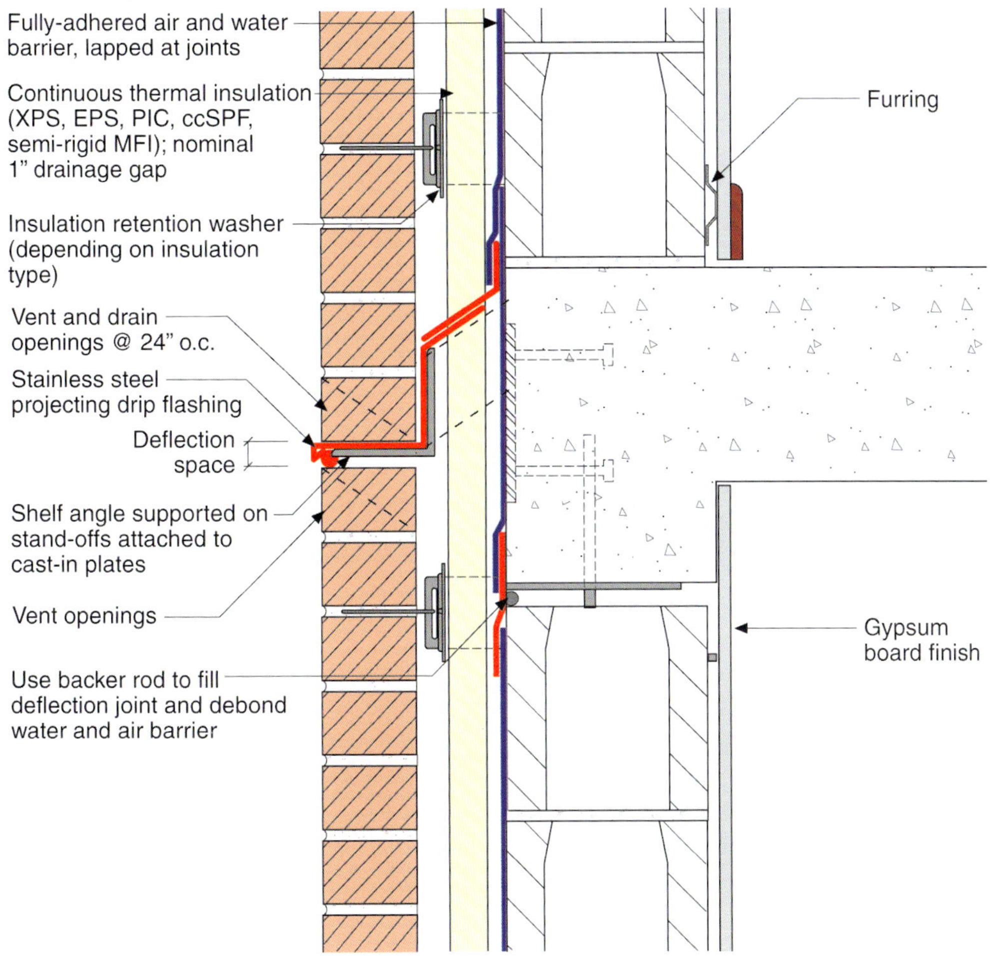

Figure 5.25: A thermally-broken shelf angle detail. By using widely-spaced (typically 36"/1 m or greater) steel knife-edge supports, thermal bridging can be very significantly reduced.

Figure 5.26: A brick relieving / shelf angle supported away from the backup wall by a steel tube at 4' o.c. (1.2 m).

Figure 5.27: Hangers used to accommodate changing substructure levels.

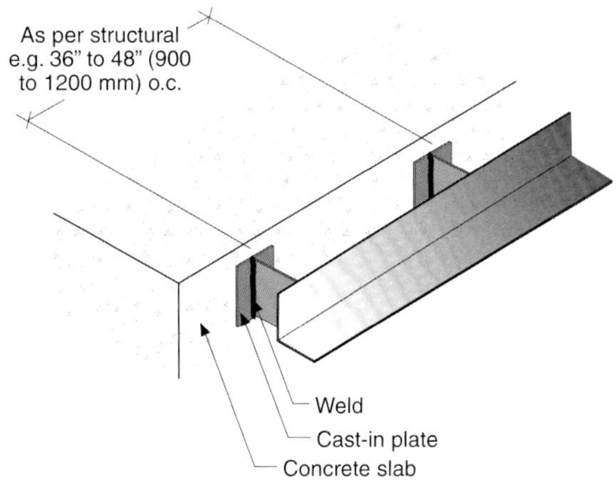

Figure 5.28: Schematic isometric of shelf angle standoffs

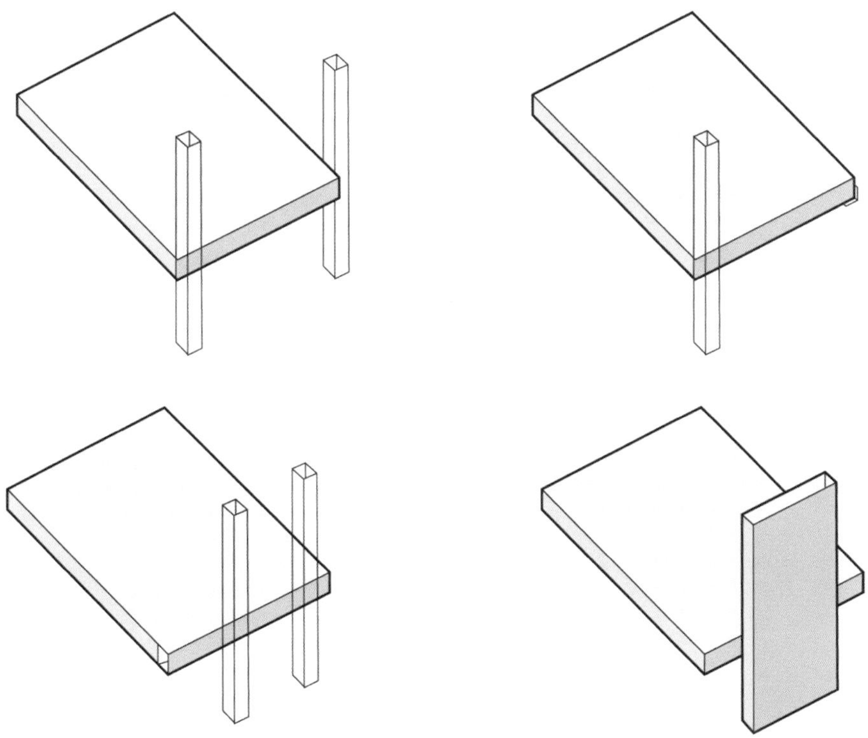

Figure 5.29: Cantilevered balconies are huge heat loss components that are very expensive to solve. Numerous alternate support arrangements avoid the need for a moment connection through the control layers.

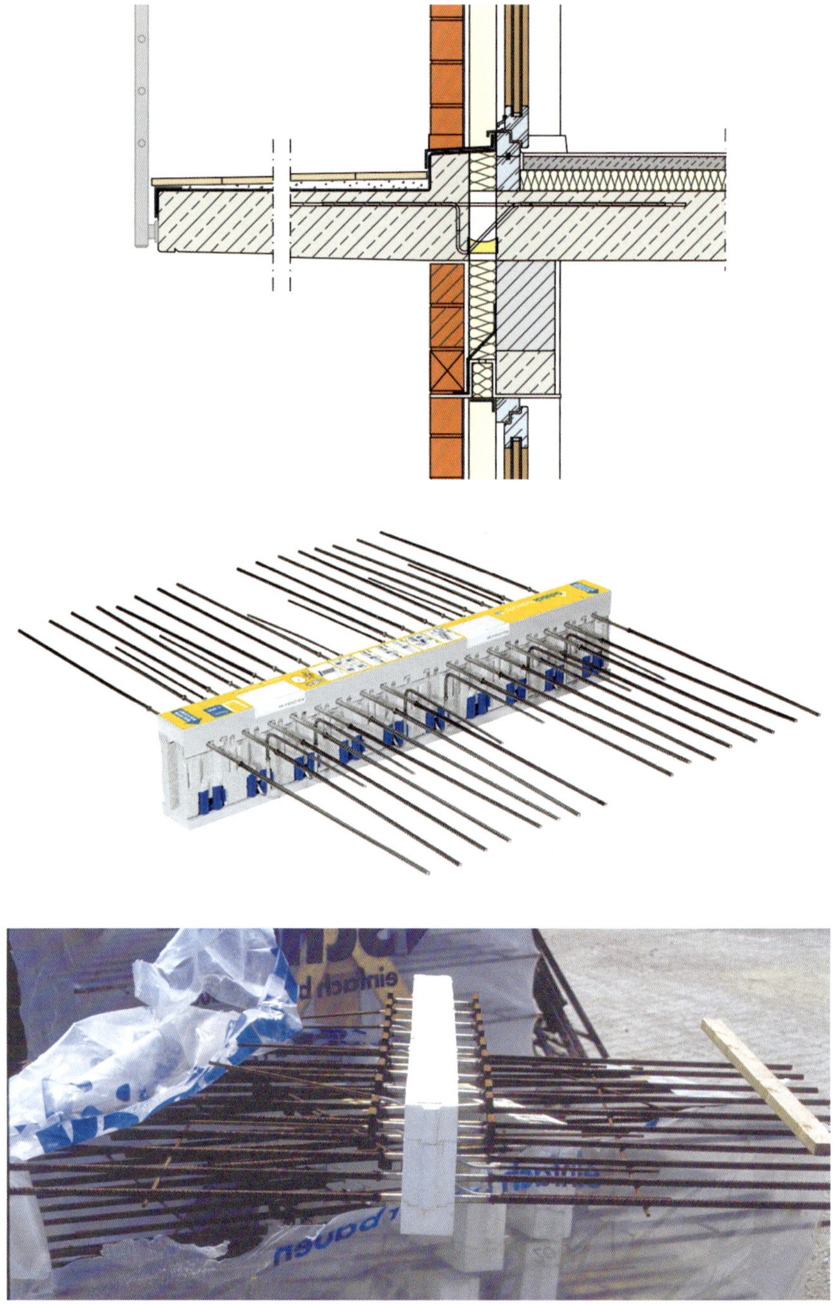

Figure 5.30: Special structural thermal break products are available that allow for cantilevered concrete and steel elements to penetrate the thermal control layer. Heat loss reductions of up to 90% are possible. (Courtesy Schoeck)

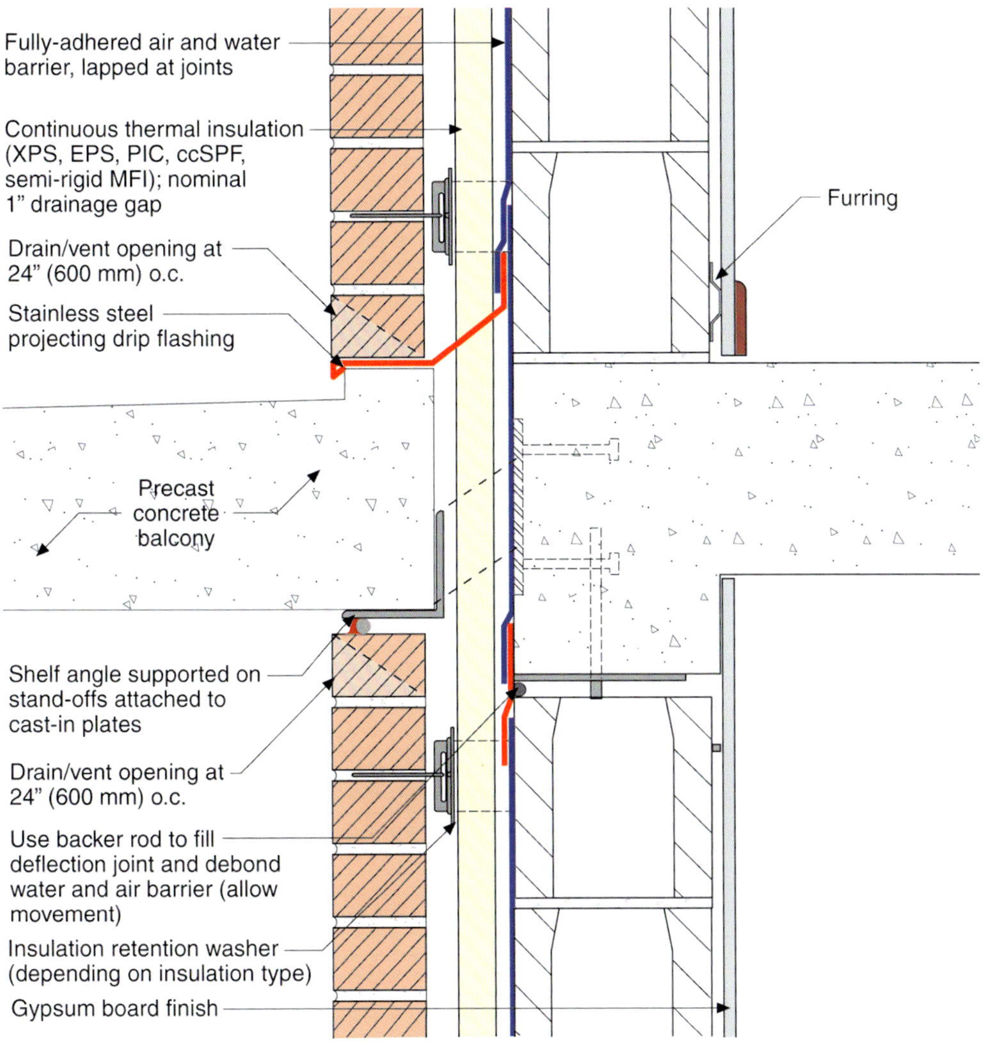

Figure 5.31: Heat loss, air and water leaks around propped balconies are relatively easy to solve by the same means as offset shelf angles. Designing the building with propped balconies rather than thermally-broken balconies is always less expensive, more durable, and higher performance.

Figure 5.32: Precast balcony supported on shelf angle and standoffs. Note upturned lip at back of precast element to ensure water is directed outward.

Figure 5.33: Using a diagonal metal brace to support the outer edge of a balcony reduces the thermal bridge to a total of four knife-edge supports per balcony.

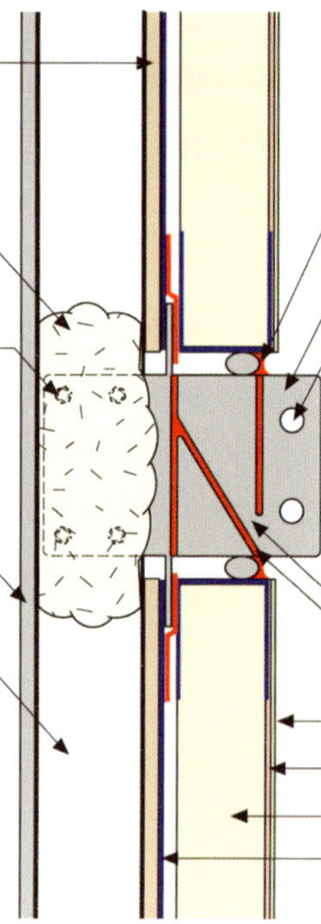

Figure 5.34: A structural penetration for a moderate load can be formed through a system like EIFS or architectural precast by use of a knife-edge and drained, two-stage joint.

Figure 5.35: A knife-edge support penetrating through a fiber-cement drained cladding. Sealant will be provided around the knife edge to limit water penetration, but the sealant can fail with little consequence to performance.

Chapter 6 Closure

As society and the building industry changes to demand higher performance of buildings and building enclosures, change in enclosure design and specification is inevitable. In short, design will need to become based more on performance and rely less on "standard practice," which is already becoming a marginal basis of design.

What has been presented throughout the book is merely an introduction to best practices, not an in-depth study of the scientific principles. Identifying the control layers, ensuring continuity, and properly specifying their performance is necessary for functional enclosures everywhere. This book's focus has been cold-climate, low-energy, high-performance commercial, multi-unit residential, industrial, and institutional buildings. However, the fundamentals presented are valid for many different building occupancies, climates, and materials.

Of course the details presented are merely demonstrative. Specific details need to be developed, as they do for all building projects, depending on trade skill and preferences, material availability, owner and designer preferences, and performance goals and budgets.

Designing buildings that consume much less operational energy than current designs while remaining affordable, healthy, buildable, and durable will require additional effort from architects, engineers, technologists, builders, and even code and fire officials. Hopefully this book will help make this very worthwhile task easier.

Chapter 7 References

ASHRAE/IESNA Standard 90.1-2007 ASHRAE Standard Energy Standard for Buildings Except Low-Rise Residential Buildings, American Society of Heating, Refrigerating and Air Conditioning Engineers, Inc., Atlanta, 2007.

ASHRAE. *Handbook of Fundamentals.* American Society of Heat Refrigeration Air-conditioning Engineers, Atlanta, 2009.

ASTM C1199 - 09e1 *Standard Test Method for Measuring the Steady-State Thermal Transmittance of Fenestration Systems Using Hot Box Methods.*

ASTM E1827 *Standard Test Methods for Determining Airtightness of Buildings Using an Orifice Blower Door*)

ASTM E1677 - 05 *Standard Specification for an Air Retarder (AR) Material or System for Low-Rise Framed Building Walls*

ASTM E2178 - 03 *Standard Test Method for Air Permeance of Building Materials*

ASTM E2357 - 05 *Standard Test Method for Determining Air Leakage of Air Barrier Assemblies*

ASTM E283 - 04 *Standard Test Method for Determining Rate of Air Leakage Through Exterior Windows, Curtain Walls, and Doors Under Specified Pressure Differences Across the Specimen*

ASTM E783 - 02 *Standard Test Method for Field Measurement of Air Leakage Through Installed Exterior Windows and Doors*

Baker, M., *Roofs.* Multi-Science Publications Ltd., Montreal, 1980.

Bodart, M, De Herde, A., "Study of Parameters and Criteria Influencing the Choice of Glazing in Office Buildings in Belgium", *Proc of DOE/ASHRAE Buildings VIII*, Clearwater, FL, December 2001.

Bombino, R., Finch, G. "Reconsidering the Approach towards Determining Overall Building Enclosure Thermal Performance for Code Compliance", *Proc. of ASHRAE Buildings XI*, Clearwater Florida, 2010.

Brand, Ronald, E. *Architectural Details for Insulated Buildings.* Van Nostrand Reinhold 1990.

Brown, W.C., Bomberg, M.T., Ullet, J.M. and Rasmussen, J. "Measured Thermal Resistance of Frame Walls with Defects in the Installation of Mineral Fibre Insulation", *J. of Thermal Insulation and Building Envelopes*, Vol 16, April 1993, pp. 318-339.

Buelow-Huebe, Helena. *Energy Efficient Window Systems.* PhD Thesis, Dept. of Construction and Architecture, Lund University, Sweden, 2001.

Canadian Building Digest 50 *Principles Applied to an Insulated Masonry Wall.* Div of Building Research, NRCC, Ottawa, February, 1964.

Canadian Building Digest 94. *Precast Concrete Walls – A New Basis for Design.* Div of Building Research, NRCC, Ottawa, October, 1967.

Carmody, J., Selkowitz, S., Arasteh, D., Lee, E. S., Willmert, T. *Window Systems for High-Performance Buildings.* Norton & Co., 2004

CEC. *California Commercial Building End-use Survey.* by Intron Inc. Report CEC-400-2006-005. March 2006.

CGSB 149.15 *Determination of the Overall Envelope Airtightness of Buildings by the Fan Pressurization Method Using the Building's Air Handling Systems.*

Cole, R., Kernan, P., "Life-Cycle Energy Use in Office Buildings", *Building & Environment*, Vol. 31, No. 4, pp. 307-317, 1996

Desjarlais, A. "Self-drying roofs: What?! No dripping!" *Thermal Performance of the Exterior Envelopes of Buildings VI,*. Atlanta: American Society of Heating, Refrigerating and Air-Conditioning Engineers, Inc., pp. 763-773, 1995.

Enermodal Engineering Limited. 1996. *Building insulation thermal anomalies.* ASHRAE Research Project 785, 1996.

EN ISO 14683: 2007. *Thermal bridges in building construction - Linear thermal transmittance - Simplified methods and default values.* Geneva Switzerland.

EN ISO 10211: 2007. *Thermal bridges in building construction - Heat flows and surface temperatures - Detailed calculations.* Geneva Switzerland.

Gratia, E., De Herde, A. "Design of low energy office buildings". *Energy & Buildings*, Vol. 35 pp.473-491, 2003.

H. Hens, A. Janssens, W. Depraetere, J. Carmeliet and J. Lecompte, "Brick Cavity Walls: A Performance Analysis Based on Measurements and Simulations", *Journal of Building Physics*, Vol 31, pp. 95-124, October 2007.

Johnson, R, Sullivan, R., Selkowitz, S., Nozaki, S., Conner, C., Arasteh, D. "Glazing energy performance and design optimization with daylighting.", *Energy & Buildings*, Vol. 6, No. 4, pp. 305-317, 1984.

Lawton, M., Roppel, P., Fookes, D. Teasdale St Hilaire, A., Schoonhoven, D., "Real R-Value of Exterior Insulated Wall Assemblies", *Proc. of BEST2 Conference*, Portland, OR, 2009.

Leach, M., Lobato, C., Hirsch, A., Pless, S., and Torcellini, P. *Technical Support Document: Strategies for 50% Energy Savings in Large Of-*

fice Buildings. National Renewable Energy Laboratory, NREL, DOE Technical Report , September, 2010.

Lee, I. *High Performance Window Systems and their Effect on Commercial Building Energy Use*. M. A. Sc. Thesis, Faculty of Engineering, University of Waterloo, 2010.

Lerum, Vidar. *High Performance Building*. Wiley, New York, 2007.

Love J., Tian W., Tian Z, "Window-to-Wall Ratios and Commercial Building Environmental Control in Cold Climates", *Proc. 3rd SBRN and SESCI 33rd Joint Conference*, Fredericton, 8 pp., 2008.

Lstiburek, Joseph and Carmody, John. *Moisture control handbook : principles and practices for residential and small commercial buildings*. New York : Van Nostrand Reinhold, 1993.

Natural Resources Canada. *Energy Use Data Handbook 1990 to 2007*. Office of Energy Efficiency, Ottawa, Canada, 2007.

NFRC 100-2010. *Procedure for Determining Fenestration Product U-Factors*. National Fenestration Rating Council, Greenbelt, Maryland, 2010.

Poirazis, H., Blomsterberg, A., Wall, M., "Energy simulations for glazed office buildings in Sweden", *Energy & Buildings*, Vol 40, pp. 1161-1170, 2008.

Posey, J.B and Dalgliesh, W.A. "Thermal bridges — heat flow models with Heat2, Heat3, and a general purpose 3-D solver", *Proc. 10th Canadian Building Science & Technology Conference* Ottawa , May 12, 2005.

Reppe, P., and Blanchard, S. *Life Cycle Analysis of a Residential Home*, Report 1998-5, Center for Sustainable Systems, University of Michigan, 1998.

Ross, B., *Design with Energy in Mind*. M. Arch. Thesis, School of Architecture, University of Waterloo, 2009.

Straube, J.F., and Burnett, E.F.P., *Building Science for Building Enclosures*, Bulding Science Press, Westford, 2005.

Straube, J.F. and Burnett, E.F.P., "Rain Control and Design , Strategies". *J. Of Thermal Insulation and Building Envelopes*, pp. 41-56, July 1999.

Timusk, J. and Lischkoff, J., "Moisture and Thermal Aspects of Insulated Sheathings", *Proc. Second Conference on Building Science & Technology*, London, Ont., 1983, pp.71-91.

Uvslokk, S., "The importance of wind barriers for insulated timber frame construction", *J. Thermal Insul. and Bldg. Envs.*, V. 20, July, p.40-62, 1996.

Appendices

Appendix A Heat Flow

Heat always flows from hot cold. Within air conditioners and heat pumps this process is manipulated to make it appear that the opposite occurs.

Temperature: a measure of thermal energy, units of Kelvin (K) or Celsius (°C), Rankine (R), Fahrenheit (°F)

Thermal conductivity: a material property, heat flow per unit area per unit thickness per unit temperature, symbol k (W/m K)

Thermal conductance: a property of a material layer, heat flow per unit area and temperature, symbol C (W/m^2 K) or (Btu/hr ft^2 °F)

Thermal resistance: a property of a material layer or wall usually measured from surface to surface, of the resistance to heat flow per unit area and temperature difference, symbol RSI (m^{2}K/W) or R (ft^2•°F•hr/Btu)

R = 1/C

Imperial value R (ft^2•°F•hr/Btu) = 5.678 RSI

Enclosure/assembly heat transfer coefficient: a property of an enclosure assembly (the thermal conductance of an assembly), specifically heat flow per unit area and temperature, symbol U, units (W/m^{2}K (Btu/hr ft^2 °F)

$U = 1/R_{total}$ or U_{SI}

Building overall heat transfer coefficient: a property of a building (the overall thermal conductance), specifically heat flow per temperature, symbol U, units W/K (Btu/hr °F)

Example:
Given that the thermal conductivity of Type 4 extruded polystyrene (for example, Styrofoam SM) is 0.029 W/mK, find the conductance and resistance of a layer 50 mm thick in both imperial and metric units.

Answer:
Conductance = conductivity k / thickness l (in meters). 50 mm = 0.050 m so ...

C = 0.029 / 0.050 = 0.58 W / m^2•K

this means, for example, that a one square meter panel of 50 mm thick Styrofoam SM will allow 0.58 watts of energy to pass through it under a one degree Kelvin (or Celsius) temperature change.

Resistance, RSI = 1 / conductance, so

C = 1 / 0.58 = 1.724 m^2•K / W

Hence, a layer of SM would have a thermal resistance of RSI 1.724. If one were to look in a building supply store, a sheet of 2” SM would be stamped with RSI1.76, since 2” sheets are slightly thicker than 50 mm. By the way, a layer of 100 mm would have a thermal resistance of RSI3.45, e.g. twice the thickness, twice the resistance (this does ***not*** work with conductance).

The thermal resistance in imperial R-value would be:

Imperial value R (ft^2•°F•r/Btu) = 5.678 RSI

$$R_{imp} = 5.678 * RSI1.724 = R9.8$$

Again, a full 2” thick sheet would have a slightly higher R-value of R-10. Thus, one often speaks of an “insulation value” of R-5 per inch. Four inches, R-20 for this layer, 1.5”, R-7.5 for the layer.

Most building enclosures include more than just materials – they also include air spaces, which are insulating. The heat transfer from the air next to the enclosure to the surface of the enclosure is also not perfect. Hence, one needs to calculate the impact of this effect. These complications are discussed next.

A.1 SURFACE FILMS

To account for both the radiative and convective heat transfer modes at the exterior and interior surfaces of building components, the radiative and convective heat transfer coefficients are used in the form of an *equivalent conductance* or, alternatively, equivalent resistances. These equivalent coefficients are usually called *surface film coefficients*. It is important to recognize that a surface film does not exist in reality. The term film is used simply so that a layer (of indefinite thickness) can be added to a typical conductive heat flow analysis.

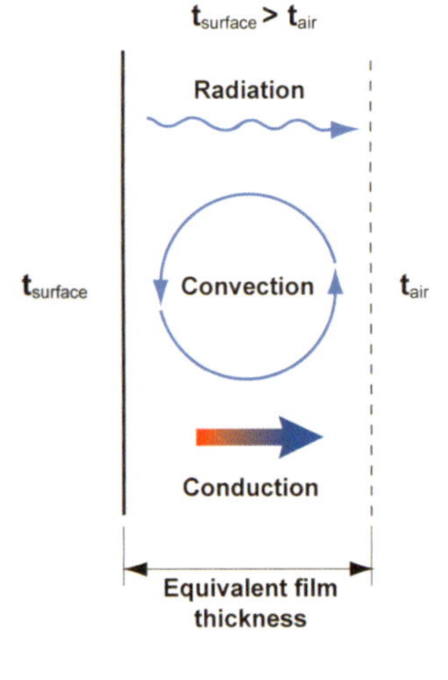

Figure A.1: Heat transfer modes at building surfaces.

Surface Position Still air	Flow Direction (e.g. indoors)	Resistance RSI [m^{2}K/W]	 R (imperial)	Conductance [W/m^{2}K]
Horizontal (i.e. ceilings and floors)	Upward Downward	0.11 0.16	1.65 1.10	9.3 6.1
Vertical (i.e. walls)	Horizontal	0.12	1.45	8.3
Moving air	(e.g. outdoors)			
Stormy – 6.7 m/s (winter)	Any	0.03	1/6	34
Breeze – 3.4 m/s (summer)	Any	0.04	1/4	23
Average conditions	Any	0.06	1/3	17

Table A.1: Equivalent Total Surface Film Conductances (h_o)

The resistance to heat flow at a surface is small relative to the heat flow resistance of most modern wall assemblies and therefore need not be accurately estimated for most purposes. Poorly insulated walls and windows have a lower overall thermal resistance and thus surface effects are more important. Hence, a more precise calculation of surface films is justified for these types of enclosures. The overall equivalent surface conductances, h_o, or resistances in Table A.1 can be used to find heat flow without further modification.

A.2 AIR SPACES

Air spaces are commonly used in building assemblies. Heat is transferred across air spaces by a combination of conduction through still air, convection flows, and by net radiation from the warm side to the cold. The modes of heat transfer vary in importance depending on: the emissivities of the surfaces, the thickness of the air space, and the absolute and relative temperatures of the two surfaces.

The heat flow across a plane air space can be found with a reasonable degree of accuracy by using detailed correlations of convection and radiation. However, a high degree of accuracy is rarely necessary or justified in light of the many poorly known variables (e.g. variable cavity widths, blocked cavities, etc.) and the relatively small influence that the airspace has on the thermal resistance of modern enclosure assemblies. Simplified values for most practically encountered situations are presented in Table A.2. The values in Table A.2 are also given in terms of resistances so that they can be directly used in the heat flow equation.

Note that the influence of airgap size is relatively unimportant in the range of 3/4" to 4" (20-100 mm). The R-value of a gap will drop rapidly as its width drops below 1/2" (13 mm), until it is negligible for gaps of 1/8" (3 mm) or less. Such gaps can make a large relative difference when inserted between two highly conductive materials (like steel, aluminum, or even concrete) but the absolute value of the thermal resistance of a very small gap is very small.

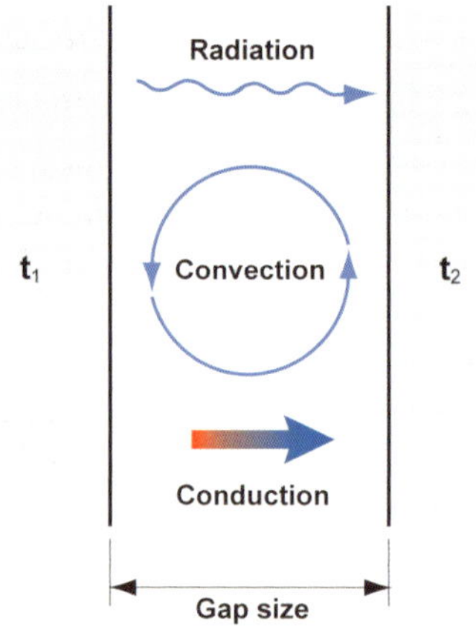

Figure A.2: Heat transfer across plane airspaces

Table A.2: Thermal resistance for non-reflective enclosed airspaces

Situation (non-reflective surfaces)	RSI value	R-value
Heat flow down (20-100 mm)	0.18	1.0
Heat flow across (20-100 mm)	0.17	0.96
Heat flow up (20-100 mm)	0.15	0.85

In many practical situations an air space is either intentionally or accidentally vented. Air flow through an air space can change the heat flow characteristics, although significant flows are required to modify the equivalent conductances listed above. In most cases, the effect of venting enclosure assemblies can be ignored, and only in extreme cases does it need to be accounted for. Extreme cases includes highly ventilated attics (at least 1% venting area) and wall panels with at least 2% vent area and cavities over 50 mm in size.

A.3 HOW TO CALCULATE THE "R-VALUE" OF AN ENCLOSURE ASSEMBLY

Remember: one cannot add conductances, you must add resistances

1. List each material in the wall or roof, its conductivity (k) and its thickness (l) in m
2. Calculate the conductance of each layer (C) using C = k / l
3. Calculate the thermal resistance of each layer using RSI = 1 / C
4. Sum the individual thermal resistances to get the answer.

Usually the thermal resistance of the air films that exist on both the interior and exterior surfaces of an assembly are added to the wall as virtual layers. This makes our calculations more accurate.

Example:
Calculate the total thermal resistance (R) and overall heat transfer coefficient (U) of the wall shown below. Use conductivity values from tabulated values.

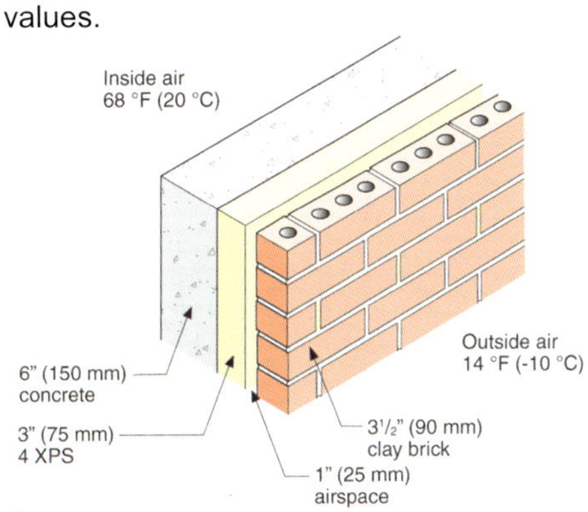

Answer:

Layer	Conductivity	Thickness	Conductance	Resistance
Interior film[Note 1]	N.A.	N.A.	8.3	0.120
Concrete	1.8	0.150	12	0.083
Type 4 XPS	0.029	0.075	0.39	2.56
Air space[Note 2]	N.A.	25	N.A.	0.17
Brick	1.3	0.090	14.4	0.069
Exterior film[Note 1]	N.A.	N.A.	34	0.029
			RSI Total	**3.04**
			Overall Heat Transfer, U	**0.33**

Note 1: Table 1 – Since the interior and exterior films are fictitious, they do not have a thickness, and so no conductivity. Hence, tables typically contain only conductances or resistances for the layer. These values can be quite variable, but as can be seen, the effect of the value of the film resistance on the total resistance of a wall or roof is small if the wall is a modern insulated assembly.

Note 2: Table 2 – The flow of heat through an air space is complicated by convection (air flows) and radiation and so tabulated values of conductance are used instead. Like surface films, the values are variable but not important to accuracy in the calculation in most modern walls or roofs.

The total resistance is RSI3.04 (or imperial R-17.2), 84% of it provided by the insulation, and heat flow will be 0.33 W/m^2 K.

This method can be used to predict heat flow through walls, roofs, or foundations comprised of uninterrupted or "continuous insulation." If insulation is installed between studs (wood or steel), is penetrated by floor slabs, or extensive steel structure, thermal bridging must be accounted for (which can reduce the resistance very significantly).

A.4 HOW TO CALCULATE THE STEADY-STATE HEAT FLOW THROUGH AN ENCLOSURE SYSTEM

Heat flow across an assembly is simply the temperature difference divided by the true R-value times the area. The temperature difference is usually just ($t_{inside} - t_{outside}$).

1. Find the total thermal resistance of the enclosure as described earlier.
2. Find the overall heat transfer coefficient U, using $U = 1/R_{total}$
3. Multiply the temperature difference across the assembly by U, i.e., $Q = U * (t_{inside} - t_{outside})$

Of course if the sun is shining on the wall, the outdoor ***air*** temperature is not the correct one to use, (the actual solar heated surface temperature is more accurate, but difficult to find sometimes – use the table provided later for guidance).

The effect of heat storage, or thermal mass, can be important for some walls since heavy mineral-based materials can store a lot of heat: see the section on thermal mass later.

Example:
Calculate the amount of heat flow through the wall of the previous example when it's -10°C outside with no sun and 20°C inside.

$$\text{Heat flow} = U * (t_{inside} - t_{outside})$$
$$= 0.33\ (20\ \text{-}10) = 0.33\ (30)$$
$$= 10\ \text{W/m}^2$$

Answer:
Heat flow outward would be 10 W/m^2.

Example:
Calculate the amount of heat flow through a wall with an R-value of 15 when it's 10°F outside with no sun and 70°F inside.

$$\text{Heat flow} = U * (t_{inside} - t_{outside})$$
$$= 1/R_{total} (t_{inside} - t_{outside})$$
$$= 1/15 * (70 - 10)$$
$$= 4 \text{ Btu/hr/ft}^2$$

Answer:
Heat flow outward would be 4 Btu/hr/ft^2.

A.5 ASSESSING ENERGY LOSS DUE TO AIR LEAKAGE

If warm/cold air leaks out of the building in winter/summer, it is of course replaced with exterior cold/hot air. This cold/hot air must be heated/cooled to make it comfortable indoors. The energy impact of air leakage is significant and must be considered since it is often an important heat loss/gain component of modern buildings. It can be included by calculating an *equivalent building heat loss coefficient* for air leakage. This can be found from:

U_{air} = 0.3 * n * V (metric units only)

Where U_{air} is the heat loss coefficient due to air leakage (W/ °C)

N is the number of complete air changes per hour (ACH)

V is the total air volume of the building (in m^3)

U_{air} = 0.018 * n * V (imperial units only)

Where U_{air} is the heat loss coefficient due to air leakage (Btu/hr °F)

N is the number of complete air changes per hour (ACH)

V is the total air volume of the building (in ft^3)

U_{air} can be included in building load calculations just like U_{wall} or U_{window}.

Example:
Calculate the amount the heat loss coefficient of a building 75 by 150 feet in plan dimension, and 12 feet high if a blower door test shows it is expected to leak at 0.2 ACH in service.

Volume V= 75 * 150 * 12 = 135,000 ft^3

U_{air} = 0.018 * n * V

= 0.018 * 0.2 * 135,000

= 486 Btu/hr °F

Answer:
Therefore the building heat loss coefficient due to air leakage is 486 Btu/hr °F. Thus, on a 10°F night, the heat loss of the building due to air leakage only would be 486 * (70 – 10) = 29,160 Btu/hr.

A.6 HOW TO CALCULATE THE STEADY-STATE HEAT FLOW THROUGH A BUILDING SYSTEM

The easiest means of estimating heat flow through an entire building is to:

1. Calculate and then list the U-value for each element (wall, roof, window, door) along with the area of that element.
2. The product of each elements' area and its U-value is the heat loss coefficient for that enclosure element per unit temperature difference (SI units: W/°C).
4. The sum of these products is the overall heat loss coefficient for the building.
5. To find the overall heat loss (or gain) multiply the overall heat loss coefficient for the building by the temperature difference across the assembly.

Example:
An industrial "big box" store is 8 m high, 32 m long and 64 m wide. The enclosure has a lightweight roof (U=0.36 W/m²/C) and precast concrete walls (U=0.4 W/m²/°C). The front of the store (which faces west) has a 6 m high by 24 m long glass curtainwall with a U-value of 2.0 W/m²/°C. Six loading doors are at the back, each 5 m high and 3.5 m wide with a U-value of 1.0. The building is estimated to leak at 0.5 ACH under normal conditions. Ignoring the effect of the floors, sun, and any other doors, find the heating requirements when it is -10 °C.

Answer:
See the sketch of the building for details of the areas.

Air leakage heat loss: 0.3 n V = 0.3 (0.50) (16384) = 2458 W/°C.

Enclosure Component	Area (m^2)	U-value ($W/m^2/°C$)	Q Heat Loss (W/°C)	% Total Heat Loss
Walls	1287	0.40	515	**13%**
Roof	2048	0.36	737	**18%**
Doors	105	1.0	105	**3%**
Curtain Wall	144	2.0	288	**7%**
Air Leakage		*from above*	2458	**60%**
Overall Building Heat Loss Coefficient (U_o)			**4103**	**W/C**

Table A.3: Enclosure component

Plan

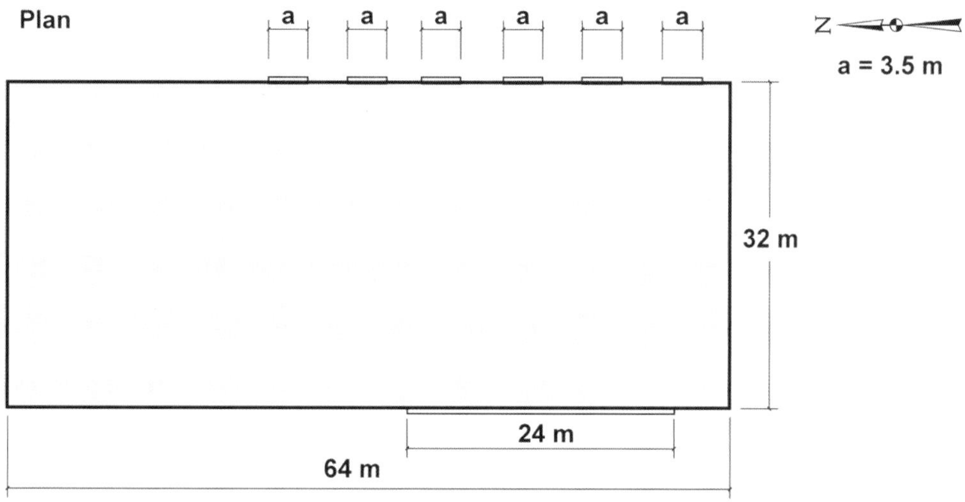

East Elevation

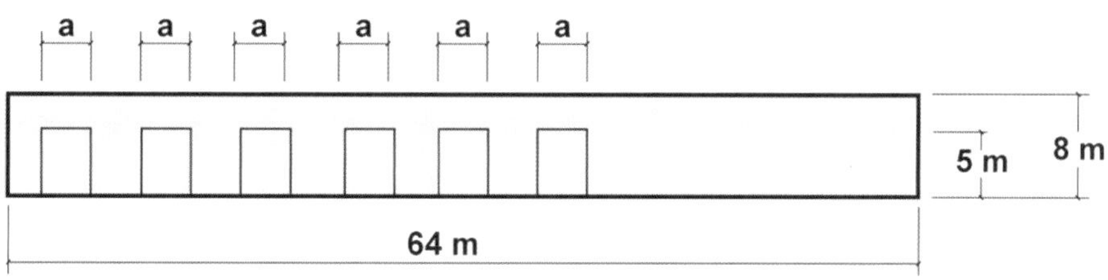

West Elevation

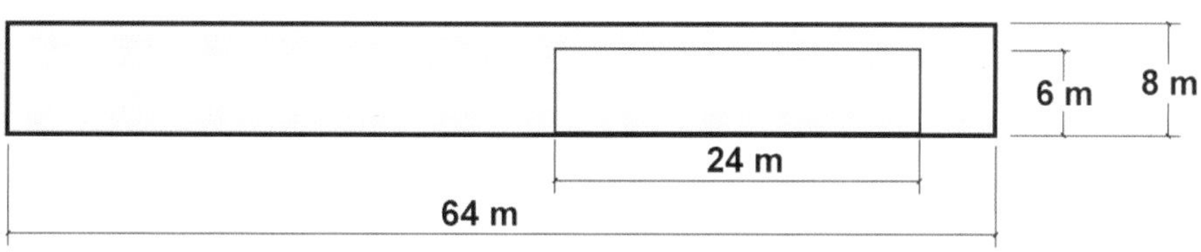

Building Volume:
= 32 x 64 x 8
= 16,384

Roof Surface Area:
= 32 x 64
= 2,048 m^2

Curtainwall Surface Area:
= 24 x 6
= 144 m^2

Door Surface Area:
= 6 (3.5 x 5)
= 105 m^2

Wall Surface Area:
= (64 x 8)2 + (32 x 8)2 – (105 + 144)
= 1,024 + 512 – 105 – 144
= 1,287 m^2

Building Surface Area:
= 2,048 + 144 + 105 + 1,287
= 3, 584 m^2

Therefore, the total heat loss can be seen to be 4103 W per degree Celsius difference (7778 Btu/hr °F). For an outdoor temperature of -10° C, indoors of 20 °C, the difference is 30 °C, and total heat loss is 123 kW.

The relative impact of each building component can also be seen. For this building type, the U-value of the enclosure is unimportant relative to the impact of air leakage. It should also be noted that the lights and equipment and people in the store may consume 30 W/m^2 of electricity in retail applications, all of which is converted to heat. Hence, if we assume an average energy use of 30 W/m^2, the interior heat gains would be:

$$2048\ m^2 \times 30\ W/m^2 = 61440\ W$$

This energy will offset the losses of 123 kW, but a heating system of some type would be needed to make up the remaining need (123-61) = 62 kW. In normal practice, engineers assume none of the energy inside a building is available to provide heating. This was acceptable in the days of poorly insulated enclosures, with modest interior heat gains, but becomes a questionable assumption in highly insulated buildings.

Example:
Consider a standard classroom 40 feet long and 30 feet wide along a double-loaded corridor. What would the heat loss coefficient be of only this classroom, recognizing that little or no heat gain/loss will occur across the partition walls (as the temperature is the same on both sides). The 12 foot high classroom has six 5 feet by 6 feet tall windows (U=0.35), opaque wall insulated to true R-value of 15, and a roof insulated to R-30. Assume ventilation and air leakage amount to 1.0 ACH.

$$A_{window} = 6 * 5 * 6 = 180\ ft^2$$

$$A_{wall} = (40 * 12) – A_{window} = 480 – 180 = 300\ ft^2$$

$$A_{roof} = (40 * 30) = 1200\ ft^2$$

$$U_{air} = 0.018 * n * V = 0.018 * 1 * (30 * 40 * 12) = 259$$

$$U_{wall} = Area/R\text{-}value = 300/15 = 20$$

$$U_{window} = U * Area = 0.35 * 180 = 63$$

$$U_{roof} = Area/R\text{-}value = 1200/30 = 40$$

$$U_{total} = 259 + 20 + 63 + 40 = 382\ Btu/hr\ °F$$

Answer:
Therefore the classroom heat loss coefficient is 382 Btu/hr °F ignoring heat loss through the slab. Thus, on a 10 °F night, when no solar or internal

gains are present the heat loss of the classroom would be 382 * (70 - 10) = 22,920 Btu/hr. As the major source of heat loss is the ventilation and air leakage ($^2/_3$ of the total) the addition of a heat recovery ventilator, demand controlled ventilation, and modest air tightening measures would be able to reduce the heat loss coeffecient to well under 250 Btu/hr °F.

A.7 ESTIMATE ANNUAL HEATING ENERGY

An approximate method, widely used before hourly computer simulations were available, is the Heating Degree Day method. This approach assumes that heating must be provided to a building below some temperature, called the balanced temperature. Above the balance temperature, heat from occupants, lighting, equipment, etc., makes up for heat loss. In the 50's and 60's, the balance temperature of poorly insulated houses was assumed to be 18 °C (65 °F), and that of offices (with the very high lighting use) was assumed to be 10 °C. Today well-insulated houses and offices will both have a balance temperature of around 10°C (50 °F). Climate data is widely available for many locations which identifies Heating Degree Days, a number that is the sum of each day of the year of the difference between the balance temperature and the average outdoor temperature for that day.

To estimate annual space heating energy then,

$$E_{annual} = U_o * 24 \text{ hrs/day} * HDD$$

If the units of U_o are in W/°C, then the result will be in Watt-hours per year. If U_o is in Btu/°F, then the results will be in Btu per year.

Example:
Estimate the annual energy demand for heating of a well-insulated building with an overall heat loss coefficient of 7778 Btu/hr °F, located in a climate with a HDD 50 °F value of 5000.

Answer:

$$E_{annual} = U_o * 24 \text{ hrs/day} * HDD$$

$$= 7778 * 24 \text{ hrs/day} * 5000$$

$$= 933{,}360{,}000 \text{ Btu}$$

$$= 933.4 \text{ MMBtu/yr}$$

Therefore the building has a very approximate heating demand of around 900 to 1000 MMBtu/yr.

Material	Density (kg/m³)	Conductivity Range (W/m K)	Conductance Range (Wm² K)
Board / sheet products			
Plywood	400–600	0.08–0.11	
OSB	575–725	0.09–0.12	
Waferboard		0.1	
Hardboard		0.105	
Vegetable Fiberboard	270–300	0.045–0.07	
Particleboard	590	0.102	
Particleboard	1000	0.17	
Strawslab	260–350	0.085–0.11	
Corrugated Metal Deck			negligible
Finishes			
Ceramic Tiles		1	
Acoustic Tiles – fiberboard		0.065	
Acoustic Tiles – glassfiber		0.036	
Gypsum Board	800–900	0.16	
Sand Plaster / Lath		0.71	
Gypsum plaster / Lath		0.16–0.35	
Sand :Cement plaster	1570	0.53	
Gypsum plaster w/perlite	720	0.22	
Gypsum plaster w/sand	1680	0.8	
Carpet Fibrous Underlay			2.73
Carpet Rubber Underlay			4.42
Terrazzo		1.8	
Hardwood Flooring		0.16	
Siding / Cladding			
Hardboard siding	640	0.094	
Wood Siding – lap		0.1–0.12	
Plywood Siding		0.09	
Face Brick – clay	2000	1.3	
Face Brick – concrete	2200	1.9	
Metamorphic Stone	2600–3000	2–2.8	
Sedimentary Stone	2200–2600	2.1–2.3	

Metal vinyl clapboard/V–groove			8
Metal – flush installed		40–80	negligible
Cement Stucco	1800	0.7–1.4	

Structural Materials

Softwood lumber	510	0.1–0.14	
Hardwood Lumber	720	0.15–0.18	
Cedar Logs and Lumber		0.098–0.12	
Concrete	2400	1.4–2.6	
Concrete (limestone)	1920	1.1–1.3	
Concrete (light)	1300	0.5–0.7	
Aerated Concrete	400	0.12–0.15	
Aerated Concrete	600	0.18–0.2	
Carbon Steel	7680	40–80	
Aluminum	2800	160–200	
Cement Mortar	1800	0.8	
Concrete Block 200 mm			5.1
Lt. Wt. Concrete Block, 200 mm			2.84
Concrete Block, 100 mm			8
Hollow tile, 100 mm			5.5
Durisol	400–500	0.072–0.085	
Adobe	1400–1800	0.4–0.8	
Clay Straw (function of density)	600–1400	0.15–0.5	
Cement–bonded rice husk	720	0.15	
Wood fiber and cement	1550	0.32	

Insulations

EPS Type 1	16	0.039	
EPS Type 2	24–32	0.034	
EXPS Type 3 and 4		0.029	
Batt Insulation		0.036–0.048	
Rigid Mineral Fiber		0.03–0.04	
Rigid Fibrous Roof Insulation		0.036	
Rigid Polyurethane		0.024	
Rigid Polyisocyanurate		0.02–0.024	
Phenolic Foam (closed cell)		0.017–0.02	
Urea Formaldehyde		0.031–0.032	

Fiberboard	270	0.052	
Cellulose Fiber	37–51	0.039–0.046	
Sawdust	145–160	0.05–0.08	
Strawbale	120–200	0.06–0.075	
Corkboard	145	0.042	
Sprayed Asbestos		0.05	
Vermiculite, exfoliated	64	130	0.06–0.07
Perlite, expanded	800	0.2–0.26	
Perlite, expanded	320	0.07–0.08	
Perlite, bonded/expanded	16	0.052	
Eel Grass batt	145–215	0.043–0.049	
Jute Resin bonded	420	0.065	
Peat slab	240–480	0.058–0.101	
Sheep's Wool, fluffy	50	0.045	

Roof Materials

Built–up Bitumen Roofing		0.17	
Asphalt Shingles			12.9
Wood Shingles			6
Crushed Stone		1.60	
Thatch–Straw	240	0.07	
Thatch–Reed	270	0.09	

Other Materials

Fresh Snow	190	0.19	
Compacted Snow	400	0.43	
Ice at –1 and –20 C	920	2.24–2.45	
Water at 20 C	1000	0.60	
Earth, dry to damp	1400–2000	0.80–2.00	
Sand dry	1400	0.30–0.80	
Air, still (conduction only)	1.2	0.03	
Glass, soda lime	2500	0.80–1.00	
Copper		380	
Lead		35	
Brass		120	
Nickel		60	
Rubber		0.2	

A.8 SOLAR HEAT GAIN THROUGH WINDOWS

When analyzing buildings for solar heat gain (beneficial in winter, usually not in summer), solar heat gain through glazing is important to assess. The peak cooling load of most buildings is controlled by direct solar gain through the windows.

The fraction of incident solar radiation that passes through a window as heat is termed the Solar Heat Gain Coefficient (SHGC). The SHGC is a property of windows that should be specified by the designer, and is readily available from most manufacturers. By definition the solar heat gain through all similar windows facing one direction is:

$Q_{solar} = A * I_{solar} * SHGC$

Where A is the area of glazing (m^2 or ft^2), and

I_{solar} is the intensity of incident solar radiation on the surface (W/m^2 or $Btu/hr/ft^2$).

Table A.4 provides an estimate of the incident solar radiation on a vertical surface at 44 °N for four orientations and four months: detailed values are available in numerous engineering textbooks. It can be seen that during warm months, the west/east sides of a building can expect around 230 Btu/hr ft^2 (750 W/m^2) and the south side will be exposed to at least 260 Btu/hr ft^2 (825 W/m^2) in fall (October).

Month	South	North	West
January 21	870 (275)	70 (22)	493 (156)
April 21	622 (197)	131 (41)	753 (239)
July 21	494 (157)	167 (53)	722 (229)
October 21	829 (263)	92 (29)	603 (191)

Table A.4: Representative "clear day" solar radiation on vertical surfaces at their peak hour for 44° N Lattitude in W/m^2 ($Btu/hr/ft^2$)

Example:
Provide an estimate for the peak solar gain that could enter through the six west-facing classroom windows, each with 1.5 m (5 ft) wide by 1.8 m (6 ft) high glazing units rated with a SHGC of 0.50.

A solar intensity of about 750 W/m^2 (240 $Btu/hr/ft^2$) would seem to be appropriate for an April day.

$Q_{solar} = A * I_{solar} * SHGC$

$= (6 * 1.5 * 1.8) * 750 * 0.50$

= 6075 Watts

or, in Imperial units

= (6 * 5 * 6) * 750 * 0.50

= 21,600 Btu/hr

Answer:

In this rather typical scenario, on a clear, sunny April or September day, over 6 kW or 21 kBtu/hr of heat would be added to the classroom by solar gain. In a well-insulated classroom (see the earlier heat loss example), this would usually provide more heat than needed even if the temperatures are well below zero, and thus require solar thermal energy storage (thermal mass), cooling or effective shading. Lower SHGC windows and/or reducing window area could also be used in conjunction with these solar gain control strategies.

A.9 THERMAL MASS

Thermal mass acts to store and release thermal energy (heat). Thermal mass is almost always beneficial for buildings with near constant interior temperature. For buildings with aggressive setback temperatures or long periods of unoccupied hours, thermal mass can cause control difficulties and may even increase energy consumption.

Thermal mass is often desirable to store solar heat gains and internal heat generation from occupancy during the day, and release it during the evening. This is one of the basic tenents of passive solar design. The limit to this strategy in all types of buildings is the rate at which heat gain can be transferred to the thermal mass during sunny periods (without overheating) not the amount of thermal mass. Because the heat loss of high performance buildings is so small, increasing thermal mass is often not useful unless the rate of heat transfer to the thermal mass is increased. This latter goal can be achieved by increasing the surface area of mass (e.g. mass more than 2 to 4” thick takes too long to charge), by ensuring that the mass is not covered by insulating or thermally lightweight finishes, and by encouraging airflow from the heated zone to remote thermal mass in other zones.

Quantifying the benefits of thermal mass to comfort and energy consumption is difficult. Hourly computer models are very helpful, but many do not give realistic results, and the coupling of gains to storage is highly sensitive to occupants and use.

For high performance buildings, the overall heat loss coefficient is low enough that the thermal mass of typical construction is quite substantial: if heating is removed it can take hours for the temperature to drop by one degree. In many high-performance buildings setback schedules will not work because the building does not cool down to the setback temperature before the temperature is set up again. This has the major advantage that a super insulated building can survive many times longer than normal buildings in the event of a power outage.

For ultra-low energy use buildings, thermal mass can be a powerful and often necessary strategy. However, it is a design strategy that is difficult to use because of the complexity of assessing actual dynamic building use, heat transfer to and from mass, the temperature swing acceptable to the occupants and other poorly characterized variables.

A.10 RADIATION-INDUCED EXTERIOR SURFACE TEMPERATURES

Radiation, either solar heating by the sun or cooling by radiation to the night-sky, can have a very significant influence on the thermal exposure of the building envelope. Although rarely warranted, detailed procedures can be used to relatively accurately calculate, in terms of energy transfer rates (e.g. W/m^2), the effect of solar heating and night-sky cooling on enclosure surface temperatures.

It is often convenient to have simple equivalent temperatures that can be entered directly into a standard conductive heat flow analysis. When an equivalent air temperature is used, the temperature is termed the *sol-air temperature*. This procedure has the drawback that air temperature is no longer realistic and the advantage that surface films play a role in the definition of the temperature. The value for the overall surface film coefficient is often assumed to be 17 W/m^2 °C although this coefficient is quite variable, especially if the surface becomes hot. In fact, under solar-heating, the coefficient is likely to be as much as two times larger.

A simple and equally realistic approach is to use *approximate surface temperatures*. Table A.5 is a compilation of the most useful approximate surface temperatures based on calculation (using refined surface heat transfer coefficients) and supported by several years of field measurements.

If the example wall in Section 3 were exposed to bright sun on a -10 °C day, the temperature of a very dark, lightweight, south-facing surface would be expected to be as much as 48 * 1 = 48 °C above the air temperature (see Table A.5). For this example, we will assume dark thermally-massive red brick (absorptance =0.85) and thus estimate a surface temperature of about 35 * 0.85 = 30 °C above the air temperature. This means:

$$\text{Heat flow} = U * (t_{inside} - t_{outside})$$

$$= 0.33\ [20 - (-10+30)] = 0.33\ (0)$$

$$= 0\ \text{W/m}^2$$

Answer:
If the wall were facing south, dark red, thermally massive, and exposed to bright sunshine, the heat flow would be approximately zero.

Table A.5: Approximate radiation-induced surface temperatures (°C)

Situation	Thermally Massive	Thermally Lightweight
Roof: direct sun	$t_a + 42\ \alpha$	$t_a + 55\ \alpha$
Roof: sun plus reflected/ emitted radiation	$t_a + 55\ \alpha$	$t_a + 72\ \alpha$
Roof: exposed to night-sky	$t_a - 5\ \varepsilon$	$t_a - 10\ \varepsilon$
Walls: winter sun	$t_a + 35\ \alpha$	$t_a + 48\ \alpha$
Walls: summer sun	$t_a + 28\ \alpha$	$t_a + 40\ \alpha$
Walls: exposed to night-sky	$t_a - 2\ \varepsilon$	$t_a - 4\ \varepsilon$

Notes: t_a refers to the ambient air temperature, **ε** is the surface emmittance, and **α** is the solar absorptance. The emmittance is about 0.90 for most materials. The absorptance varies with color from about 0.3 for white/beige objects to 0.65 for forest green to 0.95 for flat black.

- Thermally massive refers to walls with a significant amount of thermal storage capacity (e.g. brick veneer or equivalent) outside of a low conductance material (e.g. insulation). Walls with significantly more mass (e.g. multi-wythe brick, rubble) or less insulation (U > 1 W/m²/°C) will be less affected. Vinyl, metal and EIFS are lightweight walls; other walls can be interpolated based on heat capacity.
- All values are for approximately 35°-55° North.
- The values for are for likely clear night or sunny day maximums and apply to east/west orientations in summer and south orientations in the winter.

Appendix B Supporting Articles

B.1 Historical Development of the Building Enclosure

B.2 The Building Enclosure

B.3 Why Energy Matters

B.4 Building Energy Performance Metrics

B.5 Understanding Primary/Source and Site Energy

B.6 Deciding on Energy Priorities When Building New

B.7 The Perfect Wall

B.8 Thermal Control in Buildings

B.9 Can Highly Glazed Building Facades be Green?

B.10 Energy Flow Across Enclosures

B.11 Airflow Control in Buildings

B.12 Building in Extreme Cold

B.13 The Perfect HVAC

Appendix B.1

Insight
Historical Development of the Building Enclosure

By John Straube, Ph.D., P.Eng.

In traditional mass walls, e.g. a wall of solid masonry or earth, the resistance to rain penetration was only one aspect of enclosure performance (**Photograph 1**). Heat flow was also controlled by the thermal storage capacity of the massive walls, not just by virtue of the materials' thermal conductivity like the specialized insulation layers commonly used in modern building assemblies. The sun's heat was absorbed, stored, and slowly released to the interior and exterior, effectively damping typical daily fluctuations and thus increasing comfort. Vapor and airflow were also controlled by the mass of the wall. It is little wonder that such walls were used for thousands of years. Built of only brick and mortar, the wall carried all structural loads as well as performing as an acceptable enclosure. The small unit size of the brick allowed for planning flexibility so that such walls could be used for most purposes. Because mass brick walls allow a considerable amount of heat to pass through, the exterior surface temperature remained elevated throughout the winter and thus freeze-thaw durability and interstitial condensation problems were avoided. Compared to the poor control of airflow through windows and doors, the walls seemed airtight to the occupants. If the wall was sheltered by topography, other buildings, and roof overhangs, the amount of rainwater reaching the surface was so little that the wall could control this water before it reached the inner surface and caused damage. The biggest drawback to such wall systems was the large amounts of material and labor needed to construct them and the poor thermal control.

Photograph 1: Traditional mass walls

With the change from low-rise buildings with solid load-bearing walls to taller framed buildings, the dead weight and cost of traditional mass wall systems became prohibitive. Chicago's 16-storey Monadnock Building, constructed with 6-foot thick base walls between 1889 and 1891, pushed to the limit the load-bearing mass masonry wall (**Photograph 2**). Taller buildings with mass walls were practically impossible with the combination of high dead weight and low compressive strength. A large percentage of valuable ground floor area was lost to load-bearing walls and the resistance to seismic loads was poor. Today, poor control of rain penetration, heat, air, and vapor flow can be added to the list of drawbacks.

Photograph 2: Monadnock Building
(www.monadnockbuilding.com)

The industrial revolution and the scientific knowledge and technical confidence it provided resulted in attempts to produce perfect barrier wall systems. These systems very often fail to be perfect barriers because of defects in design, construction, or materials although they may still perform as required. While a unit of sealed glazing will not fail to resist rain (unless the glass cracks) the joint between the glazing and the window frame may. Similarly, metal panel systems developed in the post-war period rarely failed, but the joints and interfaces did. These examples reinforce the importance of considering the wall as a three-dimensional assemblage including joints. In many manufactured curtain walls, a small amount of rain penetration will cause no harm and either goes unnoticed or a drainage system is incorporated to deal with these small failures.

Corbusier is largely credited with popularizing the idea of separating the primary structural system from the enclosure system. Although the concept itself was well-developed by his time, the Domino house project made this approach desirable (**Photograph 3**). However, it is only in recent decades that the separation of the enclosure into layers and sub-systems for specific functions (support and control) has become more widely accepted and actually applied to building enclosures.

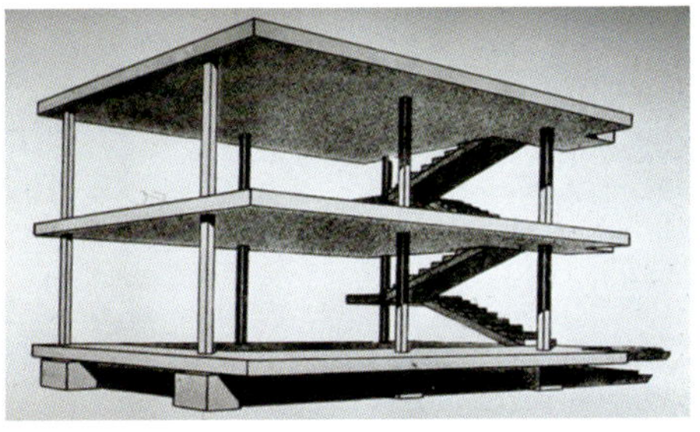

Photograph 3: Le Corbusier's Domino House
(www.usc.edu)

The current best practice in building enclosure design emphasizes the use of drainage as a rain control strategy, and demands a well-defined rain control layer, air control layer, and unbroken thermal control layer. Building science research and field experience over the last two decades have demonstrated how powerful the drained approach to rain control can be. However, other changes have also occurred over this time, specifically the use of air barriers, and steadily increasing insulation requirements. The increase in airtightness and thermal control (insulation, white roofs, radiant barriers) reduced the energy flow across the enclosure available to dry this remaining moisture. Hence, the potential duration of wetting for materials in high-performance enclosures is increasing, and this can cause durability problems.

Drainage does not remove all water that penetrates the cladding, as any rainwater absorbed by materials or clinging to surfaces can only be removed by evaporation. Similarly, air leakage condensation, which is now more likely in frequency, and severe in intensity because of higher levels of thermal insulation and higher cold-weather interior humidity levels (themselves the result of increased airtightness), cannot be dried as quickly as in the past. This lack of drying capacity, when combined with changing materials (masonry to gypsum sheathing, brick veneers to metal panels) and the substitution of traditional materials with often less durable modern ones, has increased the probability of moisture-related enclosure failures.

What is needed is a re-evaluation of how we assemble enclosures, and improvements in ensuring continuity of the control functions. As we change the insulation levels, airtightness, and materials, we need to consider changes in how materials are assembled in enclosures. The steel-stud framed walls of the 80's and 90's cannot continue to be built in the same manner in the 2010's and 2020's. It is now clear that such walls did not provide continuous thermal control. Better rain control is a critical part of the needed change, as are increases in tightness, better control of thermal bridging, and a protection of moisture-sensitive materials from extreme temperatures and prolonged wetting. The "perfect wall" approach (**Figure 1**) described above provides all of these improvements. Once thought of as an ideal enclosure assembly that would rarely be built, it is becoming the new standard for durable, energy-efficient, high-performance enclosures.

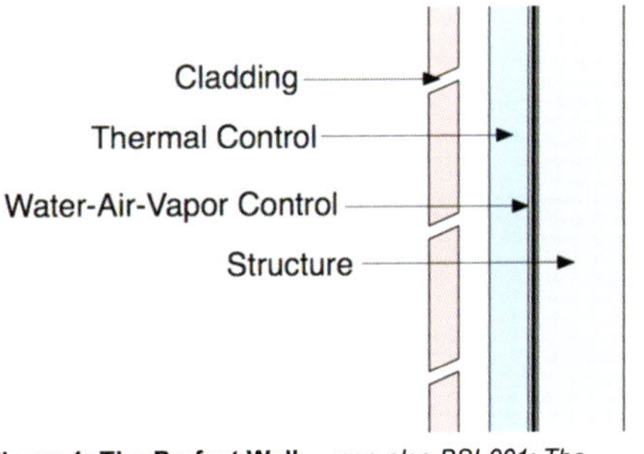

Figure 1: The Perfect Wall — *see also BSI-001: The Perfect Wall (www.buildingscience.com/documents/insights/bsi-001-the-perfect-wall/view)*

Appendix B.2

Building Science Digest 018

The Building Enclosure

2007-07-11 by John Straube

Abstract:

That part of any building that physically separates the exterior environment from the interior environment(s) is called the building enclosure or building envelope. Environmental separator is another term used to describe the enclosure, but note that this generic term also applies to separators of two different interior environments. The term building enclosure is preferred to the term building envelope largely because it is considered both more general and more precise. Also note that the building enclosure may contain, but is not the same as, the so-called thermal envelope, a term that is used to refer to the thermal insulation within the enclosure.

The enclosure, the loadings it must resist, and its functions are addressed in this digest.

The Nature of the Building Enclosure

Both the above-grade and the below-grade portions of the building enclosure are part of a physical system involving three interactive components: the exterior environment(s), the enclosure system, and the interior environment(s). The exterior environment above grade is very different from that below grade, and within any building there can be numerous interior environments. The nature of the building enclosure and its spatial relationship with the other parts of the building are shown in Figure 1.

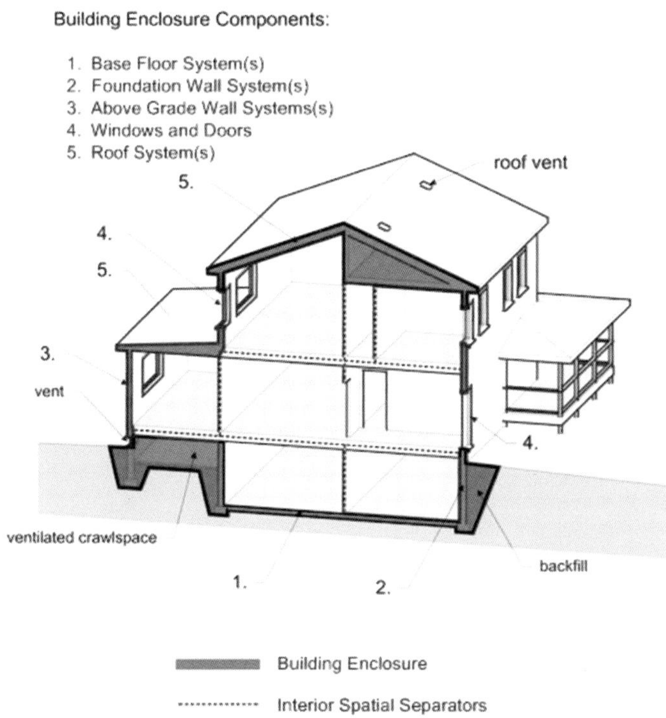

Figure 1: The building as a set of separated spaces and the as-built separators

As suggested by Figure 1, a building in general consists of a collection of spaces bounded by a set of spatial separators. There are separators between interior environments as well as separators of an interior environment and the exterior environment; collectively, the latter constitute the building enclosure. In many respects, the functions of a separator of internal environments within a building (floors and interior walls) are very similar to those of the building enclosure. Usually, however, the performance requirements of internal separators are fewer and much less onerous.

Enclosure Components

The primary *function* of the enclosure is to **separate** the interior environment from the exterior environment to which it is exposed. Physically, the typical building enclosure usually consists of the following components:

- the roof system(s)
- the above-grade wall system(s) including windows (fenestration) and doors
- the below-grade wall system(s), and
- the base floor system(s).

The building enclosure should not be thought of as a combination of numerous one-dimensional or even two-dimensional planar components. Each enclosure component is a three-dimensional, multi-layer, multi-material assembly that extends from the inside face of the innermost interior layer (e.g., the paint or wallpaper) to the outside face of the outermost layer (e.g., paint or roof shingles). The overall enclosure is made up of all the contiguous enclosure sub-assemblies.

Each enclosure component is an assemblage of layers of specified products (such as gypsum board or wallpaper) or materials (such as paint or wood). For instance, the thermal insulation could consist of a layer of blown-in glass fiber filaments. A deliberate air space or cavity is also considered to be a layer. Consider, for example, the sloped roof system shown in Figure 1; this constitutes an enclosure assembly consisting of all the contiguous layers between the finish on the ceiling and the outer face of the roof shingles. Furthermore, the compacted gravel under the basement floor (or base floor) system and the backfill to the outside of the foundation or basement wall system are each specified layers within their as–built assemblies. As such, they can be considered to be an integral part of their respective building enclosure components. By common definition each enclosure sub-assembly incorporates all the contiguous (in-contact) layers.

The climate-related loadings that a building and its enclosure actually experience are modified versions of the local climate. These microclimates are modified by adjacent buildings, the landscaping (especially trees), and other parts of the enclosure. Roof overhangs, for instance, are climate modifiers to the walls below as well as integral parts of the roof system.

The above definition of the enclosure has some limitations. First, the physical difference between site microclimate modifiers and the enclosure is not that straightforward. Second, although one can define the enclosure as the building component that separates interior from exterior environments, it is not always easy to precisely define "interior" and "exterior." For example, the backfill around a basement wall or ivy growing on a masonry wall are, by definition, considered to be part of the enclosure. They are also part of the site. What cannot be ignored is the significant impact of this soil or ivy on moisture and thermal loadings.

Consider also an attached garage. The space in the garage does not experience outdoor conditions, nor is it conditioned like the interior of the building. The enclosure could be considered to incorporate both the exterior wall of the house and the garage wall, with the space between (containing the car) being an integral part of the enclosure. Alternatively and preferably, the enclosure could be considered to be the garage wall, and the intermediate wall could be viewed as an interior separator with significantly different conditions on each side.

Similarly, two adjacent internal spaces can also have quite different environmental conditions. For example, consider the not uncommon scenario of a hotel with a warm, humid swimming pool next to an air-conditioned exercise room or a community center containing both a swimming pool and ice arena. In such cases the loadings and performance of the interior separator are more like those of a building enclosure component. In this type of condition enclosure design principles should be applied to the interior separator.

The influence of climate modifiers, such as trees, garages, overhangs, and the soil on the loadings experienced by the enclosure is not always appreciated. Especially in the case of low-rise buildings, the effect on the enclosure from the exterior environment can be intentionally reduced or moderated by the use of microclimate modifiers such as plantings, fountains, building overhangs and windbreaks. For example, a roof overhang (an integral part of the roofing system) will moderate the amount of rain on the wall below and will therefore be a microclimate moderator for the vertical portion of the building enclosure. With taller buildings it is more difficult to modify the external microclimate.

The definition of where the enclosure begins and exterior environment stops can sometimes be confusing, especially in the case of buffer spaces such as garages, screened porches, vented and storage attics, or vented crawlspaces.

The exterior environment could be considered to be a three-dimensional space with randomly varying mass and energy properties. The local climate and weather provide the major, but not the only, exterior loadings. Local climatic records   for instance, average and extreme values   are commonly available but these data are usually based on open-field, weather-station records. The building may be some distance from the weather station, and the terrain (hills, other buildings, etc.), the landscaping (trees, shrubs, etc.) and the building itself (overhangs, protrusions, etc.) can moderate the weather that each enclosure component is actually subjected to. As a matter of convention, that portion of the exterior environment that is close to, and affected by, the building is called the *external microclimate.* In fact, different parts of the enclosure are subjected to different exterior microclimates.

The interior environment is usually specified in relation to the physical needs of people and is defined in terms of temperature, relative humidity, airflow rate, and air quality. However, sensitive equipment, materials, or processes (e.g., computers, archived paper, electronics factories) may require different, often more stringent, conditioning requirements than those required by (adaptable) human occupants. Special uses, such as swimming pools, pharmaceutical plants, ice arenas, etc have significantly different conditions than normal. Because the interior spaces are usually conditioned, the state of the interior environment is often assumed to be constant, but in practice, significant variations occur over the day, between different spaces and over the seasons.

The service systems (e.g., lights, motors, stoves) and the contents of a space have a strong and time-dependent influence on the interior environment. The level of the influence depends on the controls, output capacity, and distribution effectiveness.

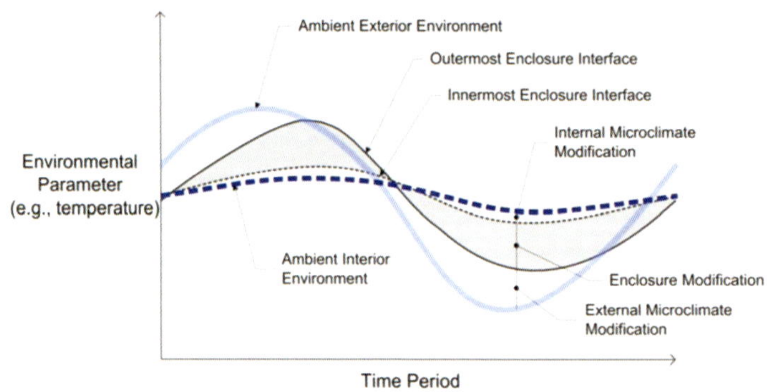

Figure 2 indicates that, while the building enclosure separates the interior and exterior environments, it really experiences several microclimates. The enclosure interacts with both environments and, in turn, affects both environments. This interaction is usually dependent on the time of day, the day of the week, and the season. There is thus a cyclical (diurnal, workday/non-workday, weekly and seasonal) aspect to the time-dependent response of the enclosure as it separates and, to some degree, modifies the influence of both environments.

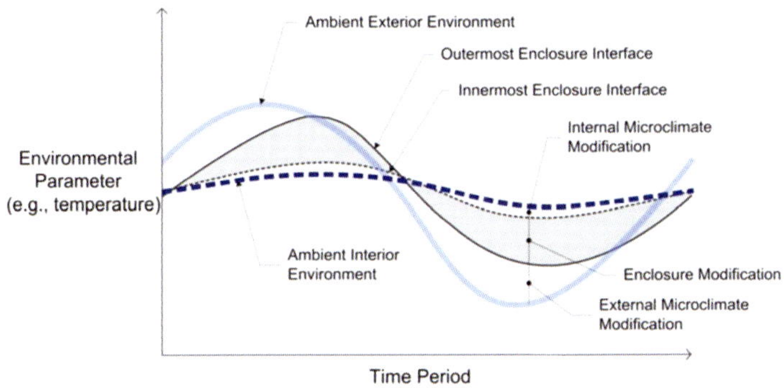

Figure 2: Contributors to environmental modification

While the enclosure may be the largest modifier, the difference or the modification between the ambient and interface conditions can also be significant. Note that the scale of

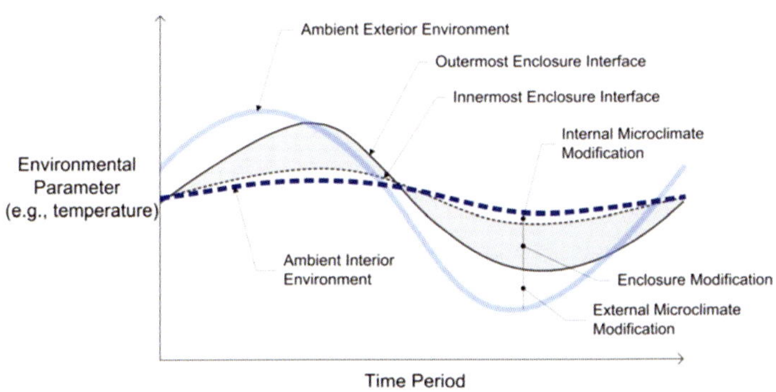

Figure 2 will be different for different parameters. Moreover, this simple representation of variation over one cycle of time for one parameter does not accurately or fully represent the complex behavior and interactions that occur.

The properties of exterior environments and interior environments are separate and major fields of study. Each environment is discussed in other Building Science Digests, Primers, and Design Guides.

Enclosure Loadings

The performance of any portion of the enclosure should be considered in relation to the various loadings generated by i) the exterior environment, ii) the interior environments, and iii) the enclosure itself. The generic term *loading* refers to any event, phenomenon or characteristic that can affect the enclosure. Each and every loading that impacts the building enclosure can be categorized into one of just five types, namely:

- gravity related
- ground related
- heat related (thermal)
- moisture related
- air related

Exterior Environment Loadings

The exterior environment described above is responsible for many different loadings. Sources of these loadings are the climate (weather over the short term), human or human-made effects, and nature or natural phenomena. Table 2.1 identifies, by type and source, those loadings due to the exterior environment that could affect the enclosure. It should be evident from this list of loadings that the design or analysis of the enclosure involves consideration of not only the usual structural loadings but also mass (air, moisture, etc.) and energy (heat, light, sound, etc.) loadings. Moreover, for design purposes, we need to have some knowledge of both average and extreme conditions for each relevant loading.

Table 2.1: Loadings from the exterior environment

Source \ Type	Heat related	Moisture related	Air related	Ground related	Gravity related
Weather or natural climate	Ambient conditions, solar	RH, fog, rain, ice, snow	Barometric pressure, wind		Water, snow, hail
Abnormal climatic effects	Reflected solar, lightning	Tornado, hurricane, flooding	Tornado, hurricane	Frost heave, landslide	Wind-borne missile
Natural phenomena	Fire, Ground water	Adfreezing, Freezing	Radon, methane, soil gas	Seismic, land- slide, settlement, termites, plants, etc.	Hydrostatic pressure, soil pressure
Human-made weather	Global warming, city effect (2-7 ^{0}C)	Smog, Acid rain	Wind related vortex/swirl		
Human-induced events	Fire	Fire (hoses, sprinklers, etc.)	Smoke, sonic boom, sound, explosion		Impact, wear and tear

Interior Environment Loadings

The **interior environment** consists of occupied, used and, often, conditioned spaces. As shown in Table 2.2, the main sources of loading are human-induced, the result of natural phenomena or due to conditioning. Variations in the state of the interior "climate" and the related extreme values tend to be smaller or less than those occurring within either the enclosure or the exterior. This occurs not only because the building enclosure modifies the effect of exterior variations but also because the interior environment is generally augmented by inputs of mass and energy, e.g., the building or parts of it may be mechanically conditioned.

For a building to successfully meet its desired performance, the required state of each of the interior environments must be ensured. The state of an interior space may either be maintained constant or may vary in time as a function of the nature and degree of space conditioning, internal gains (equipment, lighting, solar effects and human activities) and the modifying influence of the enclosure.

Table 2.2: Loadings from the interior environment

Type / Source	Heat related	Moisture related	Air-flow related	Ground related	Gravity related
Interior space	Ambient conditions, solar	RH, water (sprinklers, etc.)	Barometric pressure, wind, stack, fan induced		Water
Natural phenomena	Fire	Fungal growth, mold	Radon, methane	Settlement, termites, plants, etc.	
Human-induced events	Fire, people	People, flooding, combustion, equipment	Smoke, sound, explosion		Impact, wear and tear, dead & live loads

Loadings from the Enclosure

The enclosure itself is a source of loading and, as shown in Table 2.3, loadings arise from the enclosure element under consideration as well as adjacent elements.

To properly design buildings and their enclosures it is necessary to have some knowledge and understanding of the relevant environments as well as their interdependence. A considerable body of empirical and experimental data exists on the interaction of climate, site and building [Watson and Labs 1983, Olgyay 1963, Brown 2001, Moore 1993]. Research, notably that by Fanger [1970] and Givoni [1981], has identified and defined the physiological aspects of human response to both interior and exterior climates, and these data have been distilled into convenient forms for building designers [ASHRAE 2001].

The enclosure is only one part of the larger system that moderates or controls the environments on both sides (especially the inside) of the enclosure at some location on the building. The building itself and the enclosure both influence the exterior microclimate that interfaces with the exterior surface of the building enclosure at a specific location. Consider, for example, the local impact of wind, rain, and solar radiation.

2. **Control functions**, i.e., to control, regulate and/or moderate all the loadings due to the separation of the interior and exterior environments largely the flow of mass (air, moisture, etc.) and energy (heat, sound, etc.).

3. **Finish functions**, i.e., to finish the enclosure surfaces the interfaces of the envelope with the interior and exterior environments. Each of the two interfaces must meet the relevant visual, esthetic, wear and tear and other performance requirements.

A fourth building-related category of functions may also be imposed on the enclosure, namely:

4. **Distribute functions,** i.e., to distribute services or utilities such as power, communication, water in its various forms, gas, and conditioned air, to, from, and within the enclosure itself.

Figure 3 illustrates a representative portion of a building enclosure, its functions and the nature of the loadings. The enclosure may also serve other, usually non-physical, purposes such as advertising or as a symbol or image, e.g., as a statement about power or security or wealth, etc., but these are not considered here.

Each and every part of the enclosure must satisfy the relevant support, control, finish and distribution functions. Only the support and control functions are needed everywhere. The finish functions may not be needed in some areas (such as above suspended ceilings, in service rooms). Moreover the distribution functions, which largely service the adjacent interior spaces, need to be met only where there is a service or utility to be distributed – large segments of most enclosures do not need to fulfill this function.

To further demonstrate the relationship between category of function and loading, consider Table 2.4. All the possibly relevant loadings are listed and their functional relevance is identified. This table is important because it clearly shows that the number of loadings is considerable and that each loading affects at least one functional sub-category for the enclosure and often more than one.

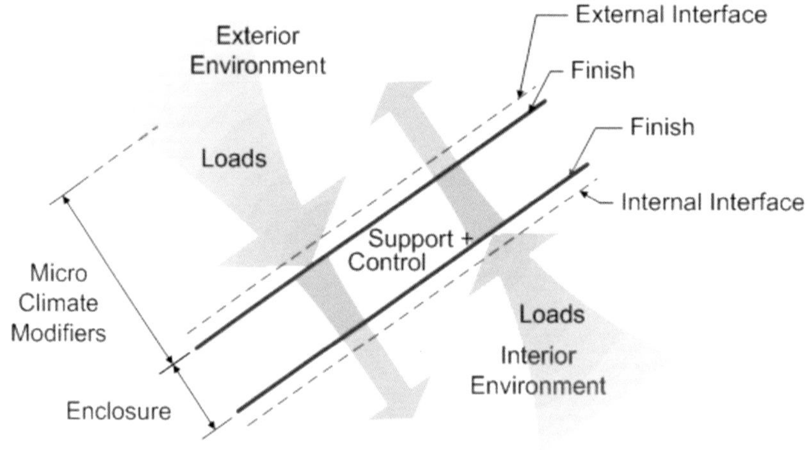

Figure 3: The enclosure and its functions

Table 2.4: General category of loadings and related functions

Causal phenomenon or loading	Specific loadings	Interior	Support	Control	Exterior
		Category of functions			
Essentially structural	Gravity – Dead (assembly, etc.)		●		
	Gravity – Live (people, snow, etc.)		●		
	Wind		●	○	
	Ground Movement (seismic, settlement, etc.)		●		
	Explosion		●		
	Rheological (creep, shrinkage, etc.)		●		○
	Impact (vehicles, missiles, people, etc.)		●		
	Fire		●		
Essentially environmental	Heat (thermal, etc.)		○	●	
	Air (pressure, movement, leakage, etc.)		○	●	
	Moisture (built-in, precipitation, etc)		○	●	○
	Smoke			●	
	Solar radiation (incident, reflected, etc.)			●	○
	Chemical attack/atmospheric (acid rain, etc.)			●	○
	Particulate matter (dust, VOC's, etc.)			●	
	People (wear & tear, etc.)	○		●	
	Insects, birds, animals, (termites, rodents, etc.)			●	
Essentially perceptual	Light (natural, incandescent, fluorescent, etc.)			●	
	Sound	○		●	○
	Visual – local	•			•
	Visual – contextual	•			•

Primary significance ●

Secondary significance ○

Tertiary significance •

Enclosure Performance Attributes

In any built facility the building enclosure must perform satisfactorily, i.e., each and every loading must be controlled or resisted (i.e., managed). There are the physical functions (previously categorized as support, control, finish and distribute functions) as well as qualitative functions or attributes. Ideally all of the attributes listed below need to be satisfied:

- constructability (or buildability)
- economic viability or cost over both the short and long term
- viewability including aesthetic, cultural and other visual expectations
- utility
- sustainability (renewable, non-toxic, etc)
- serviceability (perform in service)
- safety in relation to life, health, injury, property and economic enterprise
- productivity (in construction and installation)
- operability (for users)
- maintainability (for operators)
- repairability (for operators)
- durability
- convertibility in relation to modification, extension or conversion
- disposability.

This list identifies each of the general attributes that apply not only to the building enclosure but also to the building materials, sub-systems, building, and the building facility. Except for the first and last two, all are attributes of in-service performance.

Design

The three primary design dimensions of any building have been identified and discussed, i.e., the possible loadings, the relevant physical functions, and the attributes of performance. These apply not only to the building enclosure but also to the other physical components of the building.

Function, attribute and loading each constitute a dimension for design in the sense that the design team needs to systematically and explicitly consider each function, each attribute and each loading and all relevant combinations of these loadings. In each step a technical design criterion is set up in such a manner that the modeled or computed response is demonstrably better than or satisfactory when compared to the permissible performance threshold. In mathematical terms this process can be written as follows:

$$_{c} \quad _{w}\left\{ R^{m} \qquad R^{s} \right\}_{\mathrm{min}\,\$}^{\mathrm{max}}$$

For all (i.e., the sum of) relevant combinations of loadings (c) and for all relevant loadings (w), the modeled or computed response (R^m) must be demonstrably better (larger or smaller as the case may be) than a specific limit or threshold value (R^s) for this particular response. This conditional criterion needs to be resolved for all relevant situations and this has to be accomplished at minimum cost ($) and for maximum benefit (β).

This approach has been widely used in limit states design for structural systems, but is not yet well developed for building science applications such as building enclosure design. Such a rational approach has not been applied to enclosure design because of a lack of quantitative understanding (of building physics and material properties) and no professional and research framework to create the approach for analysis, loadings, and performance thresholds. This may change in the coming decades.

Closure

In this digest a concerted attempt has been made to develop a systematic, logical and consistent framework or context to study, analyze and design both buildings and the building enclosure. Numerous researchers [Markus and Morris 1980, Hutcheon 1963, Elder 1974, Allen, 1980] have addressed this issue, most notably and comprehensively James Marston Fitch [1960, 1975]. Fitch's writings, particularly *The Aesthetics of Form* and *American Building*, deserve to be much better known. Interested readers should review these sources.

References

AJ Handbook of Building Enclosure, Ed. A.J. Elder, The Architectural Press, London, 1974.

Allen, E., *How Buildings Work*. Oxford University Press, Oxford and New York, 1980.

ASHRAE Handbook of Fundamentals, American Society of Heating, Refrigeration and Air-Conditioning Engineers Inc., Atlanta, Georgia, USA, 2001.

Brown, G.Z. and DeKay, M., *Sun, Wind and Light. Architectural Design Strategies*, Second Edition, Wiley, 2001.

Fanger, F.O., *Thermal Comfort, Analysis, and Applications in Environmental* Engineering. McGraw-Hill Inc., New York, 1970.

Fitch, J.M., and Branch, D.P., "Primitive Architecture and Climate", *Scientific American*, Vol. 203, pp. 134-145, 1960.

Fitch, James M., *American Building: The Environmental Forces That Shape It*. Schocken Books, New York, 1975.

Givoni, B., Man, *Climate, and Architecture*. Van Nostrand Rheinhold, New York, 1981.

Hutcheon, N.B., *Requirements for Exterior Walls*. Canadian Building Digest 48. Division of Building Research, National Research Council, Ottawa, 1963.

Markus, T.A. and Morris, E.N., *Buildings, Climate, and Energy*. Pitman Publishing, London, 1980.

Moore, F., *Environmental Control Systems*. McGraw-Hill, New York, 1993.

Olgyay,V., *Design with Climate*. Princeton University Press, Princeton, New Jersey, 1963.

Watson, D., and Labs, K., *Climatic Building Design*. McGraw-Hill Inc., New York, 1983.

Appendix B.3

Insight
Why Energy Matters

An edited version of this Insight first appeared in *Perspectives*, Volume 17, Number 1. Spring 2009.

By John Straube, Ph.D., P.Eng.

The environmental crisis, and hence green building design, revolve around a wide range of issues: habitat destruction, stormwater run-off, air pollution, climate change, and resource use. However, the on-going consumption of energy to operate, condition, and light a building, as well as the energy embodied in on-going maintenance is the largest single source of environmental damage and resource consumption due to buildings. Energy security and carbon emissions have signaled an even stronger focus on energy in green buildings, particularly as the energy consumption growth rate of countries such as China, Russia, India and Brazil increase.

Reducing the operational energy use and increasing durability should be the prime concerns of architects who wish to design and build "green" buildings. I have reached this conclusion after spending years looking at actual building energy consumption, reviewing countless computer simulations, and being involved in numerous green building charrettes. It has even been suggested (Lstiburek, 2008) that 80% of a green architect's concern should be directed towards reducing energy consumption during operation.

Scientific life-cycle energy analyses have repeatedly found that the energy used in the operation and maintenance of buildings dwarf the so-called "embodied" energy of the materials. Cole and Kernan (1996) and Reepe and Blanchard (1998) for example found that the energy of operation was between 83 to 94% of the 50-year life cycle energy use (**Figure 2**).

Despite the massive amount of evidence pointing to the importance of energy consumption to green design, designers and even rating programs like LEED still seem fixated on material choices, not energy reduction. Perhaps the lack of attention to this critical issue is a result of the fact that architects are trained in the arrangement of spaces, massing, cultural influences, and the selection of finishes. Designing buildings that consume little operational energy is not

Photograph 1: Energy Matters—Building owners and designers will be increasingly concerned with energy use, energy efficiency, the choice of energy source, the distance of that source from their building, and the global impact of their building and operations.

a normal skillset: it requires much more quantitative understanding of heat transfer and radiation physics, weather and sun, and mechanical equipment operational details than architects normally possess. Nevertheless, it is the architect that makes most of the significant choices that impact energy consumption.

The trend in the last decades has not been encouraging. According to the Department of Energy's Commercial 2003 Building Energy Consumption Survey, the energy consumption is not less than buildings built before 2003 (**Figure 3**). A New Buildings Institute study has shown that LEED buildings use essentially the same amount of energy. The real savings from improved window technology, more efficient equipment, and better design tools have disguised the fact that we are wasting more energy because of over-ventilated, over-glazed, and under-insulated buildings.

If a building's orientation, massing, window area/shading, insulation arrangement, and airtightness are not properly optimized, no amount of mechanical engineering, heat pumps, chilled slabs, natural ventilation or green materials can make the building a "low-energy" building. And if it is not a low-energy building, it is not a green building.

Complex underfloor air systems, double-facades, green roofs, and heat pumps are all technology with potential, but will not save energy if the architect's design (let alone the mechanical engineer's) is not appropriate. In many cases these technologies are added to a poor building, and the result is average energy consumption despite the use of good, or even exceptional, mechanical design and equipment. Efficient and high-tech equipment and controls can moderate—but not make up for—bad building design.

The solution to this problem begins with awareness of the importance of operational energy consumption to environmental damage, resource depletion, habitat destruction, and hence, to green buildings. Solutions will take many forms, but all will involve prediction of energy consumption, and confirmation that the designed low-energy building is actually built and operated as one. Only then do we have proof it is a green building. The Architecture 2030[1] challenge is based entirely on this

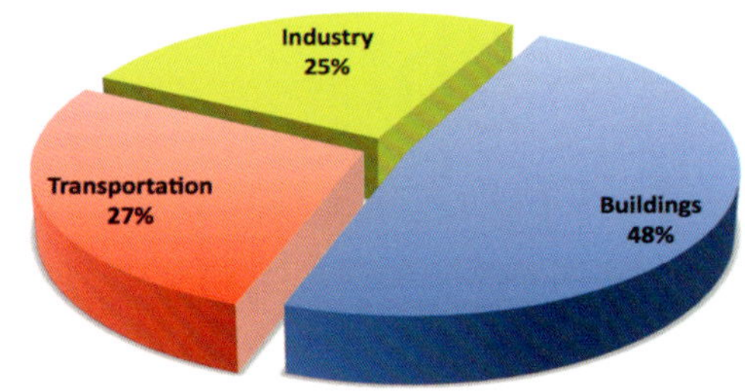

Figure 1: US Building Energy Use (from architecture2030.org)

premise. The choice of a minimum 50% reduction in energy use per building floor area has the major advantages that it is economically achievable today, with today's technology, skills and materials; is significant enough to make a real difference; and does not require any great precision (that is, one can be in error by 5% in design, operation, or measurement and the building remains a great success).

Energy consumption should be measured in terms of environmental damage, such as carbon emissions, resource depletion, or habitat destruction. None of these impacts are easily measured. However, every month a small army of people (meter readers) are deployed around the country to measure the energy used in buildings, and this energy use data is remarkably precise, condensed into easy-to-read formats, and mailed to building owners. Therefore, the most readily available and scientifically rigorous measure of energy is in units of Btu, megajoules, or kilowatt hours as provided in energy bills. One disadvantage to this approach is that all energy is not equal when it comes to environmental impact. In many places in North America, electrical energy delivered by the grid is more polluting, from a greenhouse gas perspective, than natural gas. However, there are simple correction factors available. Measuring energy in the form of dollars presumes that environmental damage is related to dollars. It is not, and this format should be avoided.

Meeting the challenge of delivering low-energy buildings will, however, require a different approach to design, the development of new skills, and a focus on new priorities. Checklists and rating programs (like LEED, BuildGreen, and GreenGlobes) are neither sufficient nor necessary,

1 Architecture 2030 a non-profit, non-partisan and independent organization, established in response to the global-warming crisis by architect Edward Mazria in 2002. Its mission is to rapidly transform the US and global Building Sector from the major contributor of greenhouse gas emissions to a central part of the solution to the global-warming crisis. Its goals have been adopted by hundreds of firms and organizations.

but could be complementary to the quantifiable end goal of low energy consumption.

Issues of recycled content, low embodied energy, and natural ventilation are not unimportant. However, if these concerns distract so much that a low-energy building does not result, then the environment is risked. LEED could take a true leadership role if Energy and Atmosphere credits were weighted far more heavily, or if the prerequisite reduction in energy intensity (based on real buildings) were raised to 30 or 50%.

The operational energy use of buildings is their biggest environmental impact. Green buildings, which must be low-energy buildings, need to be designed to respond to this reality.

References

Cole, R., and Kernan, P. "Life-cycle Energy Use in Buildings", *Building & Environment*, Vol. 31, No. 4, pp. 307-317, 1996.

Reppe, P, and Blanchard, S. *Life Cycle Analysis of a Residential Home*, Report 1998-5, Center for Sustainable Systems, University of Michigan, 1998.

Lstiburek J., *Building America*, ASHRAE Journal, Dec. 2008.

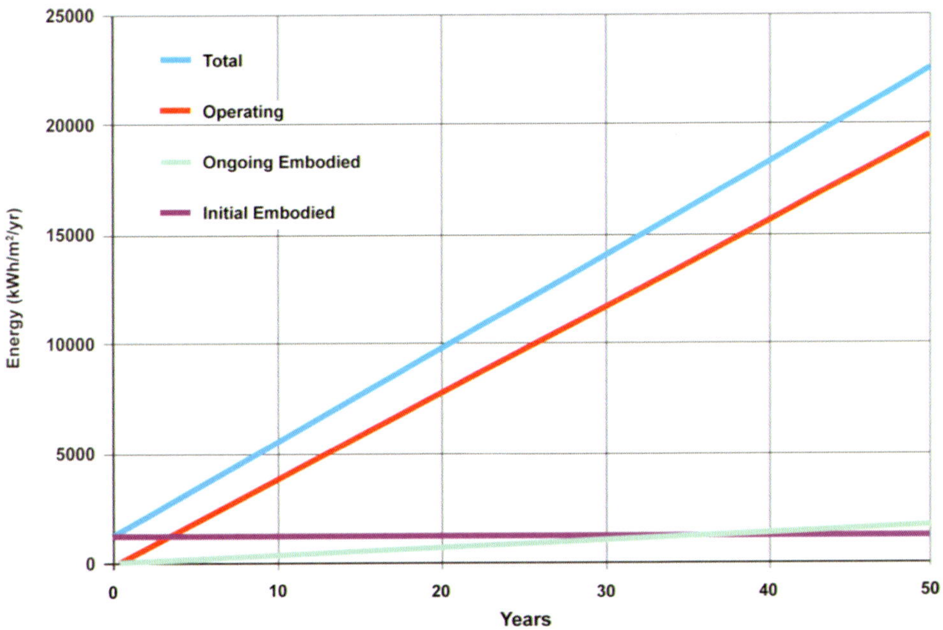

Figure 2: Embodied energy is swamped by operational energy in almost all building types (based on Cole & Kernan, Bldg & Environ, 1996)

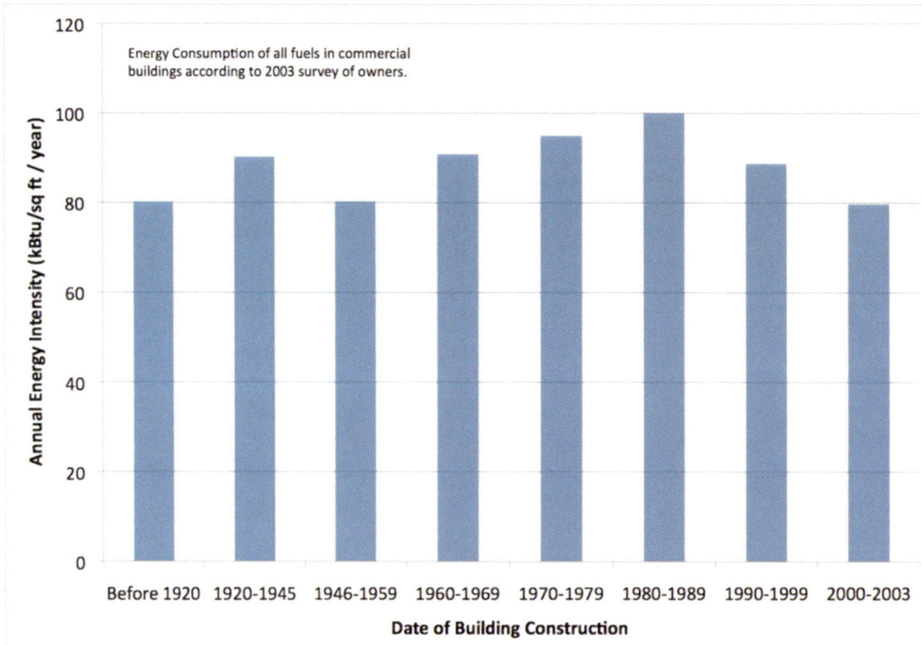

Figure 3: Actual Measured Energy Consumption versus Year Constructed for US Commercial Buildings (data from DOE Energy Information Agency CBECS 2003 Survey)

Appendix B.4

Building Science Digest 152

Building Energy Performance Metrics

2010-05 by Kohta Ueno

Abstract:

Putting metrics on building energy performance is a required step to make any progress on low-energy use and/or "green" buildings. However, there are many confusing and contradictory metrics available; to speak a common language, it is necessary to understand the topics that are behind these measurements. These topics include site vs. source energy, modeled results vs. reality, US average energy use figures, and methods of normalizing energy use. The normalization of energy use intensity (EUI), or dividing by square footage is examined; several significant problems in applying this metric to residential use are demonstrated. Various other metrics are presented, as well as a proposed method to provide all of the useful building energy information in a format that allows normalization by any chosen metric.

Introduction

"If you can not measure it, you can not improve it."

-Lord Kelvin (William Thomson)

Putting a measurement on building energy performance: you would think that it isn't that hard of a problem, right? Unfortunately, this is a topic that is simultaneously complicated, controversial (at times) ... not to mention lethally dull (possibly).

So why do we need building energy metrics? Most everyone in the building profession talks about energy savings, carbon emissions, and "green" buildings—but all of this is just smoke and mirrors until it can be quantified: "Are we saving energy—and if so, how much?" (sadly, sometimes the answer to that question is "No"—see Building Science Insight 007: "Prioritizing Green—It's the Energy Stupid").

In order to quantify savings, or create targets/goalposts, we need to speak a common language. In other words, when we throw around phrases like, "50% savings," we need to specify, "Compared to what?" These difficulties are the core of this document; the key topics to understand are:

- Site and source energy—the difference between what comes out of the gas pipe, the electric cable, and the power plant. For details, see BSD-151: "Understanding Primary/Source and Site Energy."

- Savings shown by models and simulations vs. savings shown by actual use

- Calculations of United States average usage

- Normalization methods—unless you have two houses that are identical, with the same number of identical people, and in the same climate, you need to "normalize" to compare House A with House B. But once you start doing that, you start distorting the results—understanding those distortions are vital for looking at the "normalized numbers" with a critical eye.

Organizations such as ASHRAE (American Society of Heating, Refrigerating and Air-Conditioning Engineers), Architecture 2030, and European organizations are promoting building energy rating systems and building energy metrics. They realize that with rising energy prices, a building's energy bills will be a vital component of its economic viability and its real estate value. Getting those numbers into the public domain (in a transparent form) is a great way to make market forces drive energy efficiency.

Real Buildings vs. Energy Models

When folks are throwing around phrases like, "… this building saves 40% energy relative to a code-compliant house," these numbers typically come from simulations, or building energy models. They run:

- a simulation of the building from the plans and specifications

- a simulation of a similar building, but with code-compliant features

- and then compare the energy performance (to obtain % improvement)

Examples of this method include the HERS (Home Energy Rating System)/Energy Star scale, ASHRAE 90.1 compliance (used in LEED), California Title 24, and the Building America Benchmark.

So what's wrong with that?—after all, nobody has the budget to build the two houses and wait for a year of energy bills every time they do a project. This type of energy modeling is the only way to make these predictive comparisons; it is also the way to decide (for instance) "Which makes more sense: adding more photovoltaic panels or adding more roof insulation?"

However, there are some key weaknesses that affect the usefulness of modeled results:

- Many times, the model does not—or cannot--capture the complexities and realities of the actual building. This is quite understandable: you need to enter a model in a finite amount of time, and it isn't known how every single pipe, button, and widget will actually be installed. Also, there are some characteristics that will only be revealed by doing field testing—airtightness, or duct leakage. These things can unfortunately have huge effects—which means you can only realistically predict a limited amount of information with a model, regardless of its sophistication and complexity.

- Even more importantly, the effect of the occupant of a building can be tremendous—i.e., "how the building is driven." There are many studies that look at <u>identical</u> units, and the actual usage shows a <u>factor of three</u> (or more) between the high and low users. To take the driving analogy further: imagine that a Toyota Prius (50 mpg 2010 rating) regularly got mileage between 75 to 25 mpg, depending on how you drove it. It might be obvious, but this type of variability limits how well you can predict things with energy models.

Figure 1 shows an example of the variability is seen in real houses: it plots the annual electric usage of 17 identical houses in Oklahoma, all by the same builder from the same time period. It plots simulation values against actual values—perfect correspondence would be on the diagonal line. Instead, we saw a range from 50% of the prediction to 180% of the prediction.

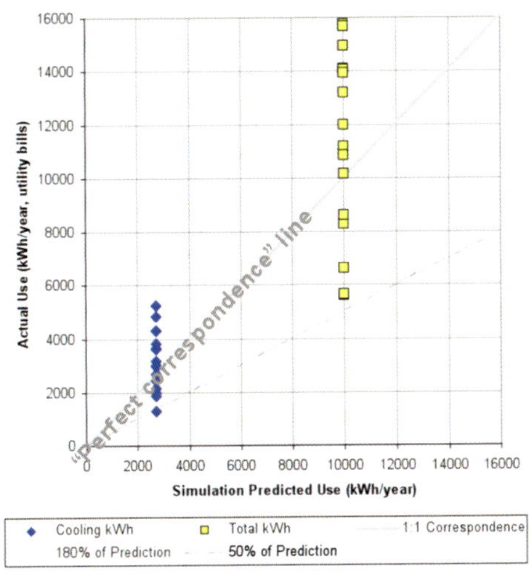

Figure 1: Simulation (horizontal) vs. actual (vertical) electric use for 17 identical houses in Oklahoma

These problems start to explain how you end up with disappointing situations where the building on paper (i.e., simulation) was "rated better than the standard," while it only achieved middling performance when the energy use was actually measured (see BSI-007: "Prioritizing Green—It's the Energy Stupid").

The difference between a simulation and energy bills is sometimes called an "Asset rating vs. Operational rating." This is the language used by ASHRAE's Building Energy Labelling effort (the "Building Energy Quotient Program"—see Figure 2)—a push to provide energy labels for commercial buildings, so that the market begins to understand (and value) energy efficiency in the commercial properties that they might buy or rent.

Figure 2: Energy Certificate Examples (ASHRAE 2008)

United States EIA Household Averages: Grading on a Curve

If using an energy model has so many problems, what standard can be used for comparison, once we have the energy bills? Much like an exam being graded on a curve, you want to compare to what the average is for a similar population. If your project is much worse than everything else out there—when you look at the energy bills—you know that you're doing something wrong.

On the residential side, one commonly cited figure is the energy use per household: the US average is 92 million Btu/year. Note that this is stated in terms of site energy. This (and the following calculations) comes from the Department of Energy/Energy Information Administration's (DOE/EIA) Annual Energy Review. Digging a little deeper, you can also find the averages broken down by census region, as shown on the map below (Figure 3), which shows usage million Btu/year per household, and how it compares with the US average (±%).

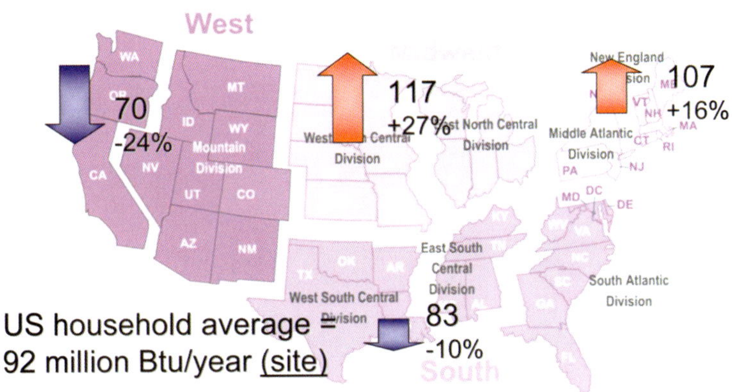

Figure 3: Site Energy: regional per household energy use (million Btu/year):

Clearly, the Northeast and Midwest have a higher per household site energy use, due to their higher heating loads. In addition, the West has lower site energy use—it has a milder climate, overall, than the other regions (especially when weighted by population location).

However, site energy does not give a full or accurate picture, as discussed in BSD-151: "Understanding Primary/Source and Site Energy." EIA data were combined with site-source conversions, and estimates of per-household source energy use are shown in the Figure 4:

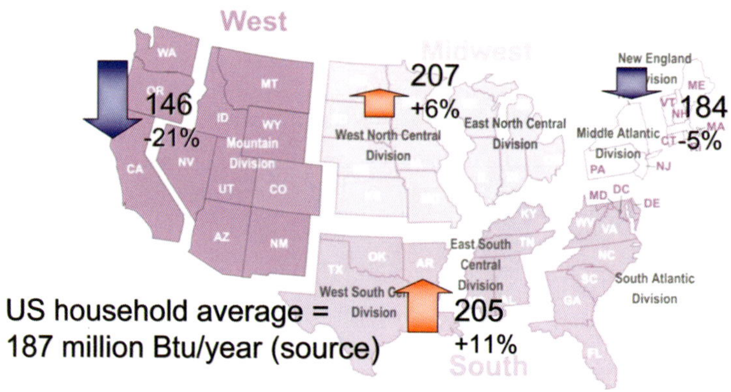

Figure 4: Source Energy: regional per household energy use (million Btu/year)

One surprise is that the South—because it is cooling-dominated—has a heavy electricity load, which makes a big difference in terms of source energy, pushing it above the average consumption. In contrast, heating-dominated climates such as the Northeast and Midwest don't have as large of a jump.

The numbers from the previous two figures (site and source per household use) are plotted in the bar graph below.

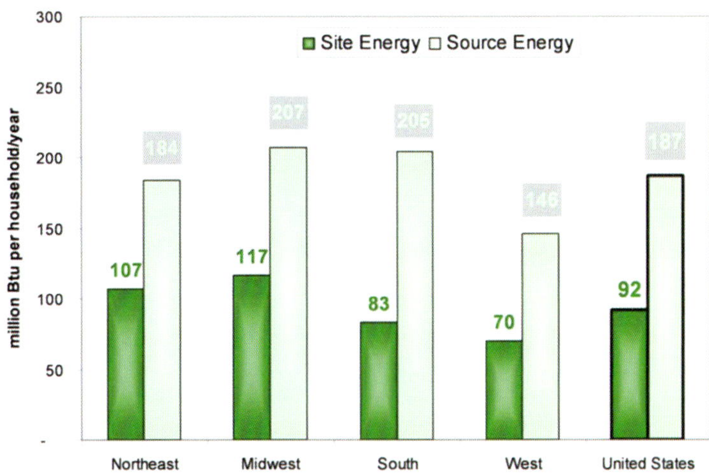

Figure 5: Site and Source Energy per household (million Btu/year): regional and US averages

Energy Use Intensity (EUI): Dividing by Square Footage

One problem with this "per household" average is, of course, that it doesn't account for house size. More importantly, it doesn't deal with non-residential buildings at all—what average value is used for a commercial building? Also, what type of commercial building is it?—this makes a big difference for the comparison number. A car dealership? Fast food restaurant? Hotel? Non-refrigerated warehouse?

The way the size problem is addressed is by dividing the energy use by the square footage. This is probably the most common energy metric out there: it is known as "energy use intensity" (EUI). It breaks down energy use (either source or site) by square foot of floor area; the common forms are kBtu/sf·year, and kWh/m^2·year. It is used, for instance, by Architecture 2030's 2030 Challenge, and the Passive House program, when they are defining their targets.

The 2030 Challenge also deals with the "usage type" question very well—they provide EUI targets for a wide variety of types of buildings. Similarly, the EPA ENERGY STAR Portfolio Manager (also online) compares an entered building with a database of similar buildings (based on usage type). That database is the Commercial Building Energy Consumption Survey (CBECS) (also from the EIA), which includes building characteristics and energy use from thousands of buildings across the United States.

Calculating EUI from energy bills and comparing it to values for similar buildings is a very useful comparison, because it looks at real numbers, as opposed to models vs. models, or models vs. reality.

Two warnings: first, this comparison is only useful when comparing site-to-site or source-to-source, when looking at EUI numbers. For reference, Architecture 2030 now publishes both site and source numbers, and by geographic region. Second, be sure not to confuse per household energy use (million Btu/year) and EUI (kBtu/sf·year).

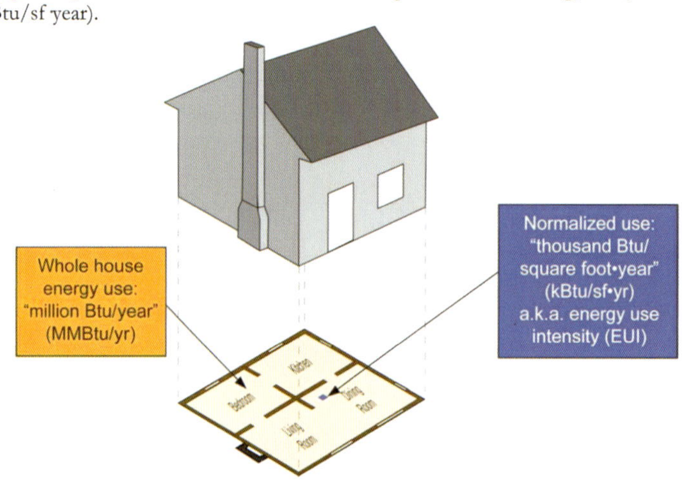

Figure 6: Household energy use vs. EUI (million Btu/year vs. kBtu/sf·year)

Figure 7 graphs site and source EUIs for single family houses, from Architecture 2030 data. It is broken down by Census Region, and includes the US average.

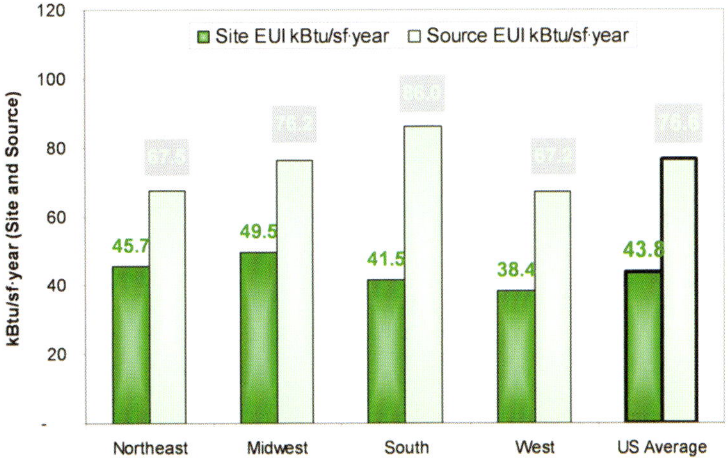

Figure 7: Site and Source Energy Use Intensity (EUI, in kBtu/sf·year): regional and US averages for single family houses.
Source: Architecture 2030

Using EUI: A Few Problems

So EUI is the most common metric out there, and it seems to handle many problems very well. You can fine tune it, based on usage type, and climate, if you have a comparison database. But there are definitely some problems with it—we're getting into those "distortions due to normalization" that were mentioned earlier.

Commercial vs. Residential: Overall, EUI is an excellent metric for commercial buildings—it costs money to build out space, so the square footage tends to get occupied and used, due to economic necessity. So EUI comparisons between similar types of commercial buildings are very reasonable.

However, residential spaces have much more variability of the occupancy of their square footage. For instance, you will see 700 square foot two bedroom apartments, but you will also see 3000 square foot houses with only two bedrooms. This might be an indication that square footage might not map as well to the energy consumption of a given house.

Climate: It is also important to remember that energy use (and thus EUI) will be affected by climate—more so in the residential realm than commercial buildings (residential buildings are "enclosure-dominated"—i.e., total energy use is affected by the weather more than commercial buildings). This can be addressed, for instance, by using regional EUI numbers.

Small House Penalty: Another problem is that EUI—just by how it is calculated—puts smaller houses at a disadvantage. This problem has also been demonstrated for the HERS system (Prahl 2000), and is true for the older version of the Building America Benchmark. Figure 8 shows the results of a set of building energy simulations, comparing household use vs. EUI.

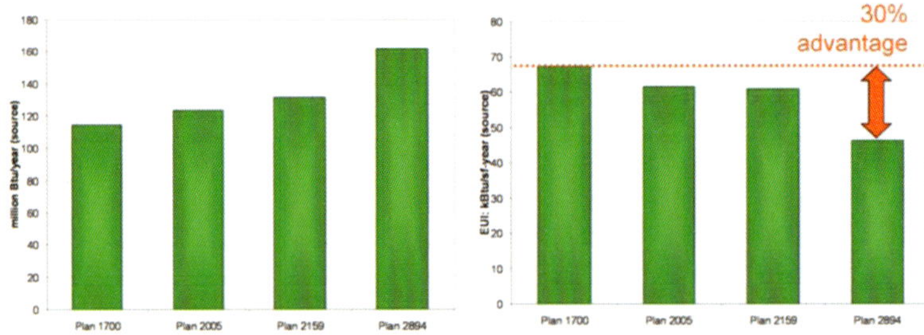

Figure 8: Square footage effects on household consumption (million Btu/year, left), and energy use intensity (kBtu/sf·year, right)

The four houses have square footage ranging from 1700 sf to 2894 sf, and the following characteristics

- identical enclosure (shell) insulation values and air leakage rates
- identical mechanical systems
- similar operating schedules
- all modeled in the same climate

Household use increases with house size, as would be expected (left hand graph). But EUI decreases with increasing house size (right hand graph), despite all other characteristics being identical. This is such an effect that the largest house has a 30% EUI advantage over the smallest, despite the houses having the same characteristics.

Another way to look at this problem is that households (regardless of size) have a refrigerator, use hot water, require ventilation, use TVs, etc.—essentially independent of square footage. These energy uses are divided by a small number in a small house, making these homes look like poor performers when they actually might be quite good on a per household basis.

Basement Effects: The previous point shows that the square footage that is used to divide energy use by has a huge effect. This gets into a grey area when you start to deal with houses with basements. The typical decision is to include the basement in the square footage if it is "conditioned"—but the definitions can sometimes be vague. Remember, adding a basement can increase the square footage by third or half again, and substantially drive down the kBtu/sf·year figure. But remember, in reality, a basement can add square footage without much of an energy penalty, because the majority of the surface area is "earth sheltered," and it regains any losses from mechanical equipment in the basement. As a result, one person has joked that the easiest way to reduce a house's EUI is to finish the basement, and include it in the square footage.

However, even if a basement is fully finished and used as everyday living space, some argue that the quality of the space is not the same as above grade space, and would apply a reduction factor when including it in square footage calculations.

Due to all of these problems, we would say that using EUI for residential comparisons has some important, perhaps critical, limitations. Note that the Architecture 2030/2030 Challenge has adopted the HERS Index as their target metric instead of EUI: this might be indicative of these problems (Architecture 2030 2008).

Other Normalizations

One normalization adds the effect of climate to EUI, for instance, in the form of kBtu/sf·year·Heating Degree Day (for heating loads). This provides some normalization based on the local climate loading, and/or year-to-year variations in climate. However, this approach assumes that heating loads are the primary energy consumer: in very efficient homes, other energy uses which do not vary with climate (refrigerator, domestic hot water) become a more significant part of the total. Furthermore, this method does not account for cooling loads unless explicitly calculated.

One final metric considered here is to normalize by occupancy (e.g., million Btu/person), instead of square footage. This is based on the belief that the ultimate value of energy use is in services to people. Therefore, energy per person should always be one of the metrics considered, in addition to indicators that divide energy use by building floor area, dollar of GDP, ton of industrial output, etc.

The number of household members would be calculated based on floor plans; a typical method uses the number of bedrooms plus one (#BRs + 1). However, US Census data provides information on the number of bedrooms and the number of occupants; which indicates that only 26% of US households had over 1 person per bedroom. The majority of the surveyed households (73%) have less than 1 occupant per bedroom. Therefore, a more reasonable calculation might be one person per bedroom.

Note that this metric would tend to reward small houses, as opposed to penalizing them.

Finally, all metrics—including this one—can and will be "gamed" by people trying to optimize their rated (rather than actual) performance. Game rooms, studies, and partitioned corners of basement might be relabelled as "bedrooms" to boost performance numbers. This is unfortunate, and perhaps unavoidable; however, given the code requirements for bedrooms (e.g., minimum square footage, and egress window area), this might limit the extent of the "gaming."

Conclusions

Overall, we are not dogmatically stating that only one building energy performance metric should be used: this document is meant to provide an understanding of the strengths and weaknesses of the options. All of them can be distorted (or "gamed"), but they can also all provide useful information, assuming that the problems are understood. For these reasons, we would recommend the use of the "mileage box" shown below, which presents all the relevant information in one place, thus allowing normalization using any preferred method.

References

American Society of Heating, Refrigerating and Air-Conditioning Engineers, Inc. "ASHRAE Building Energy Labelling Program: Promoting the Value of Energy Efficiency in the Real Estate Market," June 2008

Architecture 2030/2030 Challenge Target Table. http://www.architecture2030.org/2030_challenge/targets.html

Architecture 2030 "The 2030 Challenge Targets Embedded in the HERS Index Architecture 2030 and RESNET Agreement" Press Release, February 14, 2008

Deru, M. and Torcellini P., "Source Energy and Emission Factors for Energy Use in Buildings" Technical Report NREL/TP-550-38617, Revised June 2007

Prahl, Duncan. 2000. "Analysis of Energy Consumption, Rating Score, and House Size." Washington D.C.: U.S. Green Building Council.

U.S. Department of Energy Energy Information Administration (DOE/EIA). Annual Energy Review 2007, June 23, 2008 DOE/EIA-0384(2007)

U.S. Environmental Protection Agency (EPA). Energy Star Portfolio Manager. http://www.energystar.gov/index.cfm?c=evaluate_performance.bus_portfoliomanager

Sidebar: The "Mileage Box"

This document presents an overview of the confusing mix of building energy metrics currently available. We are not saying, "This is the only way to do the measurements"—all of these systems have pros and cons. However, many of the metrics out there—especially when normalized—don't allow conversions into the other useful metrics.

Therefore, we propose a way to present all the relevant information in one place, and allow normalization using any preferred method. We are nicknaming this graphic the "mileage box"—it is not dependent on any current rating system, energy cost, fuel supply, changing local energy code, or whether the basement square footage is counted. Also, it can be applied to either simulated performance or actual energy consumption ("asset or operational"). It shows the source energy numbers, which we believe must be provided (or at least back-calculable) in any energy metric system. It has:

- Annual source energy use (in million Btu/year)
- Annual site energy use (in million Btu/year)
- % Electric (as a percentage of site energy use)
- Square footage (if a basement is present, it is shown as above grade + basement, so it can be included or excluded as warranted)
- Number of bedrooms

The example "mileage box" shown below is for the US Average dwelling:

ESTIMATED WHOLE HOUSE ENERGY USE		
Source (MMBtu/year)	Site (MMBtu/year)	Area + Bsmt (sq ft)
187†	92	1795††
	% Electric	No. of Bedrooms
	39†	3††

So, for instance, kBtu/sf·year could be calculated by dividing either annual total (source or site) by the square footage. If carbon emissions were of interest, the percent electricity could be used together with the current fuel mix to provide an estimate. If the per person metric based on number of bedrooms (or number of bedrooms + 1) is of interest, it could be calculated, Finally, if cost is important, it can be calculated using the appropriate current or projected future cost of electricity/propane/oil/natural gas.

- †: Calculated based on *2005 DOE-EIA Annual Energy Review*, Table 2.4 Household Energy Consumption by Census Region, Selected Years, 1978-2001, and "Source Energy and Emission Factors for Energy Use in Buildings" (Deru and Torcellini 2007).

- ††: Both square footage and number of bedroom data are approximate median values, not averages. Both are taken from the *American Housing Survey for the United States: 2005* (US Census), Table 2.3. Size of Unit and Lot Occupied Units. For reference, the 2005 Residential Energy Consumption Survey Table HC1.1.2 gives a national average for "heated" housing unit square footage of 1618, and "total" of 2171. Average value for number of bedrooms appears to be in the 2.5 to 3 bedroom range.

Appendix B.5

Building Science Digest 151

Understanding Primary/Source and Site Energy

2010-07 by Kohta Ueno and John Straube

Abstract:

The difference between site and source energy is a vital concept to understand when looking at the energy performance of buildings—failing to account for the difference will result in an apples-to-oranges comparison that does not give the true picture of a building's energy consumption. This document explains how these two types of energy are accounted for differently and why.

Introduction

The difference between site and source energy is a vital concept to understand when looking at the energy performance of buildings—failing to account for the difference will result in an apples-to-oranges comparison that does not give the true picture of a building's energy consumption. Failing to report which metric (site or source) is being used is a practice that should be eliminated, given how significant the difference can be. Numerous building programs, like Building America, EPA Energy Star, Architecture 2030, and the German Passiv Haus, all use source energy metrics.

Site vs. Source

A building's energy consumption can be measured in terms of its fuel use: gallons of fuel oil, kWh of electricity, or therms of gas. Although all of them are in different units, you can ultimately measure them in terms of units of <u>energy</u>—e.g., "How much water can you boil with this energy?" For our purposes, energy is commonly measured in Btus (U.S. units) or kWh (metric units); for reference, the definition of a Btu, or British thermal unit, is the energy it takes to heat one pound of water by one degree Fahrenheit. Note that kWh, although it is <u>typically</u> used for electricity, is a unit of <u>energy</u> (not just electricity)—and can be applied to any fuel source.

If you add up all the energy (Btus) you are consuming at the meter, this is what is known as "site" energy. However, this is not the full picture. The problem is that the process of generating electricity incurs substantial losses—enough that for every unit of electricity at the plug, it might have been necessary to "burn" about 3 times that amount of energy (coal, gas, nuclear, etc.) at the power plant—see Figure 1.

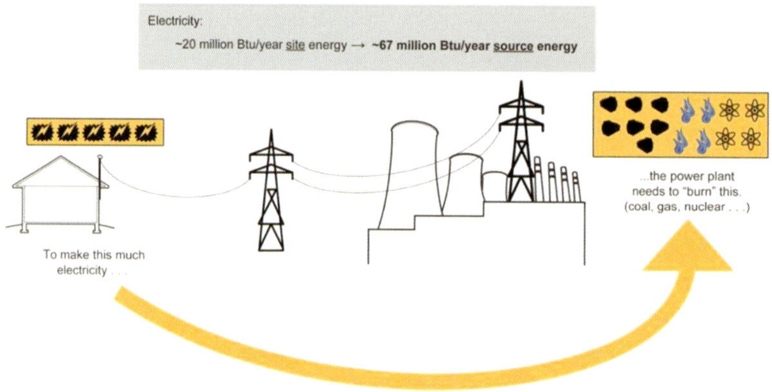

Figure 1: Source energy from power plant to electric service entrance

Although transmission losses are a part of this inefficiency, they are rather modest—roughly 10%. Transmission losses are what causes the hum of high-voltage electrical lines, and are the reason why the millions of transformers in the grid are air cooled. These losses can be reduced by better technology and will be reduced in the future by distributed power generation (i.e., the "Smart Grid"), especially if power is generated reasonably close to where it is used (e.g. in the same community for example).

In general, it can be said that "the grid" is "~30% efficient": with the majority of the losses occurring at the plant. To be specific: the power plant cooling towers that are dumping steam (i.e., waste heat) into the sky are where most of that missing 2/3 is going.

If you account for the energy consumed at the power plant, this is known as "source energy" or "primary energy." The EPA's definition is:

Source energy is a measure that accounts for the energy consumed on site in addition to the energy consumed during generation and transmission in supplying the energy to your site.

Source energy is much more important than site energy if the concern is environmental performance. Site energy is useful because it can be unambiguously measured.

A Household Example

Converting to source energy has a major effect on electricity (factor of 3), and has a small effect for fossil fuels like natural gas. So what difference does this make? Here is an example:

- A typical gas tank water heater has an efficiency level (energy factor or EF) of 0.59, or 59%.
- In comparison, a typical electric tank water heater has an EF of 0.92 or 92%.

So in terms of site energy, the electric tank has an efficiency level about 1.5 times better than that of gas tank. But if you factor in source energy:

- The same gas tank has a "source efficiency" of roughly 54%
- The same electric tank has a "source efficiency" of roughly 27%

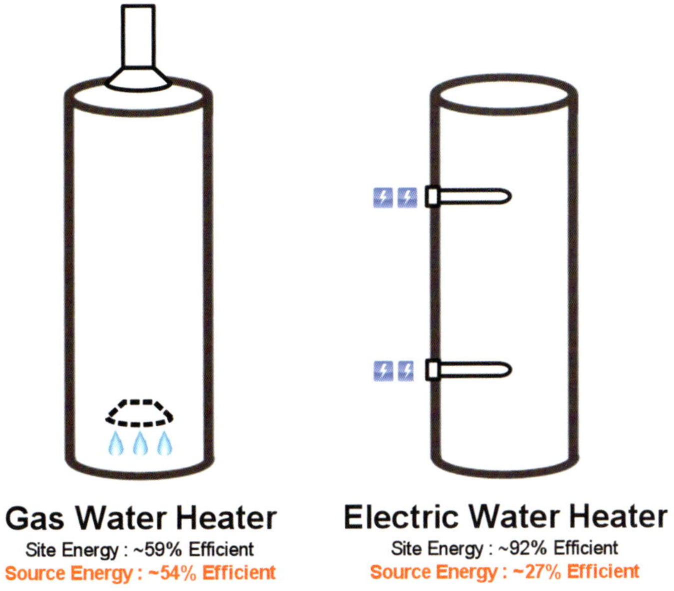

Figure 2: Site-Source Comparison for Domestic Water Heating

So the electric tank is half as efficient, in terms of source energy and, loosely, environmental impact. **Obviously, if one doesn't account for source vs. site differences, there are some pretty serious distortions when comparing electric heating with gas heating.**

As another example: some argue that superinsulated buildings are so efficient (with heating loads so low) that using electric resistance heating is a reasonable solution, instead of installing a full heating system (e.g., using point-source electric resistance heaters instead of a furnace). Based on the conversions below, this is the equivalent to installing a ~30% efficiency furnace (compared to 80-95%+ AFUE gas furnaces commonly available), when source energy is accounted for.

Site/Source Conversion Factors

A complete set of site-to-source conversions can be found in Deru & Torcellini (2007); these figures change from state to state, every day, or even hourly. This occurs because the power mix of the electric grid is constantly changing to meet load—for instance, peaking power (versus average or baseload) can have a significant effect. Also, as more and more renewable energy sources (wind, solar, etc.) are added to the grid, this "source energy penalty" will decrease. But some of the current United States national average figures are:

- Site-to-source electricity: 3.365
- Site-to-source natural gas: 1.092
- Site-to-source fuel oil: 1.158
- Site-to-source propane: 1.151

When people talk about electricity being "clean power," this typically fails to acknowledge the reality of source energy. All that's happened is that the pollution has been moved from your chimney to somewhere that you can't see it—it hasn't magically disappeared. In fact, with our current power mix, it is reasonable to argue that electricity is America's dirtiest fuel. This does not mean electricity should not be used, only that it should be used wisely.

For calculating source energy from raw "billed" fuel units (kWh of electricity, therms of gas), see the conversion factors in Conversions for Calculating Your Source Energy Use (sidebar).

But overall, when people start throwing energy numbers at you, make sure that they first answer the question, 'Are you talking about site or source?"

The Fuel Mix of the Grid

As described above, the electricity grid in the United States is powered by a mixture of fuel sources. The US power grid is divided into three main grids, which have minimal connection between them: the Western Interconnection, the Eastern Interconnection, and the Electric Reliability Council of Texas (ERCOT) Interconnection (see Figure 3).

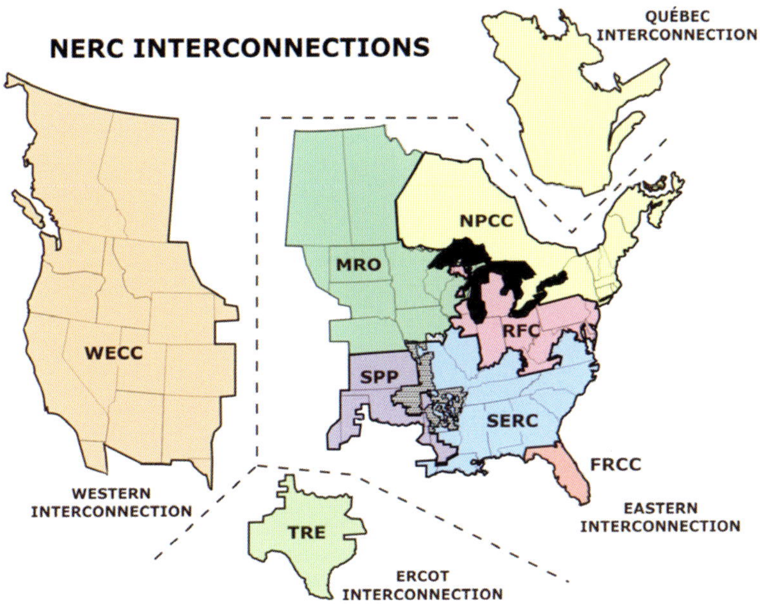

Figure 3: North American electrical grid interconnections, including the 10 NERC regional reliability councils (NERC 2007) (via Deru and Torcellini 2007)

Based on the 2004 EIA data cited by Deru and Torcellini, the fuel mix for the US average and these three grids is shown below (Figure 4). It is clear that the US grid is largely powered by fossil fuels (71% on average), and 50% by highly polluting coal plants. In comparison, only 2% of the grid is powered by renewable energy, which includes renewable fuels, geothermal, wind, and solar (PV).

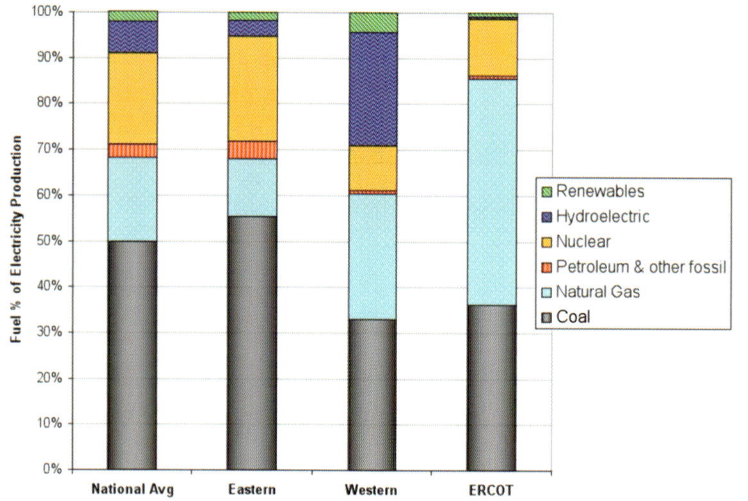

Figure 4: Electricity generation fuel mix for the continental United States (national) and three interconnections, for 2004 (EIA 2006) (via Deru and Torcellini 2007)

Of course, over time, the fraction of renewable energy is expected to increase, making the grid "greener." With the retirement of old coal, the addition of wind, high efficiency gas, biomass, tidal, or even nuclear, the carbon intensity of producing electricity will drop, and the source-site ratio will drop. Perhaps the largest change will be as a result of millions of buildings feeding energy into the grid from PV arrays or co-generation facilities, since these have a source-site ratio of nearly one.

However, this optimism should be tempered by the fact that in our current market, generating electricity with fossil fuels is still one of the cheapest options. This was elegantly pointed out by Professor Nate Lewis of Caltech, in a 2007 interview with Paul Krugman (Krugman 2007):

"So building an emissions-free energy infrastructure is not like sending a man to the moon," Lewis went on. "With the moon shot, money was no object — and all we had to do was get there. But today, we already have cheap energy from coal, gas and oil. So getting people to pay more to shift to clean fuels is like trying to get funding for NASA to build a spaceship to the moon — when Southwest Airlines already flies there and gives away free peanuts! I already have a cheap ride to the moon, and a ride is a ride. For most people, electricity is electricity, no matter how it is generated."

Global Warming Potential

Another factor to consider when looking at site-source energy conversions is not just energy or resource consumption, but also carbon emissions. Global warming potential is commonly measured in units of carbon dioxide equivalent (CDE, or $CO_{2equivalent}$), expressed as mass (e.g., the average passenger vehicle produces 5.48 metric tons $CO_{2equivalent}$ per year).

The process of treating and pumping (via pipeline) natural gas generates (before combustion) 0.088 lbs $CO_{2equivalent}$ for each $kWh_{equivalent}$ of delivered energy. Combusting one therm of natural gas, at 100% efficiency release another 11.7 lbs $CO_{2equivalent}$, which is another 0.40 lbs CO_2 for each $kWh_{equivalent}$ for a total of about 0.49.

- **Natural gas: ~0.49 lbs $CO_{2equivalent}$ per $kWh_{equivalent}$**

Electricity from the national grid, results in the emission of 1.67 lbs of $CO_{2equivalent}$ as delivered to the building's meter (Deru & Torcellini 2007, EPA eGrid).

- **Electricity: 1.67 lbs $CO_{2equivalent}$ per kWh**

Hence, providing heat by electric resistance heat releases 1.67/0.49 = 3.4 times as much greenhouse gases as heating with a perfectly efficient natural gas appliance. Top quality natural gas furnaces, and even some boilers, can achieve 95% efficiency, which changes the emission factors per unit of heat delivered to the building to 0.516 lbs $CO_{2equivalent}$ for each $kWh_{equivalent}$ (which changes the ratio to 3.2 instead of 3.4).

According to the US Energy Information Administration (EIA), 4110 billion kWh of electricity were generated in 2008 ("total net generation"). Of this energy, 1994 billion kWh was generated by coal, 248 billion kWh by hydroelectricity, and 124 billion kWh by renewables other than hydroelectricity. This consumed just over 1 billion tons of coal alone (plus other fossil fuels) and resulted in the emissions of 2.516 billion tons of CO_2, and 9.0 million tons of SO_2.

The 2005 Residential Energy Consumption Survey (RECS) notes that the 111 million households in the US consumed 1275 billion kWh over that year. If ten million homes had a 4 kW_p array, each generating 4500 kWh/yr, then 4.5 billion kWh would be produced, or about 3.5% of total electrical energy consumption. However, this relatively modest impact in overall kWh consumption reduces source energy by 3.4 times as much, or 15.1 billion kWh. This is a rather significant impact, demonstrating the disproportionate effect of offsetting electricity use.

What about using energy costs instead?

Some (notably ASHRAE 90.1 and LEED) have considered using the cost of energy as a metric, instead of bothering with site-source conversions, Btus, kWh, etc. First, costs are commonly used in economic analyses, and are what many building owners care about the most. Second, energy costs are actually a rough surrogate for/approximation of source energy. This is part of the reason why ASHRAE Standard 90.1 (the energy efficiency standard for large buildings), uses cost in its "building energy cost method" (which calculates the effectiveness of various energy conservation measures) (Jarnagin, 2010).

Unfortunately, energy costs can vary greatly by geographic region, season, and even time of day. So if someone is trying to compare buildings A and B, energy cost can easily give you a distorted picture—or one that is only accurate for the next week or two. It is better to take the energy units, and then figure out the energy costs as necessary. This allows estimation (for instance) what would happen if energy costs change over time.

Energy costs are also somewhat meaningless across time and space: a high performance building in Europe may have the same energy cost as a mediocre building in Arizona, but use less than half as much energy and emit a third as much pollution. Similarly, it is difficult to assess the relative performance of a home reported to use "$60/month" in heating energy based on a 2004 magazine article. Without knowing how much energy costs, and what proportion was gas or electric, the dollars are meaningless. Reporting the source energy consumption would allow a comparison of buildings on different continents and at different times.

Conclusions

The difference between site and source energy must be accounted for when looking at the energy performance of buildings—failing to do so will result in an apples-to-oranges comparison that does not give the true picture of a building's energy consumption. Electricity, as a fuel, requires roughly three times the energy input ("source") to produce one unit of energy at the plug ("site). In addition, the electricity grid is primarily fed by fossil fuels (71% on average): with the current power mix, it is reasonable to argue that electricity is America's dirtiest fuel. However, over time, the fraction of renewable energy is expected to increase, making the grid "greener."

References

Deru, M. and Torcellini P., "Source Energy and Emission Factors for Energy Use in Buildings" Technical Report NREL/TP-550-38617, Revised June 2007

Jarnagin, Ronald E. et al. "35 Years of Standard 90.1" ASHRAE Journal, March 2010.

Krugman, Paul, 2007. "The Power of Green" *The New York Times,* April 15, 2007.

U.S. Department of Energy Energy Information Administration (DOE/EIA). 2006. Electric Power Annual 2005, Energy Information Administration. www.eia.doe.gov/. Washington, DC: U.S. Department of Energy (accessed January, 2007).

U.S. Department of Energy Energy Information Administration (DOE/EIA). 2009. Annual Energy Review 2008, June, 2009 DOE/EIA-0384(2008)

U.S. Department of Energy Energy Information Administration (DOE/EIA). 2008. 2005 Residential Energy Consumption Survey. DOE/EIA-0632 (08)

U.S. Environmental Protection Agency (EPA). eGRID website, http://www.epa.gov/cleanenergy/energy-resources/egrid/index.html

Sidebar: Conversions for Calculating Source Energy Use

The following conversions from billed units to source energy are provided for some common fuel types. The energy use for the year should be added up, and then converted to million Btu/year (source) by type of fuel.

- Natural Gas: therms[1] × 0.1092 = million Btu (source energy)
- Electricity: kWh × 0.01148 = million Btu (source energy)
- Fuel Oil: gallons × 0.1781 = million Btu (source energy)
- Propane: gallons × 0.1187 = million Btu (source energy)

[1]: Note that natural gas use is often given in hundreds of cubic feet (CCF): this is roughly equivalent to therms, but requires a "thermal conversion factor" (typically 2-5%) to obtain therms.

As mentioned above, these conversions use today's site-source conversion factors, and assume certain fuel energy contents (Deru and Torcellini 2007), which will vary by region and time of year.

Appendix B.6

Insight
Deciding on Energy Priorities When Building New

By John Straube, Ph.D., P. Eng.

The future is uncertain. This is a truism, and yet, when we design and construct a new building, we need to make decisions in the present or very near future. In fact, this is one of the critical distinctions about designing buildings: they are expected and likely to last 50 to 100 years, but we build them now. The challenge of designing for the future is no more acute than in the current choices facing the designer of an environmentally friendly building.

The future of energy is particularly unclear at present. Will the cost of oil rocket back to $150 per barrel or languish at $40? Will the cost of clean renewable energy generated by photovoltaic's drop to a quarter of current prices? Will cost-effective technology be developed to allow coal plants to capture and sequester their emissions, and thereby provide cleaner energy at a cost competitive with today's rates? Or will a tremendous growth burst of plug-in hybrids cause an increase in electrical demands and thus push up prices? Or could drilling and production technology open up significant reserves of tight natural gas, and allow for the import of plentiful, clean LNG (liquid natural gas)? Perhaps demand growth in the developing world will swamp everything and drive energy prices to spectacular heights? These types of questions often face the designers of low-energy green buildings. In some cases, a design is dominated by only one possible future path. Following one just path may generate a design that will be a poor solution along another path. While we cannot be sure of which future will occur, some design choices leave more future options open than others.

Despite the dizzying array of possible energy scenarios, some choices are, or should be, easy. The consumption of operational energy currently comprises the vast majority of the environmental damage of a building during its life. Regardless of the future path of technological developments, it is almost inevitable that energy will not get cheaper. In fact since the early 1970s's the retail cost of energy has increased at twice the rate of inflation.[1]

Photograph 1: This boiler arrangement in this low-energy building (see Sidebar) can easily accommodate a replacement heating device of a different fuel type, be replaced or augmented with solar energy, and was even planned to service a neighboring building in the future. The reader may be interested to know that these two residential-size boilers provide space heating and domestic hot water for a 20 000 square foot 13-unit mixed-use building in a cold climate – and the second boiler is rarely needed.

Reducing a building's operational energy requirement is the best protection against an uncertain future, as regardless of where energy comes from or what its future price, using less of it will always be better than using

[1] This information comes from the US Energy Information Administration (www.eia.doe.gov); 2008. The consumer price estimates for energy in all sectors between 1970 and 2005 has increased by 6.7% compounded per year. Electricity, remarkably, is 4.5%, while natural gas is over 8% per annum. Predicting the future is hard, but the EIA can tell you what has happened in the past.

more. Efficient use of energy is therefore an obvious strategy that avoids environmental damage today, and will save money over the long term (see **Sidebar**). Generating clean energy on or near the site is another obvious approach to reducing environmental impact over a buildings lifespan but second in priority to using less.

Efficient use of energy in buildings typically entails:

1. insulation,
2. airtightness, and
3. solar control to *reduce* energy demand, and
4. efficient appliances to *meet* energy demands.

Power generation on site typically means photovoltaics (PV), or small wind turbines, but combined heat and power (fuel cells, natural gas, etc) biogas, biomass, microhydro power and other forms of generation may also be viable choices in some cases. It is not always easy to decide on what mix of these broad strategies to apply to a given project. One major factor that should be used to help decide which strategies and what mix to use is **time**.

Let me explain. During new construction, it is often relatively easy and inexpensive to insulate the opaque parts of the building enclosure. However, some parts of the enclosure are much more expensive to insulate at a later date than others.

- Walls are often the most difficult to upgrade in any building, since they are covered with expensive and durable finishes/cladding and have window and door penetrations. An owner will be very unhappy if her only upgrade path is to remove the brick veneer cladding.
- Upgrading a vented attic at a later date is both cheap and easy, especially in low-rise buildings: simply blow more loose fill insulation on top of what was built.
- A cathedral ceiling or flat roof assembly can only be cheaply upgraded by adding rigid insulation when the roof membrane is replaced in 15 to 25 years.
- The insulation level of an unfinished basement space in a building can be easily upgraded by adding insulation when the basement is finished in 5 or 10 years. It is expensive and invasive to add insulation at a later date to a finished basement.
- Insulated under a concrete slab is something that can only be done during construction. After a slab is built, insulating techniques are less effective, and cause many integration problems (door heights, head room, moisture tolerance).
- The choice of poorly insulated and/or high solar gain windows can only sensibly be corrected in 15 to 30 years when the windows need replacement.

Different "appliances" also have very different time spans. The choice of inefficient lighting (such as halogen or incandescent) can easily and cheaply be corrected by the owner within only a few years of occupancy. The technology of lighting is also advancing quickly, and so better more efficient lights are very likely going to be available.

The technology behind a $1000, 95% efficient condensing gas furnace is unlikely to improve very much in the future, as it can only become a few % more efficient. However, new technology (such as low-cost ground source heat pumps) might become available. A typical furnace or air conditioner will need to be replaced in about 20 yrs, often less, and can be replaced with little disruption at almost any time. This means that mechanical equipment choices usually do not restrict future equipment choices very much. The most future-flexible technology uses water-based heating and cooling distribution, as almost any source of low-temperature heat or cool can be used. However, air-based distribution closes few options, as low temperature sources of heat can be used economically provided the building has a low energy demand.

A heat recovery ventilator (HRV) in the North or an enthalpy recovery ventilator (ERV) in the South can be added with almost no effort at any time over the building life cycle if air distribution ducts are in place. However, a ductwork system to distribute air is very difficult expensive and annoying to add to many building types (such as housing). By contrast, a strip mall, or big box store can add ductwork easily in the future. Economizer ductwork and dampers should always be installed in most climate zones during construction because of their fast payback and low capital cost.

Controls are almost always easy to add in the future. It is currently quite easy to add occupancy controls and dimming lighting bulbs/fixtures. The situation should

Sidebar: Energy Efficiency on a Budget

This five-story, mixed-use building in a cold climate (Waterloo, Ontario) was designed provide the owners with a low energy-use building and lots of options for the future – all within a typical construction budget.

The insulated concrete form (ICF) and precast concrete plank structure was enclosed in 4" of high density spray foam insulation. Thermally broken standoffs support durable brick and fiber cement panel cladding. The windows comprise high performance glazing in fiberglass frames. A low temperature radiant heating system takes advantage of the building's thermal mass and will be compatible with future low-temperature solar hot water heating equipment. There are no fancy renewable energy systems in this building but as the future unfolds, the owners will have lots of options.

Durable, healthy, energy efficient buildings can be designed and build for the same or less than a typical building. However, to be successful, work must start with a committed owner and design team. The result, in this case, is a low-energy building that will be an asset for the owners as the price of energy increases; can accommodate different sources of energy as supply trends change, including site generated renewable sources; and will have a very long service life without major reconstruction of enclosure or distribution components.

Sidebar Photograph 1 *(left)*: A low energy mixed use building in a cold climate.

Sidebar Photograph 2 *(middle)*: Durable, inexpensive cladding over high-density spray foam insulation.

Sidebar Photograph 3 *(right)*: Thermally isolated precast concrete balcony slabs supported by building geometry. These details are an important part of low energy building design.

improve in the future. Adding variable speed motors to pumps and fans is easy to retrofit now and in the future.

Should a building be made ready for photovoltaic's or solar hot water? Absolutely. The cost of providing south-facing sloped surface can be close to zero if this requirement is included in the design phase. No more needs to be done, although adding an empty conduit through the roof to an accessible space inside is very low cost and makes the future addition of hot water pipes or electrical cables very easy.

Hence, if you doubt energy prices are going to triple over the next 25 years, and have a hunch that PVs will soon get cheap enough to compete with grid-supplied coal power, you might choose to skip the ground-source heat pump, R-75 attic insulation and R-25 basement insulation, HRV, and LED lights in your next house design. If your hunch is wrong, and energy prices double over the next ten years, or if natural gas becomes very expensive, the owner has a number of significant, low-cost, and easy-to-implement energy-saving upgrades available. For around $5000-7500 all of these changes, except the GSHP, could be implemented in a typical 2200 square foot house. And perhaps just as importantly, ensuring that the design includes a good area of sloped unshaded south-facing roof provides the future occupant of the house the potential to add PV and/or solar hot water.

It should be clear that the same sort of strategy can be applied if you have a limited budget. Focusing the available budget on both low cost items, and decisions that cannot easily be changed in the future will allow for a building that can in the future be sensibly improved as financing becomes available.

An uncertain future and/or a tight budget does not mean that a building must be designed so that it becomes a liability in a future energy- and carbon-constrained world. It does require that careful choices be made about what to focus on.

Insight The Perfect Wall

An edited version of this Insight first appeared in the ASHRAE Journal.

By Joseph W. Lstiburek, Ph.D., P.Eng., Fellow ASHRAE

The perfect wall is an environmental separator – it has to keep the outside out and the inside in. In order to do this the wall assembly has to control rain, air, vapor and heat. In the old days we had one material to do this: rocks. We would pile a bunch or rocks up and have the rocks do it all. But over time rocks lost their appeal. They were heavy and fell down a lot. Heavy means expensive and falling down is annoying. So construction evolved. Today walls need four principal control layers – especially if we don't build out of rocks. They are presented in order of importance:

- a rain control layer
- an air control layer
- a vapor control layer
- a thermal control layer

A point to this importance thing here, if you can't keep the rain out don't waste your time on the air. If you can't keep the air out don't waste your time on the vapor.

The best place for the control layers is to locate them on the outside of the structure in order to protect the structure (**Figure 1**). When we built out of rocks the rocks didn't need much protection. When we build out of steel and wood we need to protect the steel and wood. And since most of the bad stuff comes from outside the best place to control the bad stuff is on the outside of the structure before it gets to the structure.

Also, after generations of building out of rocks folks somehow got the idea that they wanted to be comfortable – and they figured out that rocks were not the best insulation. I mean rocks are not that bad compared to windows – memo to architects: you can't build an energy efficient green building out of glass, but you can get design awards and we all know which is more important. Back to rocks, they are heavy and you need a lot of them to make the wall have any decent thermal resistance so we invented thermal insulation.

But where to put the insulation? If we put the insulation on the inside of the structure the insulation does not protect the structure from heat and cold. Remember we really do want to protect that darn structure – especially for the sake of making the structural engineers life more happy. Expansion, contraction, corrosion, decay, ultra violet radiation, and almost all bad things all are functions of temperature. So all the control layers go on the outside. Keep the structure from going through temperature extremes and protect it from water in its various forms and ultra violet radiation and life is good.

What about this air control thing? Well air can carry a lot of water and water is bad for the structure. So we have to keep air out of the structure as well because of the air-water thing – or if we let it get into the structure we have to make sure it does not get cold enough to drop its water. Now, just one other thing, tends to be important if you intend on living in the building or working in the building or keeping things safe in the building, we might want to control the interior environment. We especially ought to be concerned about what is in the interior air because when we are in the interior we tend to breathe it. Well, it turns out that we can't control air until we enclose air. So we need an honest to god airtight enclosure in order to provide conditioning such as filtration and air change and temperature and humidity control. And once again the best place to control this air thing is on the outside of the structure – but under the insulation layer so the air does not change temperature. Presto: the perfect wall. A water control layer, air control layer and vapor control layer directly on the

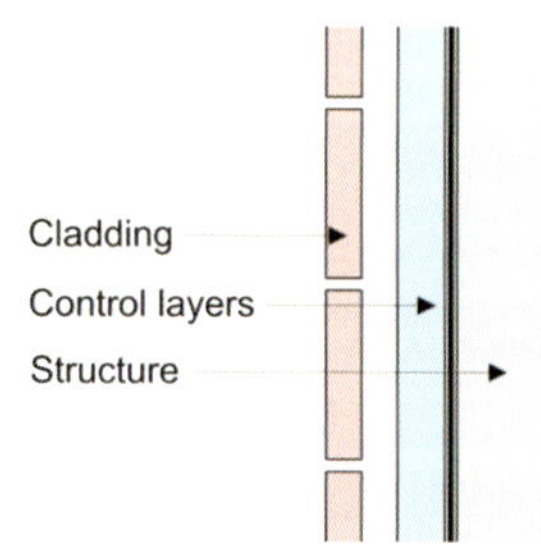

Figure 1:* *"The Perfect Wall"

In concept the perfect wall has the rainwater control layer, the air control layer, the vapor control layer and the thermal control layer on the exterior of the structure. The claddings function is principally to act a an ultra-violet screen. Oh, and architects might consider the aesthetics of the cladding to be important.

structure and a thermal control layer over the top of the other control layers (see **Figure 1** again).

This was figured out long before I was born – I think the Canadians figured it out first (1), but the Norwegians have some claims to this plus the Russians. I am going to go with the Canadians on this one because I am biased and proud of it. Also, I met Professor Hutcheon, and that is a story for the grandkids when I get some – memo to Christy and Andrew: so what's the delay here? For a more detailed discussion of the physics of all of this go to the old masters: Hutcheon and Handegord (2) and the new kids on the block Burnett and Straube (3).

In a beautiful bit of elegance and symmetry if you lie the perfect wall down you get the perfect roof (**Figure 2**) and then when you flip it the other way you get the perfect slab (**Figure 3**). The physics of walls, roofs and slabs are pretty much the same – no surprise (**Figure 4**). This insight was shone into a whole generation of practioners by Max Baker (4) when I was first getting started.

Notice in the perfect roof assembly the critical control layer or membrane for rainwater control and air control and vapor control is located under the thermal insulation layer and the stone ballast (i.e. "roof cladding") so that it is protected from the principle damage functions of water, heat and ultra violet radiation. Arrhenius* would be proud. Why we put the most critical control layers on roofs on the very, very top where they can be trashed by these damage functions never fails to amaze me. Yes, I know, they are easier to replace when they are located there. Standard answer for our disposable, unlimited resource available society.

Most problems in building enclosures occur where roofs meet walls. The classic roof-wall intersection is presented in **Figure 5** (will both credit and apologies to Max Baker). Notice that the control layer for rain on the roof is connected to the control layer for rain on the wall, the control layer for air on the roof is connected to the control layer for air on the wall…..and so it goes. Beautiful. And when it is not so…ugly.

* Dead, Nobel Prize Winner, no longer fashionable to study.

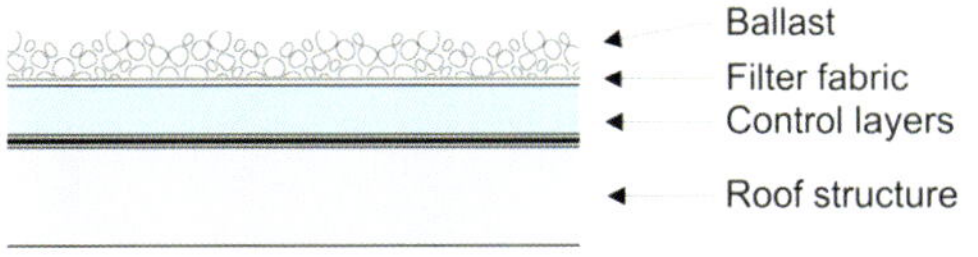

Figure 2: "The Perfect Roof"

The perfect roof is sometime referred to as an "inverted roof" since the rainwater control layer is under the insulation and ballast (i.e. roof cladding). Personally I don't view it as inverted. Those other folks got it wrong by locating the membrane exposed on the top of the insulation – it is they that are inverted.

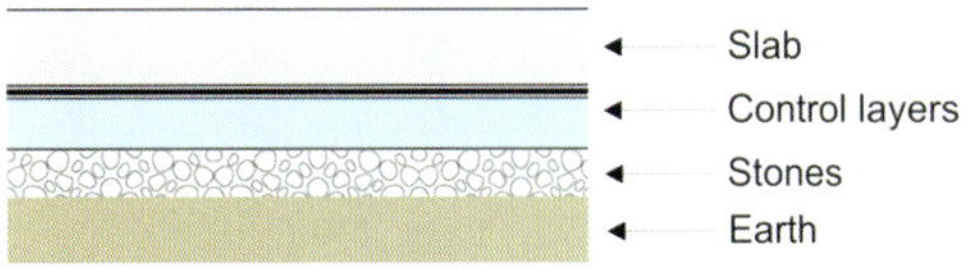

Figure 3: "The Perfect Slab"

The perfect slab has a stone layer that separates it from the earth that acts as a capillary break and a ground water control layer. This stone layer should be drained and vented to the atmosphere – just as you would drain and vent a wall cladding.

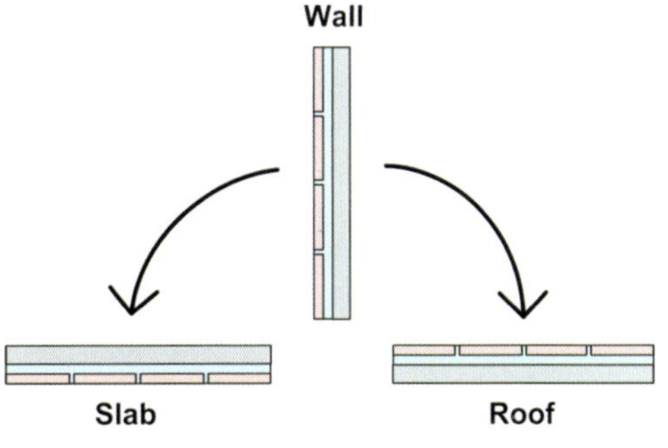

Figure 4: A Wall Is A Roof Is A Slab

The physics of walls, roofs and slabs are conceptually the same.

Time to put some meat on the bones of **Figure 1**. How should this perfect "conceptual" wall actually be built? Three ways. The best of the best of the best can be found in **Figure 6**. This is a very special wall. I refer to it as the 500-year wall for two important reasons:

- it represents 500 years of evolution; and
- it will last 500 years

It is the type of wall that you use for special buildings. Buildings that are passed down from one generation to the next. Museums, art galleries, courthouses, libraries. I call this wall the "institutional wall." It is sweet in that it can be constructed in any climate zone. The only thing that may be changed is the level of thermal insulation. My advice here is very simple: what ever you think the right amount of thermal insulation should be double it and shut up. If you love your kids don't argue with me.

A clever version of this first wall is where spray-applied closed-cell high-density foam is used to combine the four principal control layers in one material (**Figure 7**).

The second wall should be the "meat and potatoes" wall for commercial buildings. The wall every commercial building should use. The base wall that our infrastructure should depend on. So, no surprise I call it, yes you guessed it: the "commercial wall" (**Figure 8**). It has a conductive structure – metal studs. All of the insulation should – and must be located on the outside. It is a thermodynamic obscenity to insulate within a conductive structural frame. Again, you can build it

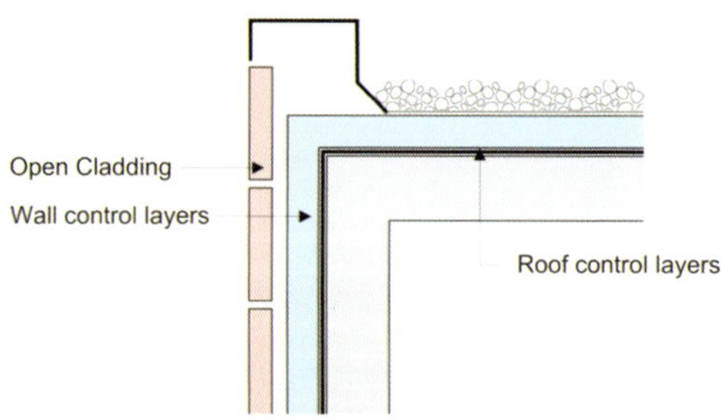

Figure 5: "The Roof-Wall Connection"

Notice that the control layer for rain on the roof is connected to the control layer for rain on the wall, the control layer for air on the roof is connected to the control layer for air on the wall… and so it goes.

anywhere in any climate location. Just consider the insulation levels (see above – particularly the part about loving your kids).

The third wall is the "residential wall" (**Figure 9**). Notice the structural cavity is insulated. That is because we are using a relatively non-conductive structural frame – the structure is wood and wood material based. Wood is not particularly conductive – that is why we do not have wood frying pans.

For this third wall to work almost everywhere (except Alaska and north of Flyn Flon** where we would not insulate even within a relatively non-conductive wood structural frame) we would split the thermal resistance of the insulation on the exterior of the structural frame with this insulation within the structural fame at least 50:50. So in an R-20 wall – at least R-10 or more on the outside of the non-conductive structural frame. And no vapor barrier on the inside of the assembly. Repeat after me, no vapor barrier on the inside of the assembly. We want the assembly to dry inwards from the control layers – and to dry outwards from the control layers. Always. Everywhere.

References

(1) Hutcheon, N. B., CBD-50 Principles Applied to an Masonry Wall, Canadian Building Digest, National Research Council Canada, Ottawa, Ontario, Canada, February 1964.

(2) Hutcheon, N.B. and Handegord, G.O.; Building Science for a Cold Climate, National Research Council of Canada, 1983.

(3) J.F. Straube and Burnett, E.F.P.; Building Science for Building Enclosures, Building Science Press, Westford, MA, 2005 (www.buildingsciencepress.com)

(4) Baker, M.; Roofs, Multi-Science Publications, Ltd., Montreal, 1980.

** Home of Bobby Clark, hockey legend, no teeth

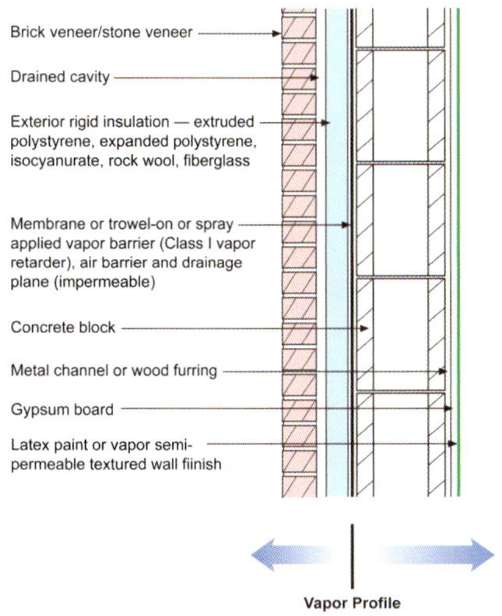

Figure 6: "The Institutional Wall"

The best wall that we know how to construct. Works everywhere in all climate zones.

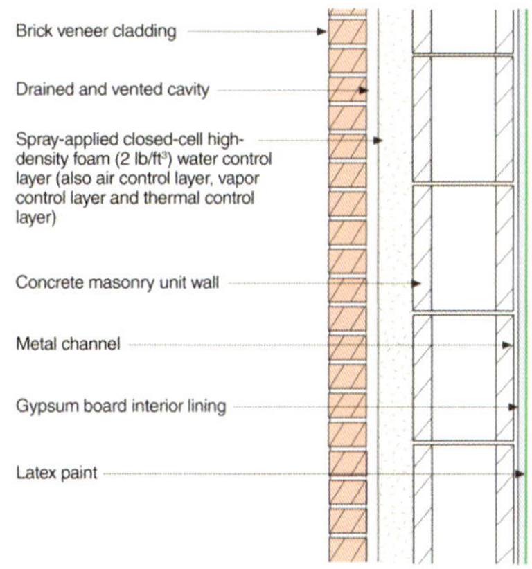

Figure 7: Clever Wall

One material combines four principal control layers.

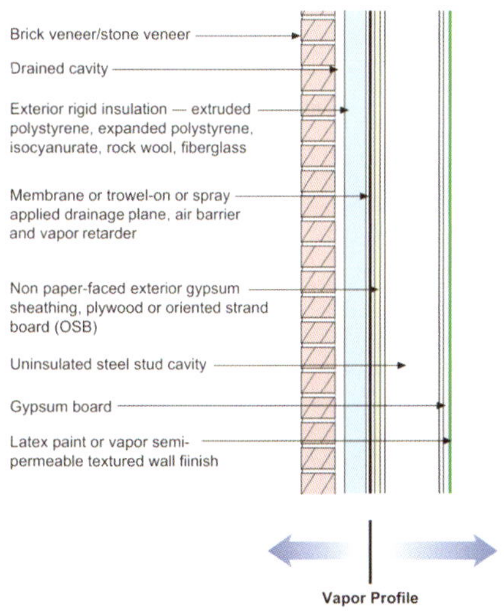

Figure 8: "The Commercial Wall"

The almost best wall we know how to construct. Affordable. Works everywhere in all climate zones.

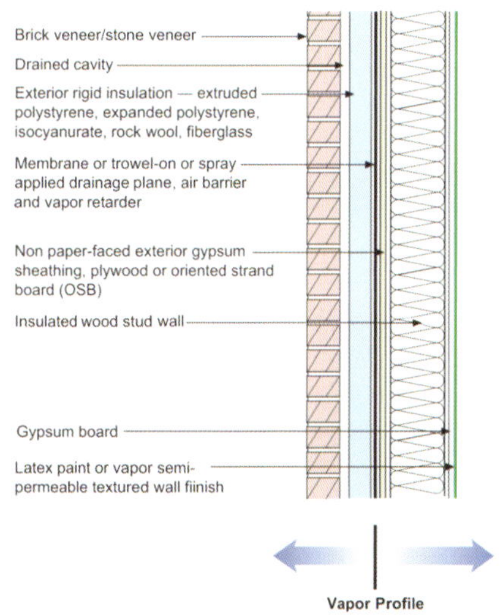

Figure 9: "The Residential Wall"

The best residential wall we know how to construct. Not cheap. Works almost everywhere – except in extreme cold climates where we would not insulate within the wood structural fame.

Appendix B.8

Building Science Digest 011

Thermal Control in Buildings

2006-11-02 by John Straube (updated 2011-12-12)

Abstract:

Providing thermal comfort without excess space conditioning costs is one of the primary requirements of buildings. Therefore, thermal control is an important aspect in almost all buildings. Understanding heat transfer and the temperature distribution through building materials and assemblies is also important for assessing energy use, thermal comfort, thermal movements, durability, and the potential for moisture problems.

Heat flow occurs through the building enclosure via opaque enclosure elements, is directly transferred into the building by solar radiation through windows, is carried along with air across the enclosure by unintentional leakage and ventilation, and can be generated within the building by occupants and their activities.

The control of heat flow in buildings requires insulation layers compromised with few thermal bridges, an effective air barrier system, good control of solar radiation, and management of interior heat generation.

Background

Heat flow can be a transient or a steady process. In the transient state, temperature and/or heat flow vary with time. Steady-state heat flow occurs when the temperature and heat flow reach a stable equilibrium condition that does not vary with time. Depending on the particular problem, the assumption of steady-state conditions may provide sufficiently accurate predictions of actual heat flow and temperature conditions. However, for some problems the assumption of steady flow can result in significant errors.

Heat flow can occur in one, two, or three dimensions. In almost all real situations, heat flow occurs in three dimensions but, from a practical point of view, it is often acceptable to simplify considerations to only one-dimensional, or series, heat flow.

Heat transfer occurs by three primary mechanisms, acting alone or in some combination:

- conduction,
- convection, and
- radiation.

Changes in moisture state, although not strictly an energy transfer mechanism, must also be considered since these state changes absorb and release heat energy, i.e., latent heat.

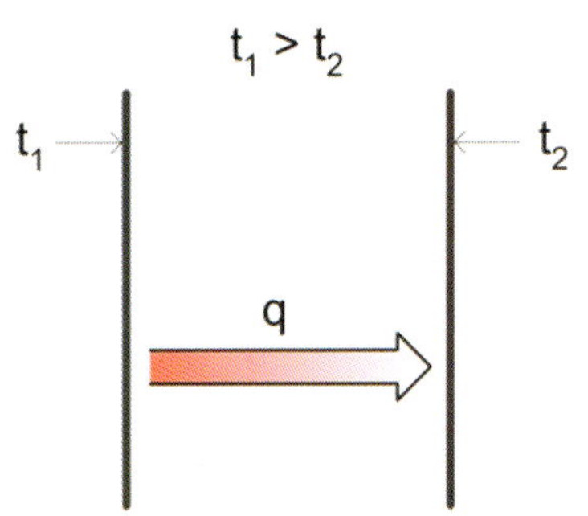

Figure 1: Conduction of heat through a solid

Conduction is the flow of heat through a material by direct molecular contact. This contact occurs within a material or through two materials in contact. It is the most important heat transport mode for solids; it is sometimes important for liquids, and it is occasionally important for gases.

Convection is the transfer of heat by the movement or flow of molecules (liquid or gas) with a change in their heat content. This is an important heat transfer mode between fluids and solids, or within fluids.

Radiation is the transfer of heat by electromagnetic waves through a gas or vacuum. Heat transfer by this mode therefore requires a line of sight connection between the surfaces involved. All objects above absolute zero radiate heat energy; it is the net radiative heat transfer that is the heat transfer of interest. Radiation is mostly of importance for heat transfer between solids and within highly porous solids, but radiation between high-temperature gases is occasionally of practical importance.

State change, sometimes called phase change, occurs at a constant temperature but still entails the movement of energy. For example, evaporation absorbs energy and condensation releases energy. This energy is sometimes called latent heat.

The mode of heat transfer often changes during the process of heat flow through and within building systems. For example, the sun transmits heat by radiation to the earth, where it can be absorbed, for example, by a brick wall. The heat is then transferred by conduction through the brick and transferred to the indoor air by convection and to the indoor surfaces by radiation.

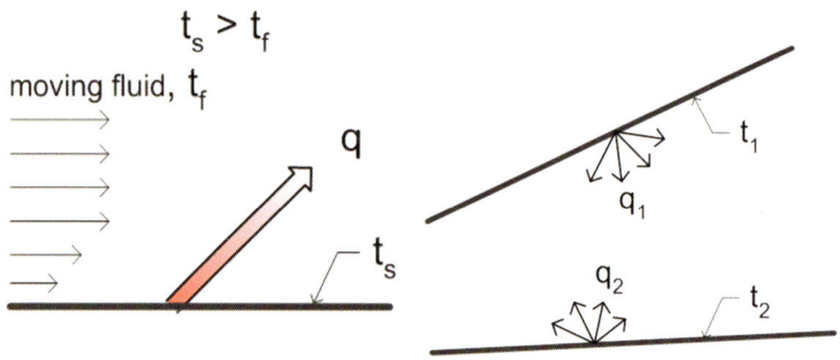

Figure 2: Convection and radiation

Insulation

All materials and layers in a building assembly have some resistance to heat flow. However, some materials with a k-value lower than about 0.05 to 0.07 W / m · K (how much?) are deliberately used in building assemblies for their ability to retard the flow of heat. These building products are called thermal insulations. Insulations are usually solid materials (so-called body insulation), but radiant barriers that control only radiation heat transfer across air spaces are also available.

Since conduction is a major mode of heat transfer and still air is a low-cost insulator, insulation products tend to be low-density materials (i.e., porous materials with a large proportion of voids filled with air) and / or made of low-conductivity elements. For example, glass of 2500 kg/m^3 density and 1.0 W/ m · K ? thermal conductivity is spun into fibers and formed into a batt of about 16 kg/m^3 density and 0.043 W/ m · K ? thermal conductivity. Fiberglass batt insulation is widely used as insulation despite the high conductivity of glass because of the very high percentage of pores filled with air. This type of thermal insulation is approximately 99.4% air.

Foam plastic insulations have a lower percentage of air voids than glass fiber batts, but are made of lower conductivity plastic material. Soft wood of 500 kg / m^3 density and 0.11 W/ m · K? thermal conductivity is produced to produce cellulose insulation of 60 kg / m^3 density and 0.042 W/ m · K ? thermal conductivity.

Most materials with high strength have relatively high density, and the strength of most building materials (e.g., concrete, wood, plastic) drops along with drops in their density. Hence, the need for low density (or more accurately, high porosity) reduces the structural capacity of most insulation. Accordingly, low-density insulation layers such as glass fiber batt, and foamed plastics are used to control heat flow in most modern building enclosures, while high-density, high-strength, high-conductivity materials such as steel studs, and concrete are used to support structural loads. In the past, building materials such as adobe, log, and low-density brick were used in a manner that combined both moderate insulating and acceptable load-bearing functions. Buildings constructed of these materials had thick walls, both to provide a reasonable level of resistance to heat flow and to provide sufficient strength.

Within porous insulations like fibers and foams, all three modes of heat transfer actually occur simultaneously. At low densities, the effective conductivity is high since convection and radiation can move heat through the relatively open space. At high densities, convection and radiation are suppressed, but conduction through the increasing proportion of solid material becomes important. Therefore, an optimum density can be chosen: one that varies with the type of material. For glass fiber A (a very fine glass fiber), it can be seen that the optimum density is about 2 pounds per cubic foot (30 kg/m^3.) However, since the cost increases as more material is used, the density of glass fiber batt is more commonly less than 1 pcf (15 kg/m^3). If higher strength is required (as it is for low slope roof, curtainwall, and exterior basement applications), higher density fibrous products of 3 to 8 pcf (50 to 125 kg/m^3 density) are available. Foam plastic insulations provide better R-value if higher strengths are needed. For example, extruded polystyrene with a density of only 2 pounds per cubic foot can easily resist pressures of 10 psi (or 1440 psf).

The wide range of values for thermal conductivity, and the inverse relationship of strength to thermal resistance, may be appreciated from the thickness required to achieve a certain level of thermal resistance. Figure 3 is a plot of the thickness of various building materials required to achieve a thermal resistance of RSI3.5 (R20).

Rockwool fiber is thicker than glass (since the spun rock contains more impurities than glass), and hence conduction plays a larger role at lower densities. Rockwool products therefore tend to use higher densities to achieve the same thermal performance as spun glass. This extra density provides these products with greater strength and more resistance to convection and radiation effects. Even though more material is used for the same thermal resistance, rockwool products compete in applications that require these properties.

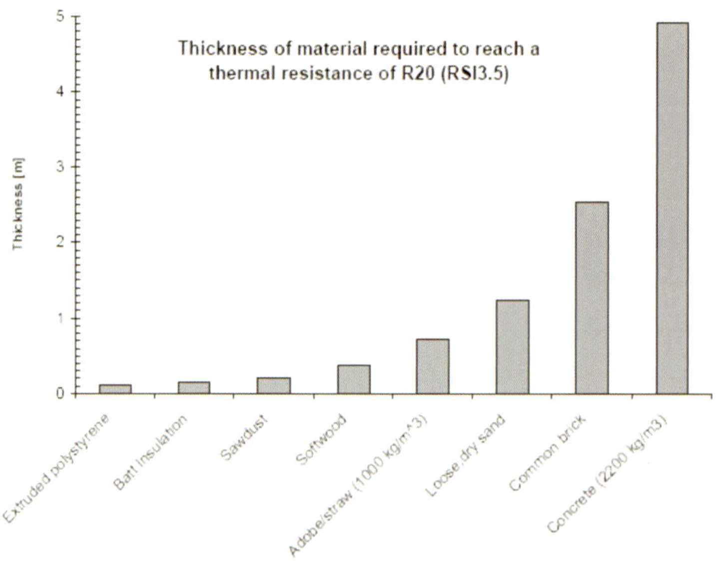

Figure 3: Comparison of the thickness of various materials required to achieve R20 RSI3.5

Thermal Bridging

Heat flow deviates from one-dimensional at corners, parapets, intersections between different assemblies, etc. When heat flows at a much higher rate through one part of an assembly than another, the term thermal bridge is used to reflect the fact that the heat has bridged over / around the thermal insulation. Thermal bridges become important when:

- they cause cold spots within an assembly that might cause performance (e.g., surface condensation), durability or comfort problems

- they are either large enough or intense enough (highly conductive) that they affect the total heat loss through the enclosure

Thermal bridging can severely compromise thermal control and comfort in some building types. Heat flow through steel stud walls and metal curtainwalls is dominated by heat flow through the metal components. Failure to break these thermal bridges can reduce the R-value of the insulating components (the insulated glazing unit or batt insulation respectively) by 50 to 80%. Filling the voids in concrete block masonry with insulation is not very effective: adding R15 insulation to a 12" block with increase the R-value of the wall by about R2. Wood framed walls are not as badly affected, but reductions of 10 to 20% are common. A separate BSD discusses thermal bridging in more detail.

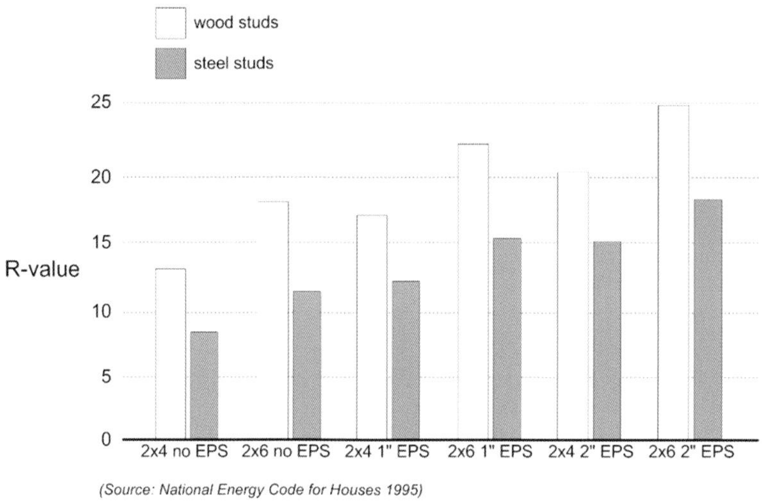

Figure 4: Best Case R-values for Walls with no extra framing for windows, floors or partitions

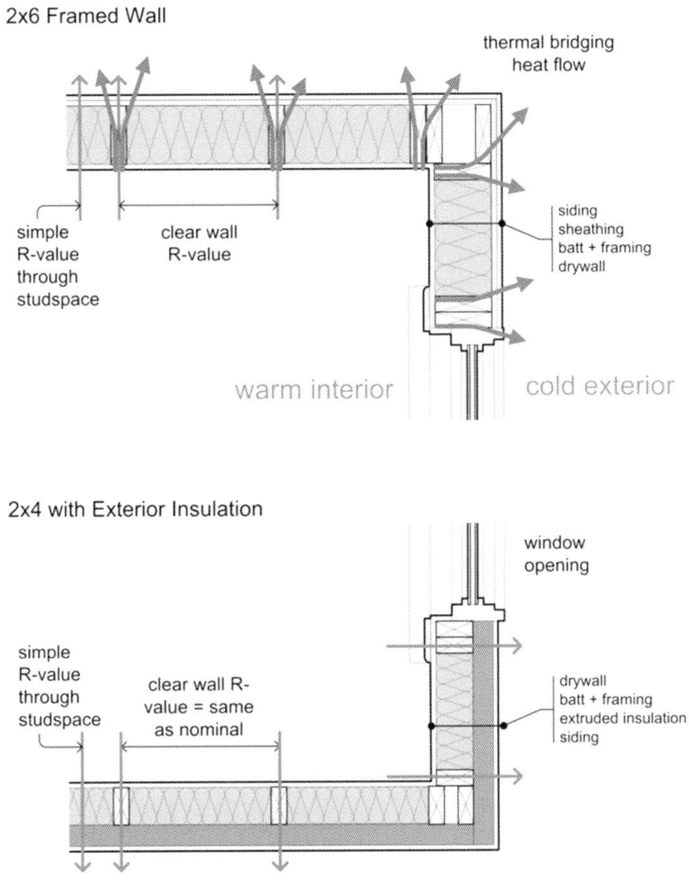

Figure 5: Solving thermal bridging through studs by using insulating sheathing

Heat Loss to the Ground

The ground temperature under and next to a building is generally very close to the annual average temperature. This means that the temperature difference between the inside of most buildings and the ground is not that large, although it is much steadier. Hence, less insulation is needed to control heat flow to or from the ground. Nevertheless, some insulation is still required in many climates zones (DOE Zones 3 and higher). In many cases, heat flow control for slabs, crawlspaces, and basements is limited by that needed for control of moisture and comfort problems, not energy.

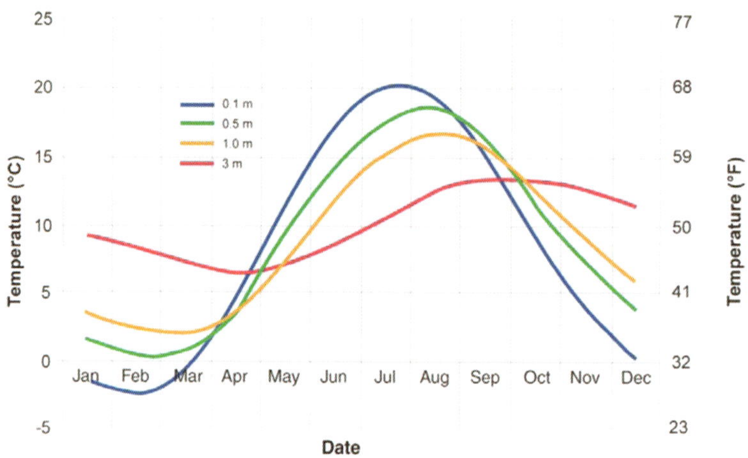

Figure 6: Temperature profile through the ground over the year in a cold climate location

Air Leakage

If cold air leaks out of the building during cold weather, it is of course replaced with cold air. This cold air must be heated up to make it comfortable. In warm weather leaking air is replaced with hot air that needs to be cooled and dehumidified. The energy impact of air leakage is significant and must be considered since it is often an important heat loss/gain component of modern buildings. For example, air leakage can account for 30% of the thermal flow across the enclosure in a well-insulated modern home.

The use of a complete air barrier system is required to prevent unintentional air leakage. Several BSD's and Guides address this critical issue.

Airflow can reduce or bypass thermal insulation in other ways than by just flowing across the enclosure.

Convective loops can form within highly air permeable insulation (low-density fibrous insulations) or small gaps around insulation (possible with rigid board insulation or improperly installed batts.

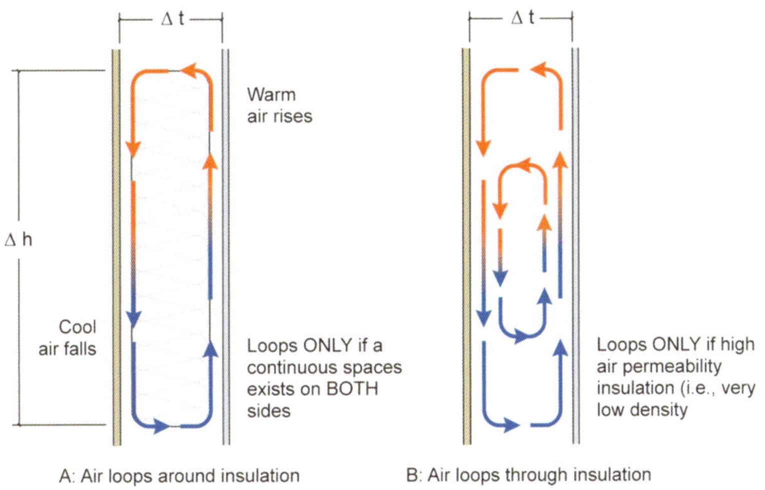

Figure 7:Convective air loops that reduce thermal control of insulation

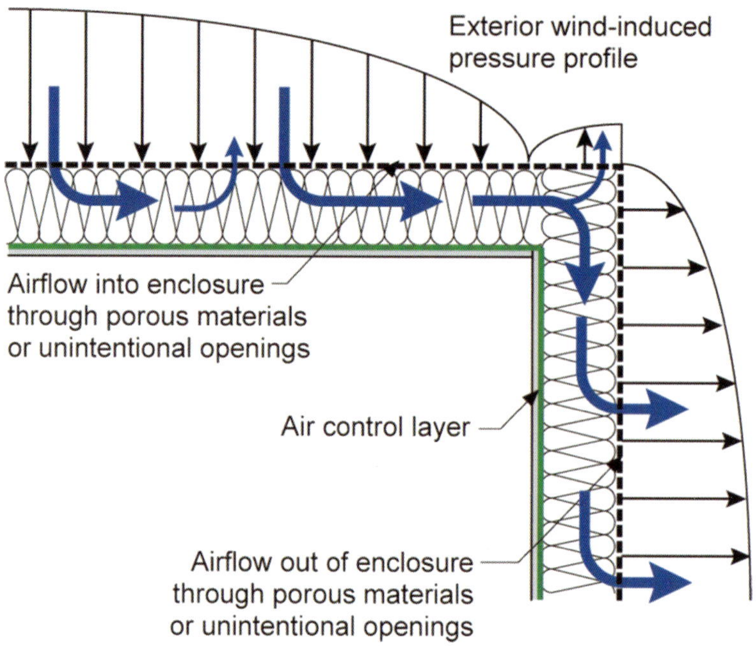

Figure 8: Wind-washing, the flow of wind through air permeable insulations, can reduce the thermal performance of insulations

Intentional ventilation has the same energy penalty as the same quantity of unintentional air leakage. Hence, the amount of ventilation should be no more than needed (see ASHRAE standards for guidance). In extreme climates, much of the energy required to heat/cool ventilation air can be recovered in heat recovery ventilator. Energy recovery ventilators also reduce the impact of humidity.

Solar Radiation Through Windows

Solar gain through windows exposed to either the direct sun, or reflected sun (reflected off the particles in the sky, creating diffuse radiation, or reflected off a surface) can dramatically affect the heat flow in a building. Hence, the building energy flows must account for the solar gain through windows. This amount of heat can dominate the performance of a modern building with relatively high window coverage (i.e., above 20 to 30% window to wall ratio).

The Solar Heat Gain Coefficient (SHGC) is the window property used to rate the amount of energy allowed through windows. The SHGC is the fraction of incident solar radiation that passes through a window and becomes heat inside the building. For example, if the SHGC of a glazing unit is 0.50, and the sun is shining on the window with an intensity of 500 W/m^2, 250 W/m^2 will enter the building.

The lower the SHGC, the less solar heat that the window transmits through and the greater its shading ability. In general, south-facing windows in houses designed for passive solar heating (with a roof overhang to shade them in the summer) should have windows with a high SHGC to allow in beneficial solar heat gain in the winter. East or West facing windows that receive large amounts of undesirable sun in mornings and afternoons, and windows in houses in hot climates, should have a low SHGC.

Solutions to control this form of thermal control include reduced window area, projecting horizontal shading (most effective on the south), exterior operable vertical shade, and solar control coatings on windows. Interior shades have a relatively small impact, but have the important role of controlling glare and providing privacy.

Passive solar heating design is used to capture the heat of the sun in a beneficial manner—this requires that most of the windows face south, and that window area be limited to collect only as much energy as needed for heating and to warm storage. Modern passive buildings have better control of the thermal losses in cold weather and hence have almost normal ratios of window to wall area.

Interior Heat Gains

In a well-insulated building, the interior heat generated by occupants and activities can be quite important. In cold weather, this interior heat offsets the heat required to warm the space. In warm climates this heat adds to the cooling load. In smaller buildings (or buildings with a large enclosure surface area to interior floor area ratio) such as housing, interior heat gains do not play a large role in most cases. Only in very well insulated homes or mild heating weather (i.e., around 10 ºC or 50 ºF) do interior heat gains form a significant proportion of heat flows in a small building.

Large boxy buildings (that is, those with a small ratio of enclosure surface area to floor area) are often dominated by internal heat gain. Thermal flow in properly insulated commercial office buildings generally is dominated by heat gain and loss through windows at the perimeter (that is, within about 30 feet of the perimeter) and by interior heat gains in the core. By employing moderate areas of high performance ($U<0.3$ or 1.25 $W/m^2{}^{o}C$) windows in a well insulated opaque enclosure, many commercial buildings will require little or no heating in below freezing weather when occupied.

Conclusions

This is just a summary introduction to the large topic of thermal control. Various Building Science Digests discuss specific topics of insulation, air barriers, ventilation, passive solar design, glazing, and thermal bridging. Despite the many specific situations that will arise in the building industry, it can be generally concluded that the control of heat flow in buildings requires insulation layers penetrated with few thermal bridges, an effective air barrier system, good control of solar radiation, and management of interior heat generation.

Appendix B.9

Insight

Can Highly Glazed Building Façades Be Green?

By John Straube, Ph.D., P.Eng.

When I see a fully glazed, floor-to-ceiling commercial or institutional building, I see an energy-consuming nightmare of a building that requires lots of heating and cooling at the perimeter just to maintain comfort. The result, on a cold winter day, is that offices exposed to the sun require cooling, while those in the shade need heat. Unless the control system is highly tuned, too many of the occupants will also be uncomfortable. Although it is well accepted that "green" buildings are above all low energy consumption buildings, there is a mistaken belief, almost a myth, that buildings with large expanses of glass are somehow green.

***Photo* 1:** An all-glass "high-performance" thermally-broken curtainwall with low-e, argon-filled glazing units and R-12 spandrel insulation. Sleek appearance but dismal performance by any normal standard. The R-value of the combined spandrel and glazing is less than R-4, and the solar gain is high enough to require air-conditioning on cold sunny winter days.

Glazing systems, including almost all modern high-performance ones, have very little ability to control heat flow and solar radiation. Older windows also did little to control heat loss and solar gain. Hence, most older buildings had restrained window-to-wall area ratios. Most of the tremendous performance gains in glazing technology over the past 25 years have been squandered on increased window area, not improved performance.

Normal thermally-broken commercial aluminum windows and curtainwalls typically have U-values of about 0.5, corresponding to an R-value of two. By using the best available low-e coated, argon-filled double-glazed units (with a center-of-glass R-value of four) in thermally-broken aluminum frames, overall window R-values of three can be approached. If the goal is low-energy buildings, why cover large portions of any building with such a low R-value system, particularly in cold climates? Spandrel panels are not much better in this regard. Although often packed with three or more inches of insulation (providing a nominal R-12), the metal back pans and aluminum framing short-circuit the insulation so that their overall R-value is typically closer to only four or five. Most so-called "high-performance" curtainwalls with a mix of vision glass and spandrel panels have overall R-values of less than 4. It is difficult to understand how such a choice of exterior wall could be considered a "green" system, when a simple low-cost wall with only an inch of continuous rigid insulation will provide significantly better control of heat flow!

But the thermal qualities are only part of the story. Glazing lets light in. That, after all, is the primary reason we use glazing. The solar heat gain that results is the reason many buildings require air-conditioning. The size of a building's air-conditioning plant is almost always defined by the glazed area: more glazing means more chillers, ducts, coils, and fans. In the past, thermally massive offices, schools and shops with limited glazing could tolerate little or no mechanical air-conditioning. It is a testament to the miracles of modern glazing (which uses low-e, low emissivity coatings to selectively allow more visible light than infrared heat radiation), that many buildings can have such large window areas and remain comfortable in the summer. Nevertheless, even very good commercial clear glazing still allows about one-third of the sun's heat to enter.

It is true that daylight can offset the need for electric lighting and provide a psychologically healthy connection to the outdoors, but one doesn't need floor-to-ceiling glass for that. In most occupancies and building types, there is no benefit to vision glass installed at floor levels (unless the occupants spend much of their time lying on the floor near the window), but there is a substantial energy penalty. Good daylighting design can reap all the benefits of glazing using vision glass covering less than half the enclosure. Numerous studies have shown there are no daylighting or energy benefits with window-to-wall ratios over 60 percent, and in most cases an area of between 25 and 40 percent is optimum (that is, lowest energy consumption). Even at these ratios, windows in a low-energy building should generally be high performance (triple-glazed in cold climates), with large thermal breaks (over 1/2" thick) and some form of exterior shading (preferably operable).

***Photo 2**: Almost the entire façade of this condominium tower is covered with low R-value and high solar gain curtainwall. It is almost impossible to achieve a low energy, comfortable building with this approach*

How should one choose glazing for green, low energy commercial buildings? At the most basic level, there are three critical measures of performance that must be chosen in design, and specified in contract documents: The U-value (1/R-value), the Solar Heat Gain Coefficient (SHGC) and the Visible Transmittance (VT).

***Photo 3:** Residential windows usually come with an affixed sticker providing verified and useful performance values: the U-value, Solar Heat Gain Coefficient and the Visible Transmittance. Commercial window specifiers need to define these performance metrics in their construction documents*

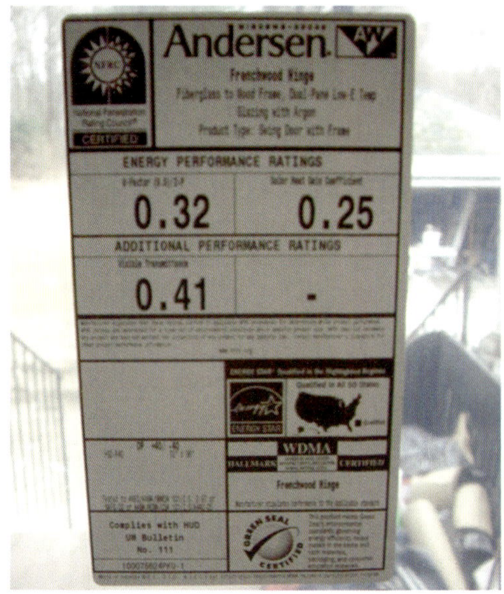

The U-value is the reciprocal of the R-value. A U-value of 0.33 is an R-value of 3 for example, (except that in the case of glazing systems the U-value includes the effect of surface films, unlike walls). As the climate becomes cold or very hot, a low U-value becomes important for energy savings and for comfort. Buildings with low U-value windows and/or curtainwalls can dispense with special, and costly, perimeter heating even in very cold climates.

The Solar Heat Gain Coefficient (SHGC) is the fraction of solar radiation that hits the glazing that passes through the glazing and becomes heat inside the room. For example, if the SHGC of a glazing unit is 0.50, and the sun is shining on the window with an intensity of 250 Btu/hr/ft^2, **125 Btu/hr/ft^2** will enter the building as heat. The lower the SHGC the greater its shading ability. Although specially-designed houses in cold climates often can benefit from solar heat gain in cold weather (ie, passive solar heating), modern offices and assembly spaces should have sufficient insulation and airtightness that the interior heat generated by occupants and activities provide most of the heat, even in cold weather. Bright sunny days often cause overheating of over-glazed south-facing spaces during cold (e.g. -10 deg F) days, whereas during unoccupied nights, when heating is needed, the low R-values result in heating demand. Operable exterior shading both obviates the need to choose a low SHGC window and provides much more control, but entails significant design effort, changes the appearance, and can be costly. Increasing opaque spandrel area, thereby reducing vision area, also reduces the total solar gain into the building of course.

The third measure, visible transmittance (VT), reports the percentage of visible light transmitted through the glazing. A typical clear, double-glazed window will have a VT of 0.60, meaning it admits 60% of the visible light. When daylight in a space is desirable, as in showrooms, high VT glazing is a logical choice. In offices and classrooms, where glare is a real issue, mid-range VT may be specified for large glazing areas, or high VT for small glazing areas. Low VT glazings such as bronze, gray, or reflective-film windows were often used in office buildings of the past as they also reduced solar heat gain. Modern spectrally selective (SS) windows allow for both a high VT with a low SHGC (and always have a lower U-value than uncoated glass). These SS windows reduce allow for significant daylighting and psychological benefits while avoiding overheating during sunny days.

In the end, glazing is a classic design problem that requires one to balance the desire for thermal comfort, energy efficiency, and light quality (all of which require small window areas) with equally important desires for view, daylight, and connectivity with the outdoors (all of which benefit from large vision-glass areas). Less window/ curtainwall area, and higher system performance is the low-cost, high performance prescription for buildings suffering from excessive glazing area. Carefully choosing, and then specifying, overall system U-value, SHGC, and VT is the best start to delivering high performance glazing on your next project.

The impact of thermal bridging through commercial wall assemblies, and heat flow through window systems can be calculated with relatively good accuracy by calculating an area-weighted average of the R-values of the windows and opaque wall sections. The equation takes the form:

$$U_{overall} = (WWR * U_{window} + (1 - WWR) * U_{wall}),$$
where $U = {}^{1}/_{R}$.

The results of a number of scenarios are plotted in the chart below.

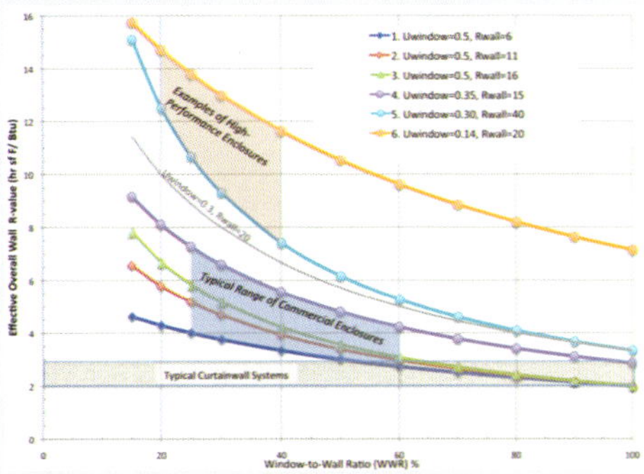

Typical curtainwall systems have an R-value of only 2 or 3, with "high performance" systems (not shown) using highly insulated (say R-12) spandrel panels and best-in-class double glazing (with low-e and Argon) may achieve R-4. Only a few systems, such as the Kawneer 7550 series or Visionwall, can achieve R-values of 6 or more.

Curve 1 above is for standard U=0.50 thermally-broken aluminum punched windows with air-filled double-glazed insulated glazing units in a R-12 batt-filled steel-stud brick veneer wall system (a true R-value of R-6 when thermal bridging is considered). The overall effective R-value of this wall is around 3 to 4 over the normal range of commercial window-to-wall (WWR) ratios of 25 to 50%.

Curve 2 shows that Increasing the R-value of the opaque wall to R-11 by adding an inch of foam on the exterior results in an increase of only R-0.5 to R-1.5 for the overall R-value for the same range of WWR.

Curve 3 shows how significant an impact window performance can make even if a good wall is provided. An externally-insulated R-16 wall, when mated with poor windows produces a vertical

enclosure with an R-value of only R-3 to R-6 for the normal range of window areas.

Curve 4 assumes a good-quality window frame with top quality glazing (low-e, argon-filled): the result for the overall vertical enclosure is still only R-4 to R-7.

These first four curves cover the performance of a wide range of commercial enclosures with a wide range of cladding types. Despite what most 10 high heat loss and heat gain. This high ratio should be avoided in both individual spaces, such as meeting rooms, as for the whole building on average.

Many designers have shown that beautiful and high-performance buildings can result from a proper balance of glazing quantity and quality. All too often, however, designers appear to choose all-glass curtain walls or floor-to-ceiling strip windows because they make it easy to create a sleek impression while leaving all the tricky details in the hands of the manufacturers. How much longer can we afford to pay the energy bills that result from that choice? It's high time to revive the craft of designing beautiful facades that don't cost the earth.

Photo 4: *A modern office building with a near-optimal window-to-wall area for energy and daylight performance in Boston, MA.*

Post Script: The very best in commercial aluminum curtainwall and windows are capable of R6 to R9. There are super windows in research labs with R-values of 10, and even 12 with dynamically operable shading capable of reducing the solar gain to less than 0.10 during bright sun and opening up to 0.60 or more during dull days. Other systems that incorporate light diffusing elements (such as Kalwall and Solera) and phase change materials (such as Glass-X) can reduce energy flows across daylight panels by a factor of ten! Eventually such technology should become more affordable and available, opening up the possibility of high glazing area and high performance.

Appendix B.10

Insight Energy Flow Across Enclosures

An edited version of this Insight first appeared in the ASHRAE Journal.

By Joseph W. Lstiburek, Ph.D., P.Eng., Fellow ASHRAE

Buildings don't work they way they used to.[1] Folks are always saying stuff like this. But let me make the case regarding just one factor and you decide. There are other factors of course, but I want to focus on only one for the moment. This one over-arching factor is the amount of energy exchange across the building enclosure.

I think that higher levels of thermal resistance and reduced heat gain across building enclosures has forever changed the performance of buildings—and not necessarily in a good way. And things are going to get worse before they get better. Sound familiar?

The lens I am going to use to look at this factor is moisture. Why? It is one of the principal damage functions acting on materials along with heat and ultra-violet radiation.[2] I think moisture is the key to understanding the performance of buildings in general—and in this particular case—in specific.

The consequence of this reduced energy exchange is beginning to be seen all over the place: mold, part load humidity problems, rot, and corrosion. Highly insulated building enclosures with reduced heat gain have low drying potentials and increased interior moisture loads. When they get wet they don't dry. Stuff is beginning to stink, rot, break and annoy.

There are two things to look at here. Levels of insulation and airtight assemblies and heat gain through glazing, roof assemblies and interior loads. Adding insulation and providing airtight assemblies reduces energy flow. That is a good thing right? Well, no—not always. Less energy flow from the inside to the outside the colder things on the outside of the building get in the winter. The colder these things, the wetter these things get and the wetter they stay. This is not good. Less energy flow from the outside to the inside in the summer and the less heat that is generated inside in the summer the less the air-conditioning system runs. This is a good thing right? Well, no—not always. We create something called "the part load humidity problem."

Higher Levels of Thermal Resistance

How do we dry wet wood? We kiln dry it. We heat the wood so that the water in the wood is warmer than its surroundings. We add energy to the water and it evaporates. There is a huge exchange of energy. When we put warm wet coffee beans in a room with very cold surfaces we call it freeze-drying. The moisture leaves the

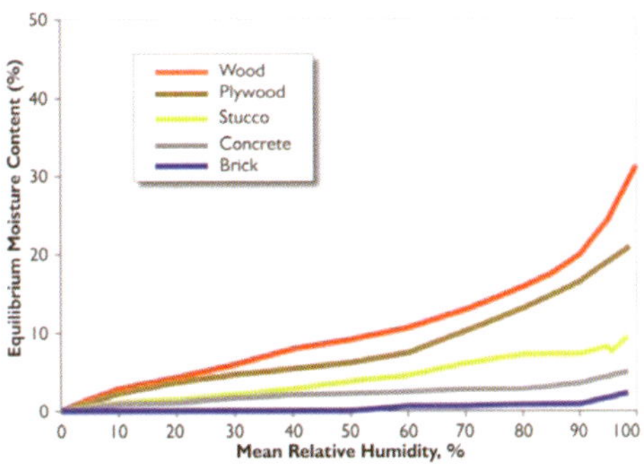

Figure 1: Sorption Curve for Common Building Materials—Note that moisture content goes up as relative humidity goes up. There is no temperature dependence or vapor pressure dependence except where temperature affects relative humidity or where vapor pressure affects relative humidity.

1 Yes, I know, pretty arrogant for a relative youngster to say. I am in my 50's, but when you get to this age you spend as much time looking back as you do looking forward. My mentors tell me that it helps with perspective. It has been said that in order to understand the future you have to understand the past. I am not so ambitious. I am only thinking of today. I think in order to understand the present you have to understand the past...

2 Someone once said that 80 percent of all building related durability problems are due to three principal damage functions (water, heat and UV radiation) with 80 percent of the 80 percent being water. I don't know who said this. Can someone help me out? I might have just made it up, but I have been saying it so long I no longer remember if I did or not. Aren't footnotes great...

coffee beans and accumulates on the cold surfaces. There is a huge exchange of energy. The greater the rate of energy exchange the greater the rate of moisture movement. Drying cannot happen with out an exchange of energy.

When my Mom and Dad bought their first house in Toronto, Canada in 1957 there was no insulation in the walls and the house was leaky to air—it had a high air change driven by a traditional chimney. We lived in a 1,200 square foot house and in January, when the outside temperature dropped to 0 degrees F., Momma cranked up the 300,000 Btu oil furnace to maintain an interior temperature of 70 degrees F. The energy flow across the building enclosure was enormous, but oil was cheap, and we were comfortable and happy. The energy flow was so enormous the building enclosure was simultaneously kiln dried and freeze-dried. In fact, the drying potential was so high, we were uncomfortably dry. As a result Poppa insisted that the furnace have a new fangled gadget attached to it—called a humidifier. How things have changed.

Well what changed? We've begun to insulate—and insulate exceptionally well—and we're getting the assemblies "tighter" to air change and convection. That results in two things—less energy exchange therefore less drying potential—and things on the exterior side of the enclosure are colder in the winter. Things being colder on the outside lead to something most folks don't consider. Many building materials are hygroscopic (**Figure 1**). This means they absorb moisture based on relative humidity. Even more strangely, they don't care about vapor pressure except if it affects relative humidity. This is a big deal. In fact it is a huge deal.

Photograph 1: The "Typical House"—Built in 1965 in Cleveland, OH with no cavity insulation in the exterior wood frame walls. No problems for twenty years. Insulation added to exterior walls in 1985 and within one year the paint begins to fail.

Photograph 2: Retrofit Cellulose Cavity Insulation—Dry blown cellulose packed into exterior wood frame walls. Injected from the exterior at top of cavity. This author, very young at the time, amazed at how well the insulation worked. Cavities were dry when opened.

Quick, snap quiz, psychrometric chart stuff....as the temperature drops, and vapor pressure is kept constant, what happens to the relative humidity? Buzz/Clang/Bell. Yes, folks, you are correct, the relative humidity goes up. The implications are staggering. Just making things on the outside of your building cold, hygroscopic things, makes them wetter. Period, end of story. But, but, the moisture content of the outside air in the wintertime is low. Yes, grasshopper, it is, but its' relative humidity is high. The amount of moisture in the absolute sense is low, but in the relative sense it is high. And hygroscopic materials don't care about the absolute sense; they only care about the relative sense. Why? Second Law stuff—isn't it always?

Check out **Photograph** 1. Nice, normal, everyday type of house in Cleveland, OH. Twenty years old, no insulation in the wall cavities, built in 1965 (yes, the photograph was taken in 1985—we did have camera's back then..). The house was perfectly happy at this point for twenty years. Then the homeowner decides to do something crazy, he decides to insulate the exterior walls to save energy and has the cavities blown with dry cellulose. What a dumb idea in 1985 eh? But, never the less, it was a beautiful job, no voids, yes it can be done, and yes it was done. No voids, no convection (**Photograph 2**—notice how young this engineer once was). Next thing you know, the paint falls off, and the siding gets smelly, moldy and begins to decay (**Photograp**h 3). Huh? The insulation did this? Yup. The drying potential is reduced due to the addition of the insulation so that the moisture entering the cladding assembly from the exterior due to rain and capillarity does not evaporate and this is further exacerbated by the colder siding having a higher local relative humidity and thus a higher equilibrium moisture content due to the sorption properties of wood. Bummer. How to fix this? Easy, wedges were used to create a gap at the overlaps of the siding reducing capillary uptake at the laps and increasing evaporation despite the cladding being colder (**Photograph 4**). This early lesson in sorption and drying potentials taught this engineer to drain and back-ventilate claddings –especially those on highly insulated assemblies (**Photograph 5**).

Photograph 3: Wet Cladding—Problems with mold and decay were worse on north and east elevations. Problems with peeling paint were worse on west and south elevations. Solar radiation cycled the cladding moisture content more on the west and south elevations stressing the paint. UV cross-linked the paint making it less flexible. Less flexibility, more movement on the south and the west led to the peeling paint. Less energy on the north and the east led to the mold and the decay.

Photograph 4: Wedgie—So simple and so obvious. After the problem is understood. Wedges created gaps that reduced capillary uptake of surface water, promoted drainage and back ventilation.

Let's extend this discussion to cavity insulation in general. The same thing happens to the cavity side of sheathing when we insulate cavities. It gets worse when the cavities also have no air flow or convection.[3] Sheathings get colder. If the sheathing is hygroscopic, its equilibrium moisture content also goes up. Why take the risk? Insulate on the exterior of the sheathing. This makes the problem go away. The risk is so high in cold climates that folks are no longer insulating the cavities in steel stud walls in places like Ottawa and Montreal—all of the insulation goes on the outside.

This is why insulating sheathing is pretty much the name of the game for highly insulated wall assemblies—or at a minimum claddings should be back ventilated and back-drained—or better, do both. This is easy for new construction. The problem is what to do with the existing building stock? What are we going to do if oil ever goes above $100/barrel? Dumb comment. Yes, but now we have a real problem. Folks cannot just go and blow/spray/fill cavities with a bunch of insulation without appreciating the consequences. This is not just a cold climate problem. It is a problem everywhere. In fact it gets worse when we go far south. Read on McDuff…

Reduced Heat Gain

I am going to let you in on a little secret that us mechanical engineers don't often share with the rest of the world—air conditioning systems only dehumidify when they run. Yup, when they don't run, they don't dehumidify. If there is no load, there is no dehumidification.[4]

Well, just make them run longer. How? They still make the air cold no matter how you run them. That is good if you need cold air. But what if you don't need cold air? What if you only need dry air? Well, you still have to make the air cold to get it to drop its water, and when

3 Air flow—air change and convection—increases energy exchange therefore increases drying potential. My Canadian friends heads are now going to explode because they have been taught that air leakage leads to "wetting" from interior sources. Yes, in extreme climates with high interior moisture loads. Think humidified buildings in Canada. Not everywhere else. Come visit the United States, the dollar exchange now works for you and air leakage improves drying potentials.

4 Sensible load. I know you knew that I knew this, but some of the email I have been getting from you folks has been quite picky, which is good, but I want to head off the obvious emails, so continue to be picky, but not on this…

you are done you still have cold air. So now what? Well, you have to heat it back up if you want to use it. This is sometimes called cooling with re-heat. There are other tricks we can do—read on—but at the end of the day we have to add energy back to the air that we cooled. There is no other practical way to do it.

It gets worse when we over-ventilate with humid air—especially when it is cloudy and not so hot. Now all you are doing is bringing in moisture. But you are doing it when there is not much of a load—not much of a "sensible load" (heat gain across the enclosure and heat gain from lights and appliances and people—the load you can "sense") to run the air conditioning system. This moisture that you are bringing in with the ventilation air is also a "load"—but it is a different kind of load that we have pretty much been ignoring up till now and getting away with it. The operative phrase to note is: "getting away with it." Mechanical engineers like me call this the "hidden load" or the "latent load" (as in "latent defect" or "hidden defect") and our heads explode when we try to explain this problem to the indoor air crazies who think that only dilution is the solution to indoor pollution. I mean how can too much outside air be bad?[5] Arrargah…

To take the water out of the air you need to use energy. To make the air comfortable/useful the air can't be too cold after you have taken the water out of it. Typically heat (energy) has to be added back to the air. Traditionally this heat (energy) was available through

Photograph 5: Drainage and Ventilation Matt Behind Cladding—This is easy for new construction and re-cladding. Existing buildings are going to be a problem. Don't say you weren't warned.

lousy glazing systems, too much glass, black roofs with no insulation, poorly insulated walls, energy inefficient lights, and crappy appliances. Well, this traditional "re-heat" due to inefficiency and poor practice is disappearing. Now we are stuck. If we don't warm up the air after we dry it we are going to get into trouble. We haven't been, and we are getting into trouble. The buildings are now getting too cold or too humid or worse, both.

Ah, this is easy you say. Just make the air conditioning systems smaller—"right size" them. Don't make me scream. I am getting tired of being told by energy weenies that my equipment is too big.[6] I still have to satisfy the "full load" requirement. If you make the equipment too small it doesn't satisfy the cooling requirements under full load. Let the people be uncomfortable you say. People should suffer you say; we have had it too good you say. Yeah, sure, another greenie weenie value judgment that also makes me scream. I don't think we should suffer. I like it being good. I don't have a problem with using energy. I have a problem with wasting energy. Until now we have gotten pretty good at dealing with this part-load problem with face-bypass, run-around coils, hot gas bypass and heat pipes but we have reached our limits.

5 Some folks don't seem to care about the energy thing either. I mean health is paramount and we should always err on the side of caution. How can we argue against health? Easy, when the "healthy" measures are "unhealthy." Over-ventilation can be unhealthy—and energy wasteful. Read on. This is leading to interesting discussions in the Standard 62 committees. Apparently the Standard 62 committees are not about energy. Apparently energy security is not a 62 problem nor is climate change nor are operating costs. Apparently, it is some other committee's problem. How about the energy committee (Standard 90.1)? Except that the energy committee does not want to touch ventilation. Cowards. My insider's perspective (on 62.2 at least) is that there is a lot of mileage to be made by scaring people about under ventilation, and folks are rising to the occasion. Unfortunately, over ventilation in hot humid climates and in mixed humid climates has led to more indoor air problems due to mold resulting from part load issues than under ventilation anywhere else—in my not so humble opinion. And speaking of health and under ventilation where are all the dead bodies? Where is the causal link between health and specific levels of specific indoor contaminants? In my not so humble opinion all of the rates have been just wild assed guesses without a sound epidemiological basis. But the resulting mold from over ventilation is real and demonstrable. Just to make it clear, I have been guessing at "rates" as well, but I have not been afraid to call them "guesses" or tried to hide the guesses behind a lot of bad science. And one more thing, this LEED stuff is making me crazy, the part where you get extra points when you ventilate at an even higher rate than what Standard 62 calls for…what a bunch of green sustainability hypocrisy…doesn't anyone at the US Green Building Council know anything about energy and part load humidity?

6 Don't go there. Just don't go there…

Building enclosures have gotten so good, glazing systems so good, interior load management so good, that we can't use the same equipment to handle the "sensible load" and the "latent load." Did I mention just how good the glazing systems have become? Did I mention the low SHGC thing? We used to just install air conditioners and only cared about making it cold inside. We used one piece of equipment to handle the sensible and latent load. The reason this worked is that the latent load was small compared to the sensible load. Not any more. The sensible load has gotten smaller big time. And, much to my chagrin, the latent load has gone up due to over ventilation. The old systems can't handle the new sensible to latent ratios. We have to separate the two loads.

We need separate systems for the sensible load and the latent load. This is a big deal and it results from the low heat gain in the modern enclosure. In other words energy conservation and good construction practice is biting us on the butt. We sometimes refer to this second separate latent load control system as supplemental dehumidification—and we need it when we have a low heat gain enclosure and efficient lighting and appliances. Others call it "preconditioning" of outside air since the largest component of the latent load is the ventilation air or the outside air. Whatever you call it, we need this second system. We don't often get it and it hurts us.

But, but these systems use energy. Yes, as the saying goes—there is no thermodynamic free lunch. You need to use energy to make air cold and you need to use energy to make air dry. We have been very good at making air cold. We now have to get good at making air dry. Think of it this way. For every 100 units of energy you save on the efficiency and on the cooling side, you are going to have to give back about 20 units of energy to be dry. You are still 80 units ahead. Problem is, if you are greedy and want the entire 100 units your building fails and your occupants get very uncomfortable and probably very annoyed.

It is pretty amazing to me but the hotel industry figured this out first. They pretty much had the problem first, so it stands to reason that they would also figure it out first. Think about the typical hotel room. The drapes are pretty much closed all the time, the room is empty during the day, and you know the unit has been sized for the full load, with the drapes open, and the room full, and everything on and you know that someone in some bureaucracy somewhere wants way too much outside air. Most of the time your latent load dwarfs your sensible load. So what do you do? You reduce the amount of outside air by installing timers on the exhaust fans and you pretty much ignore the faceless bureaucrat. Then you only run the through-wall unit to control the temperature (i.e. make it the "sensible" system) and install a dehumidifier in each unit control the humidity (the "latent" system) (**Photograph 6**). A pretty primitive but effective "fix." In new design we just "hide" the dehumidifier better—or better still make it part of the ventilation system.

Guess where this technology is now making an impact—condos and apartments. Small houses are next. We are already separating the latent from the sensible on most commercial design—after all it is the sensible thing to do.

Photograph 6: Hotel Room Fix—The through-wall unit controls the temperature (the "sensible" system). The dehumidifier controls the humidity (the "latent" system).

North versus South

So how come it is worse in the south than the north? Besides the south loosing the Civil War? In the north, it still gets cold and the outside cold air is pretty dry—and walls tend to dry to the outside in the north - so the size of the moisture drying "sink" so to speak—has not changed. But in the south, walls also dry to the inside—and if the inside is humid because of the part load problem—the walls stay wetter because the moisture drying "sink" is smaller. Additionally, in the south the moisture drive from the exterior into the wall is greater while at the same time the moisture drive from the wall into the interior is smaller as compared to the corresponding drives in opposite directions in the north.

Let me translate. It is easier to get wet from the outside in the south than it is from the inside in them north. It is also easier to dry to the outside in the north than it is to dry to the inside in the south. With less energy available, it gets worse in the south faster than in the north.

Good air conditioning design leads to dry interiors and dry interiors allow walls to dry to the inside. Part load humidity problems lead to wetter walls. Higher levels of thermal insulation and lower heat gain lead to part load humidity problems. Not good.

Now What?

OK, everybody relax. We want and we need much higher levels of thermal control (air tight, insulated assemblies) everywhere. We just have to be smart about it. We want and we need ultra efficient glazing, lights, appliances, and reflective surfaces everywhere. We just have to be smart about it. And we want and we need the right amount of outside air whenever we need it. We just have to be smart about it. This is neither expensive nor difficult, just different. We have to modify our building enclosure and mechanical system paradigms. Get used to it or be steamrolled by the change.

Insulate on the outside. Back ventilate and drain your cladding. Separate your sensible load from you latent load. Don't over ventilate. Existing buildings are going to be a bear. And have a nice day.

Appendix B.11

Building Science Digest 014

Air Flow Control in Buildings

2007-10-15 by John Straube

Abstract:

The control of air flow is important for several reasons: to control moisture damage, reduce energy losses, and to ensure occupant comfort and health. Airflow across the building enclosure is driven by wind pressures, stack effect, and mechanical air handling equipment like fans and furnaces. A continuous, strong, stiff, durable and air impermeable air barrier system is required between the exterior and conditions space to control airflow driven by these forces.

Air barrier systems should be clearly shown and labelled on all drawings, with continuity demonstrated at all penetrations, transitions, and intersections. In addition, enclosure assemblies and buildings should be vertically and horizontally compartmentalized, may require secondary planes of airtightness (such as those provided by housewraps and sealed rigid sheathing) and may need appropriately air impermeable insulations or insulated sheathing.

Introduction

It has long been recognised that the control of air flow is a crucial and intrinsic part of heat and moisture control in modern building enclosures [Wilson 1963, Garden 1965]. That this statement is true for all climates has been a more recently developed awareness [Lstiburek 1994]. A large fraction of a modern, well-insulated building's space conditioning energy load is due to uncontrolled air leakage. Wintertime condensation of water vapor in exfiltrating air (or summertime condensation of infiltrating air) within assemblies is one of the two major sources of moisture in the above-grade enclosure (driving rain being the other). Air flow through the enclosure can also carry, exhaust gases, odours, and sounds through enclosures as well as mold spores and off gassing generated within the enclosure. Uncontrolled air leakage through the enclosure is therefore often a major cause of performance (e.g. comfort, health, energy, durability, etc.) problems.

Water vapor diffusion, while amenable to simple analysis, is often (but definitely not always) an insignificant source of moisture in modern building envelopes. Wintertime exfiltration condensation is, however, acknowledged as a common building performance problem in cold climates. Warm weather infiltration condensation is often a problem in warm and humid climates (e.g. the south-eastern States) and in some cases in cool climates, especially when air conditioning or cooling (e.g. arenas) is used.

Therefore, there are three primary classes of reasons why the control of air flow is important to building performance:

1. Moisture control – water vapor in the air can be deposited within the envelope by condensation and cause serious health, durability, and performance problems

2. Energy savings –air leaking out of a building must be replaced with outdoor air which requires energy to condition it. Approximately 30% to 50% of space conditioning energy consumption in many well-insulated buildings is due to air leakage through the building enclosure. Convective circulation and wind washing both reduce the effectiveness of thermal insulation and thus increase energy transfer across the envelope.

3. Comfort and health – cold drafts and the excessively dry wintertime air that results from excessive air leakage directly affect human comfort, wind-cooled portions of the interior of the enclosure promote condensation which supports biological growth which in turn affects indoor air quality, airborne sound transmission control requires good airflow control, and odours and gases from outside and adjoining buildings often annoy or cause health problems.

There are other circumstances that require the control of air flow; for example, to control smoke and fire spread through air spaces and building voids and shafts, but these are situations that deal with extreme events, not typical service. This document will emphasise airflow control and the avoidance of related moisture problems.

Fundamentals

For air flow to occur, there must be both:

- a pressure difference between two points, and
- a continuous flow path or opening connecting the points.

Although the prerequisites are obvious and simple to state, in practical design applications it is not always clear what the pressure differences are or how to assess the existence and nature of flow paths.

In general, the approach taken to control air flow is to attempt to seal all openings at one plane in the building enclosure. This primary plane of airtightness is called the air barrier system. The word system is used since airflow control is not provided by a material, but by an assemblage of materials which includes every joint, seam, and penetration.

The following sections will present forces driving flow, air barrier systems, a discussion of flow within building enclosures, and air leakage tolerant enclosure designs.

Driving Forces

There are three primary mechanisms which generate the pressure differences required for air flow within and through buildings (see Figure 1 below):

1. wind,
2. stack effect or bouyancy, and
3. mechanical air handling equipment and appliances.

Since, it is widely acknowledged that a perfectly airtight air barrier system is unlikely to be achieved in practise, it is also desirable to control the air pressure differences driving the flow. This typically means reducing the pressure imbalance created by HVAC systems, reducing stack effect pressures by compartmentalizing buildings vertically, and reducing wind pressures by compartmentalizing building plans.

A short review of the three types of forces driving airflow is presented below.

Wind

Wind forces act on all buildings, typically creating a positive pressure on the windward face and negative (suction) pressures on the walls. Bernoulli's equation can be used to calculate the pressure imposed on a building as function of wind speed. The stagnation pressure is defined as the pressure exerted by a flow decelerated to zero speed, and is given by:

$$P_{stag} = \frac{1}{2} \cdot r \cdot V^2 \cong 0.65 \cdot V^2 \qquad [1]$$

where r is the air density [kg/m^3], approximately 1.3 kg/m^3 at 0 C and
V the wind velocity [m/s]

The pressure calculated from Equation 1 is not directly what is imposed on a building, and so a pressure coefficient is introduced to modify the stagnation pressure as:

$$P = C_p \cdot P_{stag} \qquad [2]$$

This pressure coefficient can be found for simple buildings in numerous references and ranges from a typical value of C_p = 07 to 0.8 on the windward side to C_P = -0.3 to -0.5 on the leeward. Local pressure coefficients around parapets, under glancing wind directions, and other turbulent flow conditions are often much larger, e.g., C_p = +3 to -5. Interior pressure coefficients range from -0.3 to -1.0, and the arithmetic sum is the net pressure acting across the enclosure. A concise summary of typical pressure coefficients can be found in Chapter 14 of the ASHRAE Handbook of Fundamentals [ASHRAE 1997]

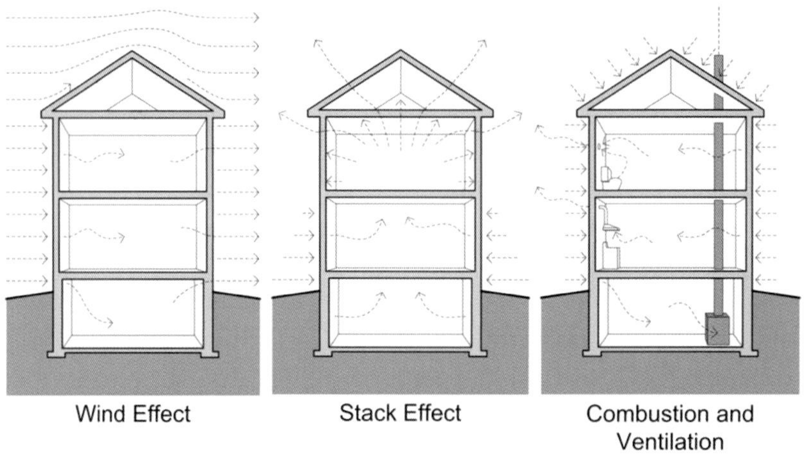

Figure 1: Forces Driving Air Flow through Building Enclosures

Low-slope roofs tend to have mostly negative (uplift) pressures, especially on the leading edge (Figure 2). Roofs with slopes above about 25 degrees experience positive pressures on the windward face, and suctions on the leeward.

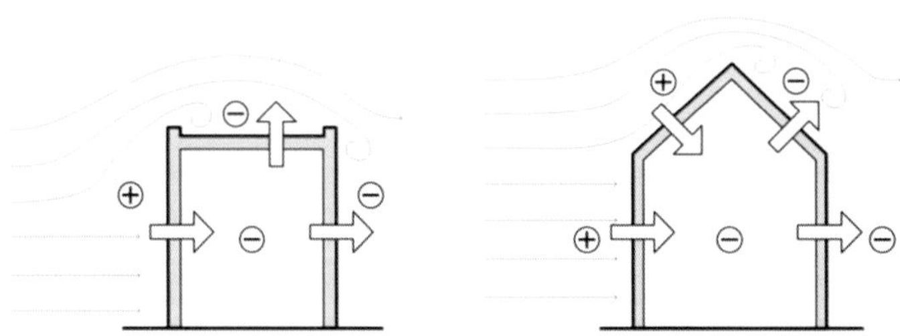

Figure 2: Wind Pressure Effects on Representative Buildings

Stack Effect

Stack effect pressures are generated by differences in air density with temperature, i.e. hot air rises and cold air sinks. The air within a building during the wintertime acts like a bubble of hot air in a sea of cold air. In the summertime the situation is reversed, although air temperature differences are usually less.

The density of dry air, r_a, varies with temperature. The greater the height of a column of air, the greater the potential difference in pressure if that column is at a different temperature. The pressure difference generated by a column of air h meters high with temperature difference between indoor and outdoor air at standard temperature and pressure is approximately:

$$P = 3465 \cdot Dh \cdot \left(\frac{1}{T_O} - \frac{1}{T_i}\right) \text{ [Pa]} \qquad [3]$$

where T_O and T_i are the outdoor and indoor temperatures respectively, (in **Kelvin** = Celsius + 273).

For example, if the air in a one meter high cylinder, open at the bottom and containing room temperature air (20 ϒC) is taken into the outdoors at a temperature of -10 ϒC, an outward pressure of 1.34 Pa would act at the top (Figure 3). The pressure at the bottom must be zero since it is connected to the outdoors. The horizontal plane at which the pressure equals the outdoor pressure (i.e. the difference is zero) is called the Neutral Pressure Plane (NPP).

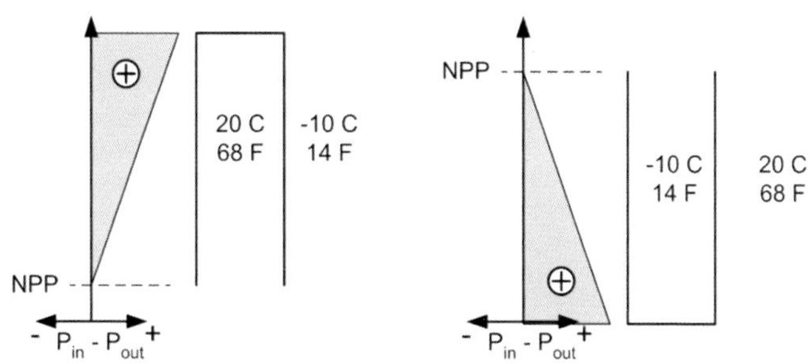

Figure 3: Pressure in Open-Ended Cylinder Due to Buoyancy

If the cylinder remained outdoors, the air it contained would slowly cool down to the exterior temperature and no pressure difference would exist. If the cylinder were then inverted and brought back indoors, the pressure at the closed end of the cylinder would again be 1.34 Pa acting outward as the cold air fell downward relative to the indoor air.

In the above examples, no flow occurred because no flow path was provided. If an open-ended cylinder containing room temperature air where used, any temperature difference between the cylinder and the surrounding air would cause flow, and the warm air would be immediately removed and replaced with cool outdoor air. However, if a heating coil were added to the cylinder to maintain the air temperature at 20 ϒC, airflow in the bottom would be heated. This is analogous to a heated building. Friction would slow air flow and result in a constant pressure drop along the height of the cylinder. Note that the NPP would now be located at mid-height and that air flow is involved (Figure 4). Obviously, the less air flowing through this cylinder the less heat energy required to maintain the interior of the cylinder at 20 ϒC.

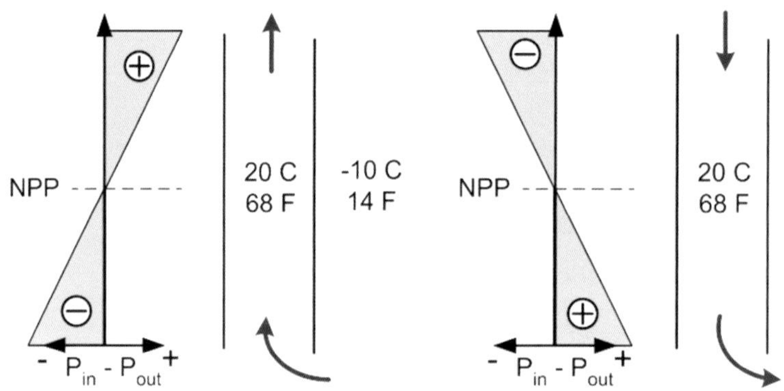

Figure 4: Flow through A Heated or Cooling Cylinder or Building

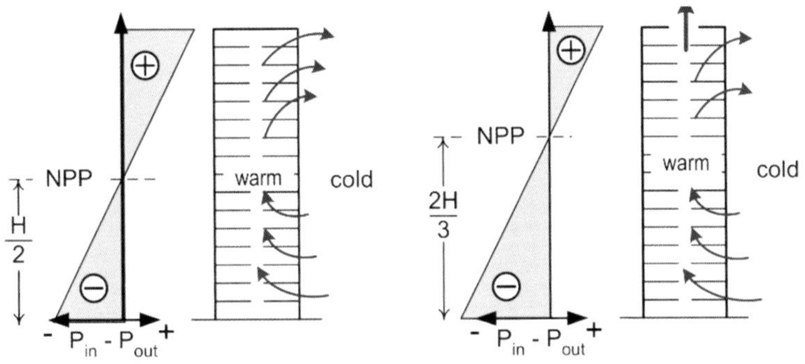

Floors leakier than walls:
-building acts like a perforated tube

Top leakier than bottom:
-NPP rise

Figure 5: Stack Effect in Real Buildings

Although houses tend to be only 3 to 9 m (9 to 27 feet) tall, stack effect can be a major force driving airflow, particularly in climates with a large temperature difference (more than 25 ΥC / 40 ΥF) across the enclosure. Often the exhaust of chimneys and fans, as well as leaky ceiling planes result in a Neutral Pressure Plane located near the ceiling level (Figure 6).

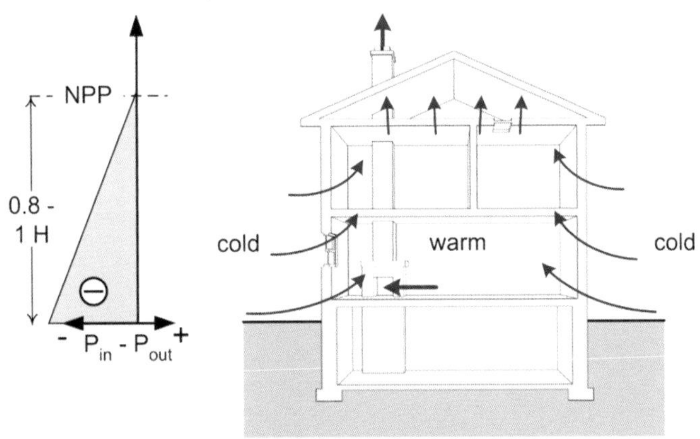

Figure 6: Stack effect during cold weather in a house

In warm climates and during warm weather, stack effect reverses and air is often drawn in at the top and pushed out at the bottom. Infiltration of warm moist air in warm weather can cause as many problems as exfiltration of warm moist air in the winter.

Mechanical Equipment

Fans and blowers cause the movement of air within buildings and through enclosures. By doing so, they can generate large pressures. If more air is exhausted from a building than is supplied, a net negative pressure is generated and vice versa.

If air is forced through the ducts that leave the building enclosure or pass outside the primary air barrier system (e.g., the very bad practise of placing ductwork in vented attics or crawlspaces) any leaks in the ductwork (and all ducts have some leakage, most ductwork is very leaky) will result in a net exhaust of air, and hence a net negative inward pressures on the building enclosure. The reverse can happen if leaky ducts outside the air barrier are under a net suction pressure.

Bathroom exhaust fans, clothes dryers, built-in vacuum cleaners, dust collection systems, and range hoods all exhaust air from a building. This creates a negative pressure inside the building. If the enclosure is airtight or the exhaust flow rate high, large negative pressures can be generated. These negative pressures have the potential to cause several problems:

- by driving inward air leakage through the enclosure, outdoor air may transport moisture into the enclosure during hot humid outdoor weather conditions
- the negative pressures can cause backdrafting of combustion appliances.
- the efficiency of most air handling devices will decrease with increasing back pressures.

Commercial HVAC systems both supply and extract air from buildings. This means both positive and negative pressure can be developed, depending on the balance of supply to exhaust flows. Dust collection systems, commercial range hoods and other industrial air handlers can move thousands of liters per second, seriously disturbing the pressure-flow relationship within a building and should be handled on an individual basis.

In design, one should aim for almost no mechanically-induced air pressure across the enclosure. This is achieved by balancing systems so that the same amount of air is supplied as is exhausted. In some case pressurization can be used to control airflow direction — buildings that are depressurised in winter will not have air leakage condensation problems, buildings that are pressurized may have winter condensation problems (if the enclosure leaks) but will exclude pollution from, for example, a parking garage.

Controlling Air Flow Through Enclosures

Controlling the flow of air across the enclosure, e.g., from the interior to the exterior or vice versa, is the most important aspect of air flow control. While no building is perfect, the goal of a design should be near zero flow and

The primary plane of air flow control in a wall is generally called the *air barrier.* Because such a plane is in practise comprised of elements and joints, the term *air barrier system* (ABS) is preferred. In framed, low-rise residential buildings, the primary air barrier system is often comprised of an inner layer of drywall (sealed around the perimeter and at all penetrations) or sealed polyethylene. However, outer layers of sheathing, (such as gypsum, waferboard, fiberboard, EXPS) and housewrap or building paper provide additional resistance to out-of-plane air flow through the enclosure assembly. In many modern building assemblies, exterior sheathing is designed and detailed to be part of an outboard air barrier system. Note that the plane of airtightness labelled by the designer (and all building sections should indicate what is intended to be the air barrier) or builder as the air barrier system may not in fact act as the ABS.

Basic Requirements of Air Barrier Systems

Typically, several different materials, joints and assemblies are combined to provide an uninterrupted plane of primary airflow control. Regardless of how air control is achieved, the following five requirements must be met by the air barrier system (ABS):

1. Continuity. This is the most important and most difficult requirement. Enclosures are *3-D systems*! ABS continuity must be ensured through doors, windows, penetrations, around corners, at floor lines, soffits, etc.

2. Strength. If the ABS is, as designed, much less air permeable than the remainder of the enclosure assembly, then it must also be designed to transfer the full design wind load (e.g., the 1-in-30 year gust) to the structural system. Fastenings can often be critical, especially for flexible non-adhered membrane systems.

3. Durability. The ABS must continue to perform for its service life. Therefore, the ease of repair and replacement, the imposed stresses and material resistance to movement, fatigue, temperature, etc. are all considerations.

4. Stiffness. The stiffness of the ABS (including fastening methods) must reduce or eliminate deflections to control air movement into the enclosure by pumping. The ABS must also be stiff enough that deformations do not change the air permeance (e.g., by stretching holes around fasteners) and/or distribute loads through unintentional load paths.

5. Impermeability. Naturally, the ABS must be impermeable to air. Typical recommended air permeability values are less than about 1.3×10^{-6} $m^3/m^2/Pa$. However air barrier materials are commonly defined as materials which pass less than Q< 0.02 lps/m² @75 Pa. Although this is an easy property to measure it is not as important as might be thought. In practise, the ability to achieve other requirements (especially continuity) are more important to performance, and the air "permeance" of joints, cracks, and penetrations outweighs the air permeance of the solid materials that make up most of the area of the ABS. Hence, a component should have an air leakage rate of less than Q< 0.2 lps/m² @75 Pa, and the whole building system should leak less than Q< 2.0 lps/m² @75 Pa.

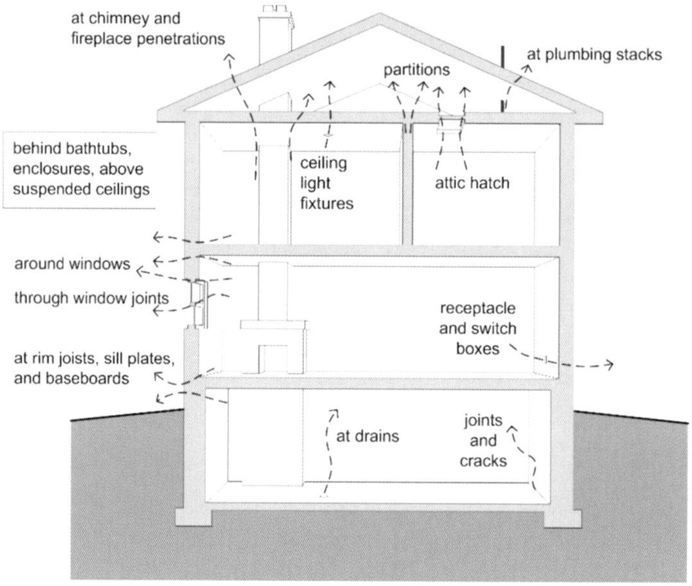

Figure 7: Typical Air leakage paths

As noted earlier, the secondary planes of air flow resistance fulfill several functions, either on their own or in conjunction with the other planes of air flow resistance. These secondary barriers not only add marginally to the overall airflow resistance of the assembly, they provide a level of redundancy if the primary air barrier is designed, built, or performs imperfectly. If the secondary barrier is of sufficient air tightness it may provide a great improvement to overall airtightness so long as compartmentalisation is provided . For example, research has shown that housewraps, sometimes called air infiltration retarders, can significantly reduce airflow through an imperfect primary air barrier even if they are not designed or built as an ABS. The satisfactory performance of many older wall systems can often be explained by the unintentional, and often synergistic, contribution to airtightness that layers such as building paper, board and panel sheathing, brickwork, etc.

Air Barrier Systems vs Vapor Barriers

The fact that many vapor barriers also retard or eliminate air flow sometimes causes confusion. In fact, much of the older literature (and a disappointing proportion of current documents) confuse or combine the function of the ABS and vapor barriers, and the difference between the two is still one of the most common building science questions. Hence, the distinction will be presented here once again.

The function of a vapor barrier is simply the control of water vapor *diffusion* to reduce the occurrence or intensity of condensation. As such, it has one performance requirement: it must have the specified level of vapor permeance and be installed to cover most of the area of an enclosure.

Many codes require the use of a vapor barrier in all enclosures. Some codes wisely require that vapor diffusion be controlled when an assembly "would be adversely affected by condensation." The need for a specific vapor barrier layer can be assessed by simple calculations, and rarely is a layer with very low permeance like polyethylene sheet justified. The 2009 IRC, IBC, and IEC will all have tables outlining science-based requirements for vapor barriers. They usually do not require the use of a very low permeance polyethylene as an air barrier or a vapor barrier. See also BSD -106 Understanding Vapor Barriers for more information.

Air barrier systems control air flow and thereby control convective vapor flow. As can be seen from the previous sections, the control of air flow provides other benefits and has at least five performance requirements to meet.

Canadian building codes require an air barrier system in all enclosures that would be adversely affected by condensation. In practise, this means air barriers are required for almost all conceivable types of building enclosures, especially since they provide more than just control of condensation. Some states (e.g., Massachusetts commercial code) and agencies (e.g., Army Corp of Engineers, General Services Administration) are beginning to mandate air barrier systems and enforce airtightness standards.

The vapor permeance of the air barrier must be considered in the same way as all other materials in an assembly should be. For example, in cold climates, a vapor barrier on the exterior is usually not acceptable (but can be designed for, as it is in an exposed membrane low-slope roof or a wall with metal cladding), whereas in hot humid climates, this location would be desirable. But the vapor permeance of the ABS is no more important than the vapor permeance of any other materials in an assembly, such as the cladding, sheathing, insulation, interior finish, etc.

Common Air Barrier Systems

A sheet of 6 mille (0.15 mm) thick polyethylene is often used as a vapor barrier in very cold climates (Zones 6 and 7). Poly is cheap to buy and install and has very low vapor permeance. However, it fails or barely meets most air barrier requirements other than air impermeability. It is difficult and relatively expensive to achieve continuity, especially since it is pierced by services and enclosure penetrations at many locations. It is likely to fail structurally when exposed to wind gust loads, and can fail through fatigue if it flaps because of varying wind pressures. Hence, it is not very durable. It is so flexible that it can deform and transfer loads through unexpected paths, deform batt insulation, tear fastenings, pump air, etc. Nevertheless, for undemanding applications such as low-rise housing, poly may act as both the vapor barrier and the air barrier. Poly on the interior should not be used in air conditioned buildings in Zones 5 or lower as its low vapor permeance poses an unacceptable risk of interstitial condensation during warm weather.

The airtight drywall approach (ADA) employs the interior layer of painted drywall as part of the ABS and poly, vinyl wall paper, or an appropriate type of paint as the vapor control layer. The drywall is stiff and strong enough for most applications and because it is visible, it is easy to inspect, repair, and to ensure continuity. Difficulty in achieving continuity is often encountered at service penetrations, wall-floor interfaces, intersection walls, transitions, partition walls, etc. Despite these difficulties, with a little care and training the ADA air barrier system is often quite successful and can be used in a wide-range of steel and wood framed roof and wall systems in both residential and commercial construction.

The vapor barrier and air barrier are often separated in enclosures employing housewraps or exterior sheathing (e.g., well sealed exterior gypsum, plywood, or OSB) as the primary air barrier. Poly, foil-backed drywall (in very cold climates), or paint is installed near the interior and acts as the vapor barrier. The advantage of such exterior air barriers is that they are often easier to install in such a manner as to span over all of the many interior service penetrations, plumbing, structural components, etc.

Another approach to exterior air barriers is the use of air impermeable, usually foam plastic or foil-faced, rigid insulation boards with sheathing tape and/or gaskets at joints. Such systems have the advantage of fewer penetrations, but the disadvantage that they are difficult to inspect and repair. The ability of these systems, including their joints, to transfer wind loads through connectors to the structural frame must be investigated for each application.

Spray foam, open-celled or closed-cell can be used as a very effective part of an air barrier system. When applied to the exterior of blockwork or sheathing, they form a continuous fully-adhered system. Transitions, movement joints, and penetrations require careful detailing to ensure airtightness. When the foam is sprayed within a stud cavity, much more effort is required to seal the joint between the floor sheathing and bottom plate, top plate to rim joist (often spray foam), etc.

Reinforced concrete is usually sufficiently airtight to form part of an ABS so long as the concrete is dense and crack size and spacing are controlled by the appropriate use of reinforcing (standard spacings used in reinforced concrete design are sufficient). Blockwork is not generally sufficiently impermeable to act as an air barrier. Small shrinkage cracks further compromise its airtightness. The application of a thick layer of block paint, or a thin layer of parging, preferably with fibres to limit crack sizes, renders blockwork walls airtight.

In commercial applications, bitumen-based air barriers, in either liquid or reinforced membrane form, are often adhered to the exterior of blockwork, concrete, or exterior gypsum sheathing. To ensure continuity, compatible membranes are used to bridge cracks and tie the ABS to windows, etc. This type of air barrier tends to be very vapor impermeable and so also acts as a vapor barrier. Therefore, the majority of the insulation must be applied outside of such an ABS in cold climates.

Polymer-modified and fibre-reinforced thin plasters are also available as air barriers. These systems require flexible joint details around penetrations, but are non-combustible and may be as vapor permeable or impermeable as specified. A variety of heavy-duty elastomeric liquid-applied air barrier membranes are also available. These products can have a range of vapor permeance values (from very impermeable to very permeable). With the appropriate analysis, they can be placed anywhere in an enclosure and used in all climates.

In framed systems, two air barrier systems, one inside and one outside of the framing, are often desirable, with framing members consciously designed as in-plane compartment separators to resist the internal lateral flows generated by wind pressures (Figure 8). Such redundancy is needed because of the susceptibility of these systems to wind washing, ABS failure, convective loops, and other airflow control problems.

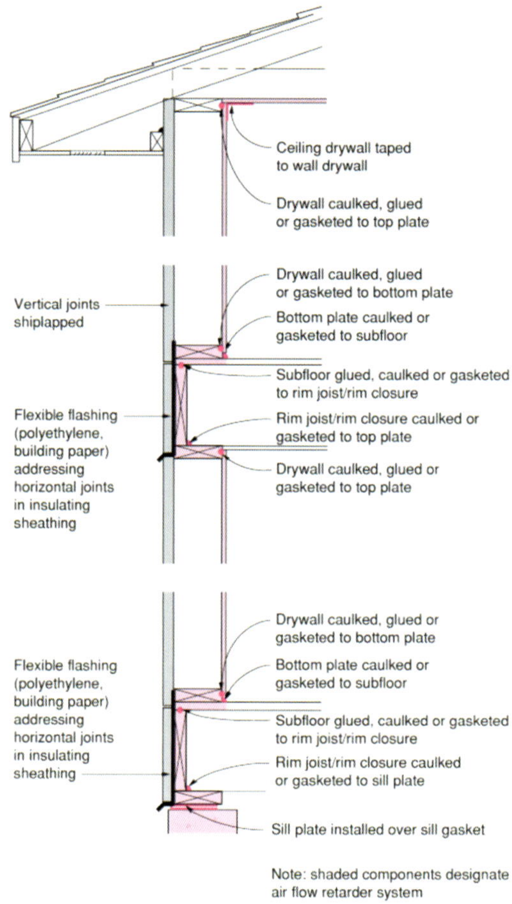

Figure 8: Airtight Drywall Approach applied to a wood-framed Enclosure

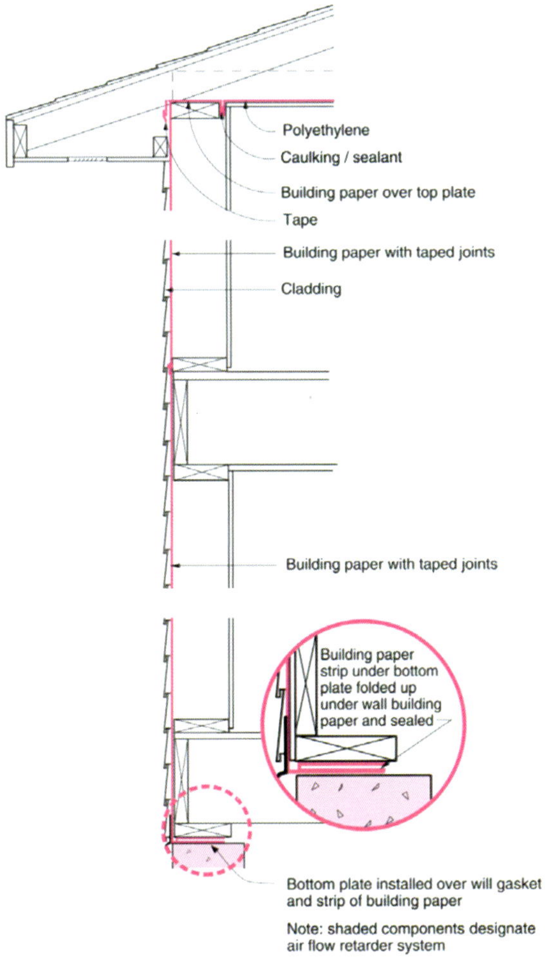

Figure 9: Housewrap air barrier system

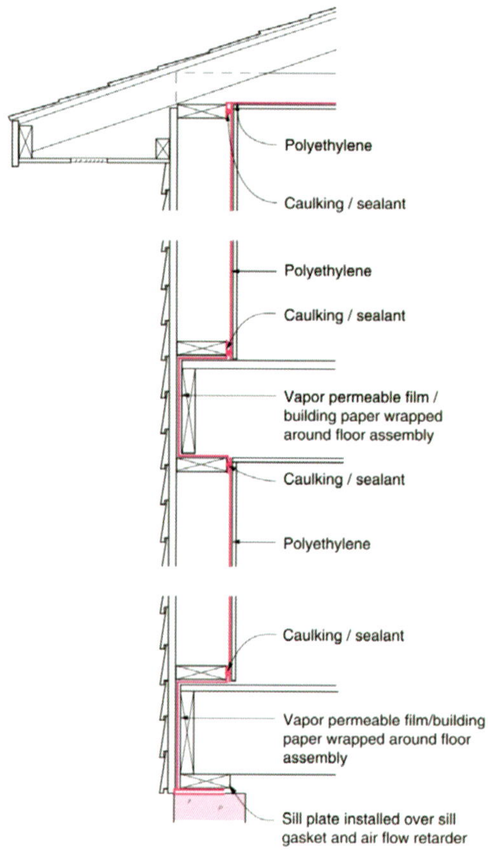

Figure 10: Polyethylene air barrier system (very cold climates only, Zones 6 or higher)

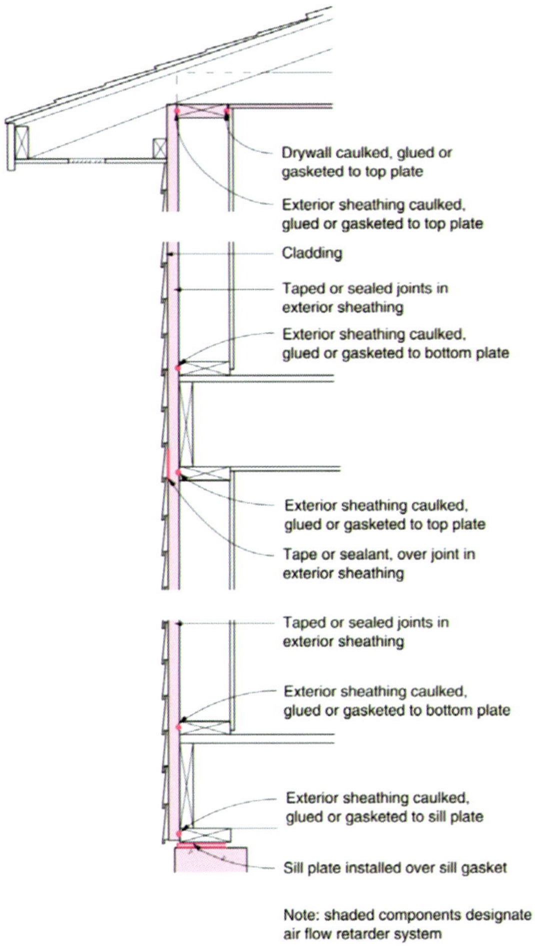

Figure 11: Exterior Sheathing (Insulated or not) Air Barrier System

Conclusions

Air flow control is important for several reasons: to control moisture damage, reduce energy losses, and to ensure occupant comfort and health. Airflow across the building enclosure is driven by wind pressures, stack effect, and mechanical air handling equipment like fans and furnaces. A continuous, strong, stiff, durable and air impermeable air barrier system is required between the exterior and conditions space to control airflow driven by these forces.

Air barrier systems should be clearly shown and labelled on all drawings, with continuity demonstrated at all penetrations, transitions, and intersections. In addition, enclosure assemblies and buildings should be vertically and horizontally compartmentalized, may require secondary planes of airtightness (such as those provided by housewraps and sealed rigid sheathing) and may need appropriately air impermeable insulations or insulated sheathing.

It must be noted that increased airtightness must be matched by an appropriate ventilation system to dilute pollutants, provide fresh air, and control cold weather humidity levels. Good airflow control through and within the building enclosure will bring many benefits: reduce moisture damage, energy savings, and increased health and comfort. However, while airflow usually causes wetting in enclosures, it also can be a powerful drying mechanism. Therefore, enclosures with increased air flow control demand greater attention to other sources of drying (diffusion is the only practical mechanism available) and the reduction or elimination of other sources of wetting (built-in, rain and diffusion).

References

Brook, M.S., "Rationalizing Wall Performance Criteria", *Proc. Sixth Conference on Building Science &Technology*, Toronto, March 5-6, 1992, pp.145-161.

Brown, W.C., Bomberg, M.T., Ullet, J.M. and Rasmussen, J. "Measured Thermal Resistance of Frame Walls with Defects in the Installation of Mineral Fibre Insulation", *J. of Thermal Insulation and Building Envelopes*, Vol 16, April 1993, pp. 318-339.

Controlling Stack Pressure in High-Rise Buildings by Compartmenting the Building. Research Report for CMHC, March, 1996.

Garden,G.K., *Control of Air Leakage is Important.* Canadian Building Digest 72, National Research Council of Canada, Ottawa, 1965.

Lecompte, J., "Influence of Natural Convection in an Insulated Cacity on the Thermal Performance of a Wall", *Insulation Materials, Testing and Applications, ASTM STP1030*. D.L. McElry and J.F. Kimpflen, Eds., Amercian Society for Tesitng and Materials, Philadelphia, 1990, pp. 397-420.

Lstiburek, Joseph and Carmody, John. *Moisture control handbook : principles and practices for residential and small commercial buildings*. New York : Van Nostrand Reinhold, 1993.

Ojanen, T. and Kohonen, R.,"Hygrothermal Performance Analysis of Wind Barrier Structures", *ASHRAE Transactions, Symposia*, Chicago, 1995, pp. 595-606.

Straube, J.F., Burnett, E.F.P.,"Rain Control and Screened Wall Systems", *Proc. Seventh Conference on Building Science &Technology*, Toronto, March, 1997, pp.18-37.

Timusk, J. and Lischkoff, J., Moisture and Thermal Aspects of Insulated Sheathings", *Proc. Second Conference on Building Science &Technology*, London, Ont., 1983, pp.71-91.

Uvsløkk, S., "The Importance of Wind Barriers for Insulated Wood Frame Constructions," *Proc. of Symposium and Day of Building Physics,* Lund University, August 24-27, 1987, Swedish Council for Building Research, 1988, pp. 262-267 and Uvslokk, S.,1996, "The importance of wind barriers for insulated timber frame construction", *J. Thermal Insul. and Bldg. Envs.*, V.20, July, p.40-62.

Wilson, A.G., "Air Leakage in Buildings", Canadian Building Digest 23, National Research Council of Canada, Ottawa, Dec 1963.

Insight
Building in Extreme Cold

An edited version of this Insight first appeared in the ASHRAE Journal.

By Joseph W. Lstiburek, Ph.D., P.Eng., Fellow ASHRAE

North of the Arctic Circle there are only two seasons—this winter and last winter. Who would ever want to live there? Being human, we can't help ourselves. We've been there a long time. And, we are going there in increasing numbers because of resources—that's where a lot of them are. We've learned a few things. We use snow to create temporary structures. The igloo is the ultimate sustainable structure. We use frozen rivers in the winter as truck interstates to provide fodder for cable TV shows. We never shut machinery off because in the extreme cold we can never start it again. So, how do you build permanent structures in such a miserable place?

The physics are straightforward, but the engineering is not. Moisture flow by air leakage and vapor diffusion from the inside to the outside is a huge concern. Even tiny gaps leaking air can lead to substantial icicles and frost boles. There is no argument that an actual air barrier is essential. The vapor drive when it is 40 degrees below zero[1] outside is formidable. This is one climate that needs a real vapor barrier, not a vapor retarder. There is no argument that an actual vapor barrier is essential (less than 0.1 perm). Seemingly insignificant thermal bridges that would be laughable in other climates are a big deal.[2] Triple glazing is not an option; it is essential. There is little room for error.

1 At minus 40 it does not matter whether it is Fahrenheit or Celsius as both scales cross at minus 40. Minus 40°F is equal to minus 40°C. Folks up north call it minus 40°FC for "freaking cold."

2 I remember asking why a blow drier is often hung outside the front door north of 66, and a second blow drier is hung inside the front door? Exfiltrating interior air carrying moisture leaking through the door hardware causes it to freeze solid. The exterior blow drier is used to melt the ice in the lock to get in. The interior blow drier allows you to get out.

Photograph 1: Extreme Enclosure—"Superinsulated" does not do justice to this house constructed by the Hydaburg Tribe in Alaska that is ultra-airtight and ultra-vapor tight (less than 0.5 ach at 50 Pa [0.20 in. w.c.] and less than 0.5 perms).

Photograph 2: **Exterior insulation**—Multiple layers of rigid insulation on the exterior of a wood-frame structure held in place with wood furring strips. Note the inset location of the windows allows the water management layer of the window to connect to the water management layer of the exterior wall (the fully adhered membrane).

We need, in essence, a "perfect" air barrier and a "perfect" vapor barrier. That means using a fully adhered sheet membrane that looks more like a refugee from the roofing industry than a building enclosure housewrap and vapor barrier.[3] The engineering problem entails constructing a six-sided cube wrapped with fully adhered sheet membrane with all of the insulation exterior to the membrane. Not just a little bit of insulation, but a lot of

3 After too many years of rips, tears, blow-offs and other annoyances, I will not specify 6 mil polyethylene in a hostile climate to serve a critical function. I took my lumps early on my own projects. Never again.

insulation. Did I mention a lot of insulation? No penetrations, none at all, are allowed in this sheet membrane. All services must be kept to the inside of the membrane (wiring, plumbing, communications, and HVAC).

How this cube touches the ground is a big deal. You don't actually want the cube to touch the ground. The ground is frozen all the time—permanently—that's why it's called "permafrost." It gets really ugly if heat from the building melts the frozen ground. That's why buildings on permafrost are elevated with ventilated airspaces between them and the ground to control heat loss from the building to the ground. An elegant solution to this foundation problem is a "space frame" with adjustable piers that thermally isolate the building from the ground. The ventilated space frame supports the cube.

The top of the cube is also a big deal. The top of the cube must be watertight but not vapor tight. It also must not blow off and it must deal with snow. Most of the time snow blows off, which is a very good thing. Contrary to popular belief, the arctic does not get all that much snow. It's just that the snow hangs around a long time. But, we can't rely on the snow always blowing off, and we can't rely that we'll always get only a little bit of snow. So, we have to deal with the potential for ice dams. Did I mention the blow-off part? It can get really windy, and the wind can whip up the snow, and blow the snow into tiny spaces. So, if you have tiny spaces connected to the outside, you should expect snow. The snow particles can get very tiny, like fine sand blowing around in the desert. The desert analogy is not crazy. The arctic is, in fact, a desert because of the small amount of precipitation. It's just a really, really cold desert.

The sides of the cube are fairly easy to deal with. The inside must be vapor tight and airtight. The outside must be watertight and vapor open. We know how to do this well in other places besides the arctic. Ventilated claddings just rock in this type of environment.

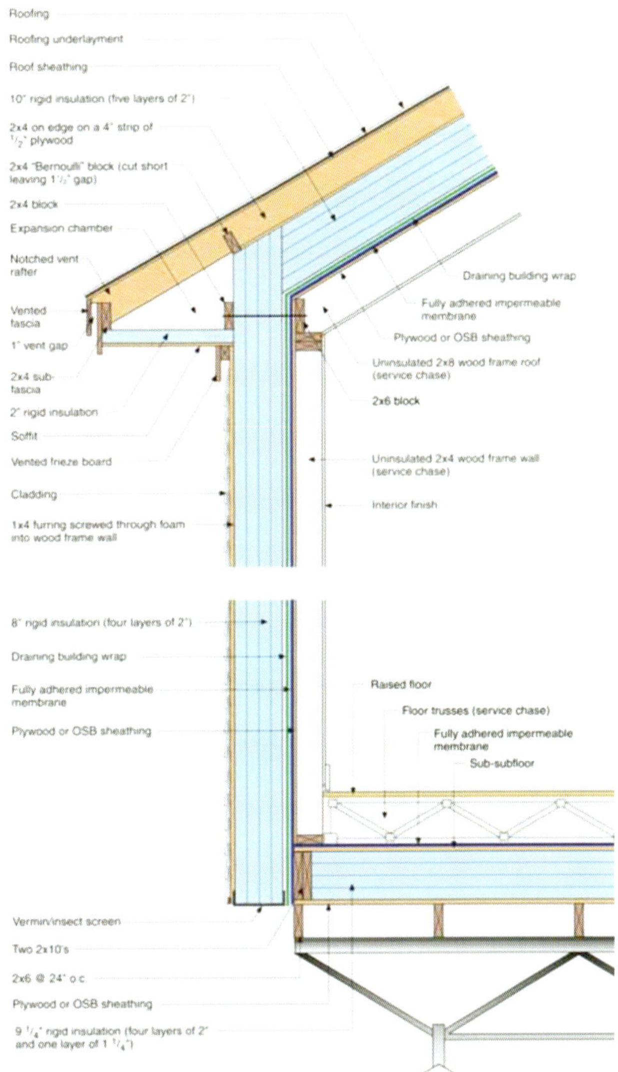

Figure 1: Extreme enclosure Section—Building supported on space frame. Note the vented frieze board, the vented facia, the insulated soffit, and the Bernoulli blocked soffit expansion chamber to control snow entry. The roof and wall assemblies are drained to the exterior between the exterior rigid insulation and the exterior (to the structure) fully adhered impermeable membrane. This membrane is the air barrier, as well as the vapor barrier for this structure.

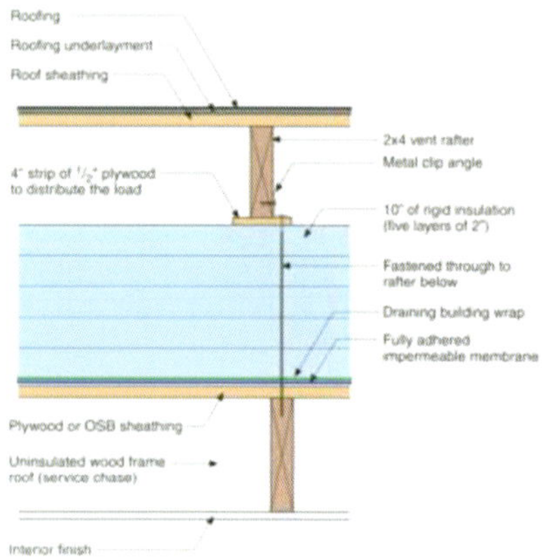

Figure 2: Roof Section—The vent rafters are connected through the roof rigid insulation with a metal bracket and a long roofing screw. Note the 4 in. (100 mm) strip of plywood under the rafter to distribute the load due to the limited compressive strength of the rigid insulation.

Photo 3: Ventilation Rafter—The 2x4 (standing proud) is connected through the roof rigid insulation with a metal bracket and a long roofing screw. You can purchase 14 in. (356 mm) long epoxy-coated steel screws. Stay away from stainless steel screws because the heads twist off. Stainless steel tends to be soft. Ask me how I know.

Putting it all together in a system that works and that is constructable is not easy. We have learned elsewhere, in high snow-load areas, how best to deal with ice dams. We need a ventilated roof cladding to deal with the thermal resistance of snow. The roof cladding needs to be at the same temperature as the exterior. When the outside temperature is below freezing, the roof cladding must be below freezing. If it is above freezing, the snow melts, runs to the edge of the roof and freezes. Then, we get big hunks of ice that can fall down and hurt you big time. Oh, they also dam up the melt water and the roof leaks, hence the term ice dam.

Because snow has an R-value[4] (Remember the igloo?), we need to flush heat away from the underside of the roof cladding using ventilation. That part is easy. What is difficult is keeping snow out of the ventilation space, and then designing an assembly that can live with the snow getting in anyway. Stuff happens. We can go to a long-dead European by the name of Daniel Bernoulli to help us out.

Bernoulli's equation tells us that if we have a fluid flowing in a pipe, and a sudden enlargement exists in the pipe, there will be a corresponding drop in pressure. If we then shrink the pipe again, we have an increase in pressure. Folks will recognize this as an expansion chamber. We can use the concept of an expansion chamber to control rain entry and snow entry into soffits. If you have a sudden drop in air pressure the air typically gives up the snow or rain droplets it is carrying. By turning the soffit into an expansion chamber, we can collect the snow in the soffit, rather than have it migrate farther up the ventilated roof cavity. Once in the soffit, we can let the snow eventually melt there and then drain the snow melt out. But just in case, because stuff happens, we've learned to design the entire ventilated roof cavity to be able to drain snow melt out.

Connecting the roof to the wall is not straightforward either. The walls can cause problems with the roof. Solar radiation heats the outside air immediately adjacent the exterior cladding, as well as the air behind the cladding in the ventilation air space. This warm air rises. We do not want this warm air to get into our ventilation space under our roof cladding as it will heat up the roof cladding The bottom of the roof overhang must be insulated to keep the solar heated air rising up the exterior walls from melting snow on roof overhangs, and vented facias. The roof ventilation air comes from the edge of the roof assembly, not from under the roof overhang.

4 It's somewhere between R-1 and R-2 per inch, depending on density.

The ventilation spaces under the roof cladding and under the exterior wall cladding also serve to compensate for flaws in the interior air barrier and vapor barrier. Yes, they need to be perfect, but sometimes air barriers and vapor barriers don't quite achieve perfection, and you had better accept that buildings move and age, and the building materials themselves age and, well you get the message. Exfiltrating moisture-laden air and vapor that is diffusing outward eventually end up in these ventilation spaces. Once in these spaces, we can remove the moisture to the exterior. Evaporation, sublimation, and drainage all play a role depending on time of year since that determines the temperature of the relevant surfaces.
Another approach is to drain the wall to the exterior between the exterior insulation and the air/vapor barrier. This works since the drainage gap is warm. Recall that frozen water does not drain. Combining roof and wall cladding ventilation with roof and wall drainage covers most of the bases.

So, we have an ultra-vapor tight and ultra-airtight enclosure that is superinsulated. Let's put people inside, along with their stuff. Presumably, they will want to live and work as much as they can inside because it is so miserable outside. Yes, you guessed it, they are going to need air. Air for them and for their stuff and for whatever they are doing. We are not going to argue about how much they need. We will leave that for another day. But, it is clear that this air is going to come from the outside.

The problem with getting air from the outside north of the Arctic Circle is that it is really, really cold and the more of it you bring in or the more of it that you need, the more of a problem it is with respect to comfort and energy. Air-to-air heat exchangers[5] are essential. In fact, it makes sense to have two of them in series, it is that ugly. They also

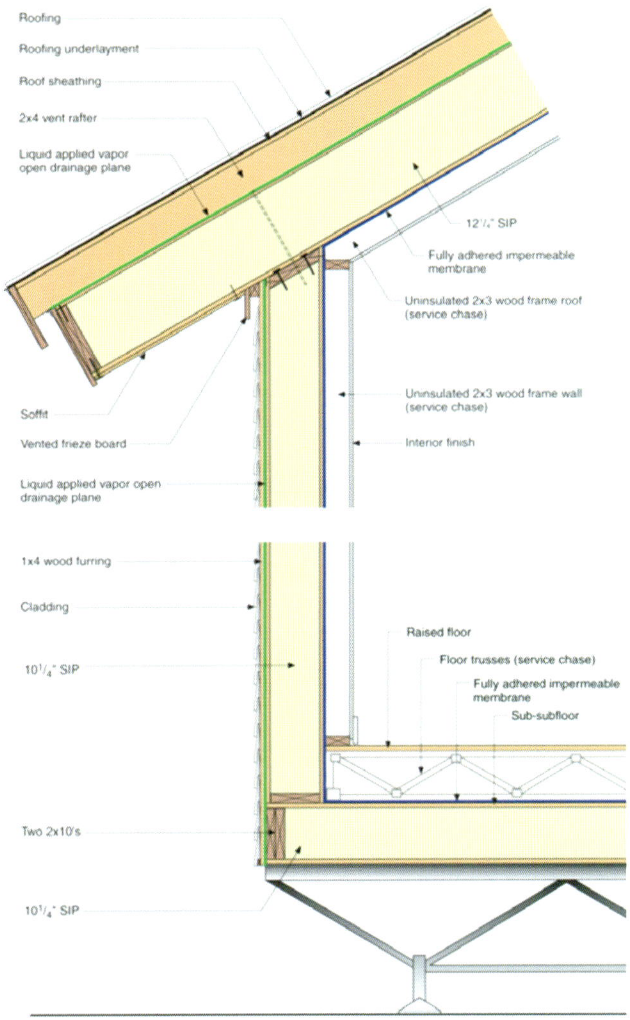

Figure 3: Extreme Structural Insulated Panel Section—A soffit expansion chamber is not used in this detail. However, the ventilated roof cavity is drained and water tight. Note the liquid applied vapor open drainage plane. Most building wraps flutter if they are not supported on both sides. This solves that problem.

5 Heat recovery ventilators or HRVs to you youngsters. Why folks insisted on changing a perfectly good name is beyond me. My daughter says it was so that we could get a better acronym—think SUV and MP3.

freeze up and must be defrosted on a regular cycle every hour when it is really cold.

How do you provide heat to an ultra-airtight enclosure? Well, if you burn something, you had better figure out how to get the products of combustion out and how to get combustion air to the thing you want to burn. Sealed combustion appliances are the only option. Folks who try other approaches usually end up as a footnote in the annual Darwin Awards. The good news is that we have lots of systems that work. They are also extremely efficient.

An even more interesting problem is how to deal with sewage. It tends to freeze. Sewage pipes, permafrost, buildings that move, and extreme cold leads to lots of crap. This is a complex problem that we will leave to another day. I will say that electrically heated waste lines are common, as well as above grade superinsulated and heated septic tanks and leach fields. Unfortunately, so are open sewage ponds and sanitation reminiscent of third-world nations.

These buildings are likely the future of housing in general: superinsulated, ultra-airtight, great windows, sealed combustion, mechanical ventilation, drained and ventilated assemblies.

We need to live and work in the arctic because of the energy resources we need. The irony is sweet. We have to build these ultra-efficient structures in the arctic because we haven't built ultra-efficient structures everywhere else.

Author's Note

Much of the fundamental research work demonstrated in the building constructed by the Hydaburg Tribe was developed under the U.S. Department of Energy's Building America Program.

Photograph 4: Mock-up Constructed First—Things never go together perfectly the first time, which is why test huts, test sections, and mock-ups are essential. Again, ask me how I know. Failure is not an option north of the Arctic Circle. Skill testing question for the geeks in the audience: what are the red dots?

Photograph 5: Interior Structural Frame—The wood-frame wall cavities and wood rafter roof cavities remain uninsulated to control cavity condensation. The cavities provide space for services. The air barrier membrane and vapor barrier membrane are exterior to the wall and roof framing.

Appendix B.13

Insight
The Perfect HVAC

By John Straube, Ph.D., P. Eng.

All space-conditioning systems are intended to provide a comfortable and healthy indoor environment. Many, even most, systems are designed in such a manner that they cannot reliably provide fresh air and comfort while at the same time being energy efficient. In fact, the most popular residential furnace/AC systems and commercial VAV systems are fundamentally flawed from their conception.

How hard any space conditioning system needs to work depends on the design of the building, particularly the enclosure and the use of the building. However, some conditioning will be needed even for a super-insulated cube occupied by a dead hermit. The space conditioning may be provided by actively powered mechanical equipment or passive systems. However, the energy efficiency has more to do with not making the system work hard than not installing one.

It is useful to know what an ideal HVAC system would look like. Although compromises sometimes have to be made, they should be made with the knowledge of how and why they are imperfect. This article defines the characteristics of a perfect HVAC system for both single-zone single-family residential and multi-zone residential, commercial and institutional applications. Together with *BSI-001 The Perfect Wall*, it provides the basic concepts for energy efficient, durable, healthy, and comfortable buildings that can work in all climate zones.

Design Goals A fundamental requirement for building design is to ensure health and safety. Hence, ventilation must always be provided when people are in the building, and some ventilation is needed to remove building-generated pollutants at all times. The required rate of outdoor air varies, from a few cfm per person to as much as 30 cfm (15 lps) per person depending mostly on politics. The challenge in practice is not the rate of ventilation, but actually ensuring that it is provided. In most cases, higher ventilation rates are used in an attempt to compensate for poor delivery reliability.

A comfortable indoor environment is about 20 to 24 °C (68 F – 75 °F) and 20-60% RH. If one widens the comfort range, a larger proportion of people will be uncomfortable. That does not mean that an indoor summer temperature of 26 °C (79 °F) will be uncomfortable for most people but it will be uncomfortable for a significant proportion (more than, say, 10%) of the North American population.

The temperature that defines comfort is *not* the air temperature, but something called the operative temperature. The operative temperature is a combination of the air temperature, the weighted average of all surface temperatures of a space (defined by the mean radiant temperature, MRT), and air velocity. At low air velocity, the operative temperature is the simple average of the MRT and air temperature.

With a high performance building enclosure (e.g. a perfect wall and exceptional glazing) the surface temperatures of a room will become very close to the air temperature, and comfort will be enhanced, even if the air temperature approaches the extremes of the comfort band. A typical modern building enclosure during cold weather will have wall and window surface temperatures that are several degrees below the indoor air temperature, and therefore to reach a comfortable operative temperature the indoor air temperature will need to be higher. The reverse is true in the summer.[1]

System Functions

Regardless of the quality of the building enclosure, some conditioning of the indoor space is required in essentially all climates. The list of required functions of any indoor environmental control system includes:

1. Make it colder
2. Make it hotter
3. Increase humidity
4. Decrease humidity
5. Provide fresh air, and
6. Filter air and remove pollutants.

1 Poor building enclosures, operative temperature, and zoning combine to explain the reasons behind the oft-heard complaint of why a building is cooled to a lower temperature in the summer than it is heated in the winter.

Practically speaking one needs systems, active or passive, to:

1. make heat or remove heat (= make cold),[2]
2. move heat or cold around the building,
3. deliver heat or cold to the space,
4. add or remove moisture from the space,
5. move ventilation air around the building,
6. filter air to remove particulates and exhaust pollutants directly, and
7. reject heat to the environment (when cooling).

The best HVAC system is one that has discrete components that individually fulfill each of these functions. This allows each sub-system to be optimized for its function, and a control system can also easily be implemented to operate each function optimally with none of the compromises or trade-offs inherent in multi-role components.

The most dysfunctional, difficult to control systems combine multiple functions into one device. Although popular, these systems also often consume the most energy.

If reliability and comfort are the design goals, the choice of ground source heat pump, electric chiller, condensing boiler, or passive solar gain is secondary in importance to the choice of how the HVAC functions will be fulfilled. Choosing an energy-efficient source of heat (such as a condensing boiler) and cool (such as a modern chiller with IPLV of under 0.6) is often wasted by systems, such as VAV, that combine the functions of ventilation, humidity control, and heating/cooling. VAV systems, controlled by thermostats only, often reheat previously chilled air, and cannot reliably deliver the quantity of fresh air required (or over ventilate, often both). As a result, a less efficient heating/cooling plant in a system with separated functions can deliver significant energy savings at the same time as improving air quality, comfort, and reliability.

Small Residential HVAC Systems

A typical single-zone residential system uses a furnace and an air conditioning unit coupled with an air handling unit (AHU).[3] This system separates the heat production, and cool production, and heat rejection functions with individual pieces of equipment. The heat and cool air generated in the AHU is transported via air in shared ductwork around the building (although hydronic and refrigerant systems are better, more expensive, alternatives they require the commercial approach described later). There is also a filter that removes airborne particles. A separate humidifier (although rarely needed in modern buildings) is sometimes included to add humidity. This sounds perfect, except for the fundamental flaw that there is almost never any provision to provide fresh air. This must be solved. There is also no ability to remove moisture (dehumidify) on demand. This also needs to be solved, especially in climates with humid weather.

To provide ventilation, a perfect system would provide separate ducted supply and return air systems. To reduce the energy used to heat or cool this ventilation air, a Heat Recovery Ventilator (HRV) can be used to transfer heat between the supply and exhaust streams as well as to provide the fans needed to move the air. In climates with significant outdoor heat and humidity, an Energy Recovery Ventilator (ERV) would be preferred to pre-cool and pre-dry incoming air using the cool and dry exhaust air.[4] Of course point sources of pollutants, such as the kitchen stove and bathroom would have dedicated exhaust fans to remove these pollutants.

Heating or cooling would ideally be delivered to the space by radiant floors and ceilings respectively, as these provide the most comfort, with no moving parts in the space, with the lowest energy penalty and sound level. In very well insulated and airtight homes, the benefit of radiant conditioning is significantly reduced as the loads are so small and hence air-based delivery becomes a practical means of approaching "perfect HVAC" performance at a fraction of the cost of radiant systems.

In practice, it is feasible in small residential systems to use an air handler and space-conditioning duct system to distribute ventilation air throughout the house. This combination of uses (distribution of conditioning air with ventilation air) requires smarter-than-average controls to work (see later) and careful design to avoid excess energy use. But it can work very well for a single zone systems like small single-family homes and apartments.

2 Recall that cold is a relative term meaning less heat. Hence, cooling means to remove heat.

3 Andy Ask honestly defines an Air Handling Unit (AHU) as "a box with a fan in it". It is common to add a coil to add or remove heat, and a filter box to filter the air passing through an AHU.

4 It is important to note that an ERV can never, under any circumstances, dry the indoor air. It only reduces the humidity *load* due to ventilation, it can never eliminate it.

The lack of humidity control provided by normal HVAC systems can be a major concern in climates with humid summer weather (e.g., Minneapolis, Miami). Only a random amount of dehumidification is provided if and when the cooling operates for long enough to allow water to collect, build up on coils, and then drain down and away. This process takes about 10-15 minutes of continuous cooling operation before dehumidification begins. Hence, a separate dehumidification device is required for the perfect HVAC system. Several suppliers of dehumidification ventilation equipment (often called an outdoor air unit in commercial applications) are now available to help solve this problem.

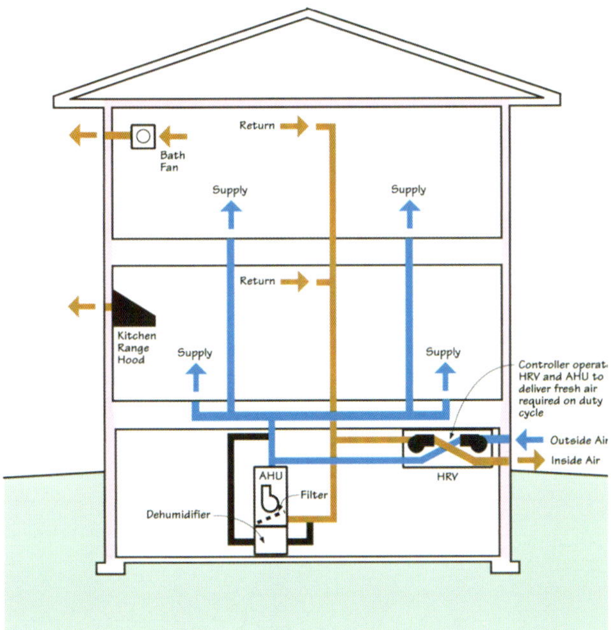

Figure 1: A small, single-zone residential heat, cooling, filtration and ventilation system that shares the ductwork between systems but separates the critical functions.

Combining these solutions provides the perfect practical manifestation of a residential HVAC system shown in **Figure 1**: energy-efficient, easy to control and trouble shoot, and guaranteeing a fresh air supply while controlling humidity. For the premium system, a radiantly heated floor and a radiantly cooled ceiling or valence could be added to this system relegating the air handling system to ventilation, mixing, filtration, and humidity control.

Larger Multi-Zone HVAC Systems

In commercial buildings (and large multi-zone residential), distributing the much larger quantities of energy around the building requires large ducts, and the associated larger inter-floor spaces to accommodate. Large ducts from central air handling units also penetrate multiple zones of occupancy and fire separation, causing all kinds of expense and performance problems. In the ultimate of madness, many multi-story MURBs actually use large ducts that penetrate through floor systems all the way to the roof, allowing flow to be regulated at the whim of stack effect pressures, and expect fresh air to find its way down corridors and under cracks below suite entrance doors! Hence the perfect commercial HVAC system uses liquids (such as refrigerant, water, or glycol) to move energy around the building, and only uses ducts to move ventilation air through the building while conditioning air temperature within each occupancy (residential unit, office tenancy, etc.).

By separating the control of temperature from the control of ventilation and from the control of dehumidification, very robust, simple to control, and energy efficient systems result.

For example, in a commercial occupancy, a central system can be used to produce chilled and heated liquid (water or refrigerant) by a range of technology. Space heating and cooling can then be delivered to the space by a range of products located in the space to be conditioned. For example, radiant heating and cooling systems embedded in the floors, walls, or ceilings. They, of course, require separate and good control of space humidity to be used in cooling mode. This type of "terminal unit" provide the quietest, most comfortable (by controlling MRT and air temperature), and lowest energy means of conditioning the temperature in a space.

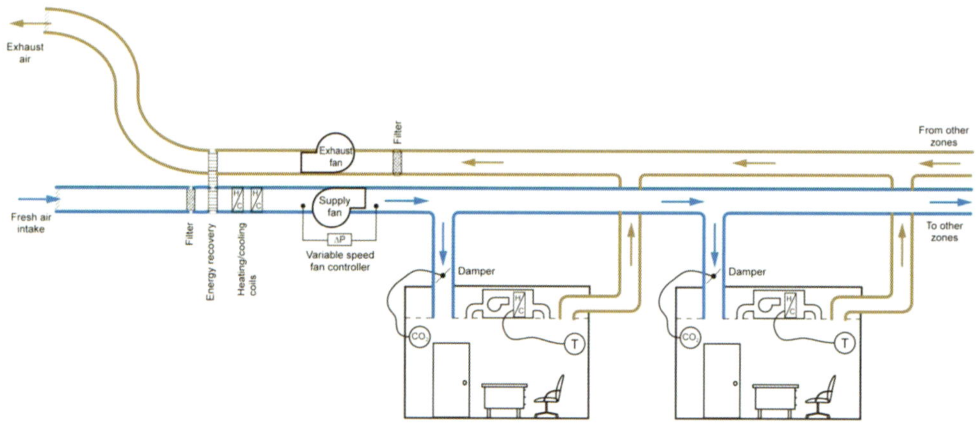

Figure 2: A multi-zone direct outdoor air system that dehumidifies supply air to offset occupancy and ventilation humidity, controls the ventilation rate dependent on occupancy, and heats or cool individual spaces with a radiant or fancoil terminal unit controlled by a space thermostat.

Other more economical approaches can be used that provide very close to the same performance. For example low-velocity fan coil units with dry coils and ECM motors that take air from the space and mix conditioned air back into the space are now available.[5] Not only are these less expensive than radiant systems, they have a faster response time. Valence cooling and baseboard heating panels are other options.

A dedicated outdoor air system (DOAS) should be used in multi-zone or large single zone spaces to provide neutral temperature ventilation air independently of heating or cooling. Such a system will ideally incorporate some form of HRV or ERV. In multi-story buildings, a separate floor-by-floor DOAS is preferred to deal with duct penetrations through floors. Variable speed air handlers that maintain a constant pressure in the supply ducts should be the standard choice for this application. The use of a DOAS avoids the too common indoor air quality problem of re-circulating polluted air from one space or occupancy to another.[6]

Ideally, separate dehumidification equipment should be provided in each space. Practically speaking, most of the humidity load comes from the ventilation air and occupancy, and hence adjusting the supply air humidity level in a DOAS ventilation system can be used to control the humidity level within reasonable limits (e.g., +/-5% RH).

Controls

Of course, none of this equipment can work without the proper controls. Heating and cooling production should be controlled by a thermostat, located in the space to be conditioned. Only if a group of spaces will always have similar thermal loads, is it acceptable to share thermostats.

[5] Parallel VAV boxes which draw return air from the space and condition temperature with dry coils are essentially less efficient versions of these fancoils.

[6] It is likely unbelievable to many lay people, but most hospital conditioning systems will actually re-circulate air from one patient room to the public hallway or the adjoining patient room, thereby encouraging airborne pathogens to move around via the HVAC system. The problems this approach generates have traditionally been partially managed by very high air change rates, but such problems can be completely eliminated with low air change rates and DOAS systems.

Despite what some may say, one cannot control humidity without the use of a humidity sensor coupled to humidity production / dehumidification equipment. These devices are called humidistats and are widely available and inexpensive. Typical cooling systems respond only to a signal from a thermostat and hence can only control temperature, not humidity, and therefore do not reliably control indoor humidity. There are no practical exceptions.

Ventilation should similarly be controlled by a "ventstat." In most applications, ventilation has been provided based on a timer system, with assumed occupancy densities and times. This obviously has severe limitations in commercial/assembly occupancies and explains many air quality complaints: the occasionally-occupied meeting room is a common ventilation problem, and over-ventilated empty classrooms waste a huge amount of energy. The good news is that economical sensors that measure carbon dioxide (which is an excellent measure of both human occupancy and activity level) are now readily available. Hence, a CO_2 sensor can act as a ventstat for a space, and control the supply of ventilation air as needed, when needed, as much as needed (of course a minimum ventilation rate is required even when no one is in the space to remove space-generated pollutants). This is generally termed "Demand Controlled Ventilation" (DCV).

A DOAS + DCV is absolutely the only "perfect" approach for large multi-zone commercial and institutional buildings, and practically the only energy efficient means of reliably delivering fresh air. Whether displacement ventilation, underfloor air, or high aspiration ceiling level air delivery and mixing is used has little impact on energy consumption, comfort or IAQ. DOAS+DCV has a tremendous positive impact. For single-zone housing, a simpler timer approach, combined with whole house mixing is more economical and usually more than sufficient[7] although a small amount of energy can be saved by using a ventstat in zero energy home designs.

[7] The FanCycler controller was developed to inexpensively provide this type of control. It provides for user settings that varies the duty cycle of the indoor air handler operation.

Complex integrated Building Automation Systems (BAS) are often touted as beneficial, even necessary, for energy efficiency and high performance. In fact, the opposite is often true. A system that requires a central controller to operate multiple pierces of equipment in concert to work is indicative of a problematic HVAC system design. Space conditions should be controlled by sensors in the space acting on equipment associated only with that space.

Supervisory systems, those that monitor but do not control, and central setpoint control, those that adjust zone target temperatures up and down after hours and on weekends, are quite useful, and fail in a tolerable fashion. But any system that needs to control remote pumps, chillers, and valves to meet humidity, temperature and air quality requirements in a space is almost always unnecessary and too complex to be properly built and commissioned, and rarely are such systems properly maintained and calibrated over the long term.

Conclusions

The perfect HVAC system concepts described here very reliably deliver, healthy, comfortable, and energy-efficient space conditioning. Many of the ideas in this article have been deployed in the past, but fallen out of favor as cheap energy, leaky buildings, and tolerant occupants allowed compromises and short-cuts. However, as the demand for better indoor air quality, comfort, and energy efficiency increase, the combination of decoupling the different required functions and the use of simple direct-coupled controls becomes imperative, not optional. Given good building enclosure design, such systems are not expensive to build, easier to commission and operate, and can save significant amounts of energy.

Appendix C Glossary

Absorption Process whereby a porous material extracts one or more substances from an atmosphere, a mixture of gases, or a mixture of liquids.

ACH **Air changes per hour**
A common unit of measure of ventilation rate for a space, or air leakage rate for a building, defined as the volumetric flow rate divided by the volume of the space considered.

Adhesion Property that describes a material's ability to bond to a surface physiochemically or chemically.

Adhesive A substance or compound used for bonding surfaces together, usually applied in the form of a liquid or paste. An adhesive and sealant or base coat may be the same material.

Adhesive Failure Loss of adhesion of a material to the surface to which it is applied. Usually applied to a sealant line, floor finish, or coating; *see Adhesion.*

Adsorption (1) Process in which fluid molecules are concentrated on a surface by chemical or physical forces or both; (2) surface adherence of a material in extracting one or more substances present in an atmosphere or mixture of gases and liquids, unaccompanied by physical or chemical change. [*from Gatley*]

Aesthetic Joint A groove cut into cladding system, often EIFS, precast, or stucco, that serves the function of decoration and/or to provide a starting or stopping point for finish coat application. ***Variant: Aesthetic Reveal***

AFUE **Annual Fuel Utilization Efficiency**
A measure of annual efficiency of conversion of primary gas or oil consumption to the production of hot water or air. Widely used in boilers and furnaces.

AHU **air handling unit**
A mechanical space conditioning device comprising an enclosure, and a fan to move air. In many cases, an AHU might also contain a filter, duct connection flanges, and/or heating or cooling coils (which, strictly, make it a fancoil unit); *see also Fancoil.*

Air Control Layer **(formerly Air Barrier)**
The complete air control layer system is comprised of materials and assemblies, each with their own performance requirements; *see also air control layer system.*

Air Control Layer Material **(formerly Air Barrier Material)**
A material that has sufficiently low air permeance and adequate strength that it can be part of an air control layer system. Recommended maximum air permeance for a material is 0.02 l/(s-m^2)@ 75 Pa (0.004 cfm/ft^2 @ 0.3" WC) when tested according to ASTM E 2178 or E 283.

Air Control Layer System **(formerly Air Barrier System)**
Air control layers are three-dimensional systems of materials designed, constructed, and/or acting to control air flow across a building enclosure, or between a conditioned space and an unconditioned space. In multi-unit/townhouse/apartment construction an air control layer system should also separate the conditioned air from any given unit and adjacent units. The pressure boundary of the enclosure should, by definition, be coincident with the plane of a functional air control layer system. In multi-unit/townhouse/apartment construction the air control layer system may also be the fire barrier and smoke barrier between units. In such assemblies the air barrier system must also meet the specific fire resistance rating requirement for the given separation.

Air control layer systems are assembled from "materials" incorporated in "assemblies" (or "components" such as windows) that are interconnected to create "enclosures." Each of these three elements has measurable resistance to air flow. The minimum recommended resistance or air permeances for the three components are:

Material 0.02 l/(s-m^2)@ 75 Pa (0.004 cfm/ft^2 @ 0.3" WC)

Assembly 0.20 l/(s-m^2)@ 75 Pa (0.04 cfm/ft^2 @ 0.3" WC)

Enclosure 2.00 l/(s-m^2)@ 75 Pa (0.4 cfm/ft^2 @ 0.3" WC)

Materials and assemblies that meet these performance requirements are said to be air control layer materials and air control layer assemblies. Air control layer materials incorporated in air control layer assemblies that in turn are interconnected to create enclosures are called air control layer systems.

Air Leakage Uncontrolled and/or unintended airflow through a building enclosure or between units of occupancy. Leakage from indoors to outdoors is known as exfiltration and leakage from outdoors to indoors is known as infiltration. Air leakage can cause indoor air quality problems, condensation, excess energy use, comfort complaints, and smoke transport.

Air Pressure Boundary The air pressure boundary is the boundary (comprised of a series of planes to form a three dimensional boundary) that generates the largest pressure drop (usually much more than half the total) when the enclosure is subjected to a pressure difference.

Air-impermeable A material or assembly that does not allow the passage of air. In practice, with respect to air barrier materials for buildings, air-impermeable materials allow less than 0.02 l (s/m^2) @ 75 Pa (0.004 cfm/ft^2 @ 0.3" WC) of airflow when tested according to ASTM E 2178 or E 283.

Air Infiltration Uncontrolled inward leakage of air (that may contain entrained water vapor) through cracks and insterstices in any building element and around windows and doors of a building, caused by the pressure effects of wind or the effect of differences in the indoor and outdoor air density.

ASHRAE American Society of Heating, Refrigerating and Air Conditioning Engineers

Back wrapping The carrying of EIFS mesh reinforcement and base coat around the ends of the insulation boards and terminating between the insulation and substrate. Typically used at system interfaces and terminations to firmly attach the base coat to the substrate and protect the edges of the insulation board at these locations.

Backer Rod A resilient foam material (typically closed cell polyethylene) of circular cross-section installed under compression in a joint to provide a backing, to control sealant joint depth, act as a bond breaker to prevent three-sided sealant adhesion, and provide an hour-glass contour of the finished sealant bead.

Barrier System Another term for a face-sealed perfect barrier rain control approach to enclosure design; *see also face-sealed.*

Base Coat A compound used to embed and to cover the reinforcing fabric of an EIFS, depending on the system type. The base coat acts as the primary weatherproofing layer.

Bead A line of sealant or adhesive. In glazing, an applied sealant in a joint such as caulking bead, heel bead, glazing bead, etc. Also a molding or stop used to hold glass or panels in position.

Beam A structural supporting member that resists loads primarily in bending. May be vertical or horizontal and made of any material or composite.

Below Grade The portion of a building that is below the line of the surrounding ground level.

BEopt Software for identifying optimal building design (BEopt is a computer program designed to find optimal building designs along the path to zero energy.).

Bond breaker A tape, sheet, wax or liquid applied treatment that prevents adhesion on a designated surface. Usually used with sealant to ensure a proper joint; *see also backer rod.*

Building Enclosure The elements of a building that act as the environmental separator between the interior environment and the exterior environment. Walls, windows roofs slabs, and basements are parts of building enclosures.

Note: The enclosure is a special type of environmental separator. Environmental separators also exist within buildings as dividers between spaces with different environmental conditions.

Building Envelope An outdated term for the building enclosure.

CAD Computer-aided design

Capillary Action The movement of water within the confined spaces of a porous material or between two adjoining hydrophilic materials due to the attractive force of surface tension. Only significant in gaps of less than about $^{1}/_{8}$" (3 mm).

Capillary Break A hydrophobic material or non porous material (such as glass, plastic, or metal) or gap between parallel layers of material (often less than $^{1}/_{16}$" or 1.5 mm) sufficient to stop capillary action.

Cap Flashing A flashing overlapping the vertical leg of base flashing to prevent water from migrating behind the base flashing.

Caulking Alternate term for sealant.

Cement Board A rigid sheathing made of cement-bonded fibre-reinforced composites (typically glass or woodfibres are used as reinforcing). Cement board is moisture resistant and non-combustible.

CEC California Energy Commission

CFL Compact fluorescent light

CFM **cubic feet per minute**
A unit of volumetric flow rate, often used as a metric of ventilation, airflow, or air leakage.

Cladding A material or assembly that forms the exterior face of a wall. Examples of cladding include stucco, EIFS, metal panels, brick/stone veneer, wood siding, and vinyl siding.

Cold joint A visible junction between to two liquid applied materials, such as concrete or EIFS coating, caused by allowing one side to cure before the application of the second side.

Commissioning A set of testing and inspections done prior to occupancy, which should note any problems that should be fixed immediately (fix and tune). Testing should include duct leakage, building enclosure leakage, air pressure relationships under all operating conditions, proper venting of all combustion appliances under all operating conditions, carbon monoxide output of all combustion appliances, and confirmation of airflow and refrigerant charge in the HVAC systems. Additionally, the homeowner/occupants should be educated about the correct operation and maintenance of the building and equipment.

Condensation The change of state from vapor to liquid. A common factor in moisture damage. Occurs on surfaces, which must be cooler than the air containing vapor next to it. Vapor supply to the condensation surface is usually by airflow but can be by diffusion.

Conditioned Space The part of the building that is designed to have controlled environmental conditions (particularly temperature) for the comfort of occupants or for other reasons particular to the use of the building.

Control joint A formed, sawed, or assembled joint acting to regulate the location of cracking, separation, and distress resulting from dimensional or positional change.

Cooling Degree Days (HDD) and cooling degree days (CDD) are climatological metrics used to express the magnitude of the heating or cooling load in a given location. These metrics are expressed in terms of a "base temperature" (e.g., "Heating Degree Days Base 65 F"). HDD are calculated by taking the daily average temperature; if it is colder than the "base temperature," the difference is calculated. This difference is summed for the period (week, month, year) in question. Note that cooling season data are often presented in terms of "cooling degree hours" (due to diurnal temperature swings), which are not directly translatable to CDD. ***Variant: HDD***

Concealed Barrier An approach to enclosure rain control that employs a single waterproof barrier to rain penetration. The barrier is not on the exposed face of the assembly but concealed behind the cladding and other material layers, which reduces the amount of rainwater reaching the barrier and the environmental loads on it. Drainage is by definition not required for good performance in this approach. A subset of perfect barrier approaches; *see also face-sealed, drained, rainscreen.*

Corrosion The deterioration of metal by chemical or electrochemical reaction (usually with oxygen, forming oxides) resulting from exposure to weathering, moisture, chemicals or other agents or media.

Crack The result of tensile or shear forces that exceed the strength of a material at a particular location.

CRI **color rendering index**
A metric of the ability of a lighting source to render colors correctly relative to an ideal light source. A CRI of 100 is the maximum value. For reference, incandescent lights are in the 95-100 CRI range, and compact fluorescents are reaching the 80s currently; *see http://www.lrc.rpi.edu/programs/nlpip/glossary.asp*

CT **color temperature**
A specification of the apparent color of a light source relative to the color appearance of an ideal incandescent source held at a particular temperature and measured on the Kelvin (K) scale; *see http://www.lrc.rpi.edu/programs/nlpip/glossary.asp*

Cure To develop the ultimate properties of a wet state material by a chemical process. Different than drying, which is not a chemical process, although drying is often a necessary part of a chemical process.

Delamination A separation along a plane parallel to the surface.

Dehumidification Removal of water vapor from air.

Dew Point The physical location along a temperature gradient at which the relative humidity of a sample of air reaches 100%, i.e, becomes satured. Condensation may not occur at the dewpoint location no condensation nuclei or surfaces exist or if a superior (lower energy) surface is present nearby (the condensation plane); *see also Dew Point Temperature.*

Dew Point Temperature The temperature at which the relative humidity of a sample of air with constant water vapor reaches 100%. The air is saturated with water when it reaches the dew point and condensation will occur on a surface.

DHW Domestic hot water

Diffusion The movement of individual molecules through a material. The movement occurs because of concentration gradients (and to a much lesser degree) thermal gradients, independent of airflow. A mode of water vapor transport in building enclosures that is much slower than airflow.

DOE U.S. Department of Energy

DOE2 Hourly building energy analysis program that can predict the energy use for all types of buildings. Being superseded by Energy Plus.

Drainage Plane Water repellent materials (building paper, housewrap, foam insulation, etc.) which are designed and constructed to drain water. They are interconnected with flashings, window and door openings, and other penetrations of the building enclosure to provide drainage of water to the exterior of the building. The materials that form the drainage plane overlap each other shingle fashion or are sealed so that water flow is downward and outward.

Drained A building enclosure rain control strategy (or ground water control) that accepts that some water will penetrate the outer surface (the cladding, which "screens" rain) and removes this water back to the exterior by gravity drainage over a drainage plane, through a drainage gap, and exiting via flashing and weep holes. Many wall systems (lap siding, brick veneer) and sloped roof systems (metal, asphalt shingles) employ drained strategies. Also called Drainscreen; *see also rainscreen, drainscreen.*

Drainscreen An enclosure system that control rain penetration using the drained approach; *see drained.*

Drip edge A geometric feature provided in an exterior building surface to ensure that flowing water will drip free rather than be pulled back toward a vertical element due to surface tension or gravity. A drip groove is commonly employed in solid materials like concrete whereas a drip edge is used for thinner sheet materials.

Dry (1) To loose moisture content (2) To develop the ultimate properties of a wet state material solely by evaporation of volatile ingredients.

Durability The capability of a building, assembly, component, product, or building to maintain serviceability over a specified time.

ECM Electronically commutated motor

EEBA Energy and Environmental Building Association

EER **Energy Efficiency Ratio**
A measure of air conditioners and heat pumps, defined in terms of Btu of energy moved divided by the watt-hours of energy input. Same as Btu/hr output divided by watt input. Commonly measured at 95 °F outdoor temperature. EER divided by 3.41 provides the Coefficient of Performance (COP).

EF **energy factor**
A water heating efficiency metric; information on water heater EFs is available at the Gas Appliance Manufacturer's Association (GAMA) website.

EFL Environments for Living®

Efflorescence The deposition of dissolved salts by evaporation on the surface of a porous material (such as concrete or brick) having been transported within liquid water (usually by capillary action).

EGUSA Energy Gauge USA software (FSEC's residential front-end user interface for DOE2.1E simulation tool).

Elevation A drawing or rendering of a face of a building.

Embed To encapsulate wihin, such as steel anchors within concrete or EIFS reinforcing mesh in the base coat.

Emseal Brand name for a pre-formed compressible foam joint filling product used to provide a seal between building components.

End dam A vertical or near vertical upstand from the end of a flashing, or window sill, used to prevent water from flowing horizontally off the end of the flashing or sill.

End wrapping The act of wrapping the reinforcement and base coat around the edges of the insulation board and terminating and bonding to the substrate at an opening in the substrate. Like back wrapping, end wrapping is a means of securely fixing the lamina where it ends at joints and penetrations.

Enthalpy Thermodynamic quantity equal to the sum of the internal energy of a system plus the product of the pressure-volume work done on the system; enthalpy cannot be directly measured; however, enthalpy differences between the initial and the final statepoints of a process can be measured. [*from Gatley*]

EPA U.S. Environmental Protection Agency

ERV Energy Recovery Ventilator or Enthalpy Recovery Ventilator

Equivalent Leakage Area
Of a building. Quantitative expression of the airtightness of a building enclosure. EqLA is the method set by the Canadian General Standards Board in which a blower door depressurizes the building enclosure to 10 Pascals and the leakiness of the enclosure is expressed as

a summary hole in square inches. ELA is set by the ASTM equivalent procedure at a pressure differential of 4 Pascals. ***Variants: EqLA, ELA***

Expanded Polystyrene Insulation (EPS)
A rigid cellular foamed plastic insulation material manufactured by expansion of polystyrene beads within a mould. This mould creates an open cell structure filled with air. EPS Type I is the most widely used insulation. Type I has a density of 1 lb /ft^3 (16 kg/m^3), Type II is a denser, more durable insulation of 1.5 lb /ft^3 (24 kg/m^3) density with higher R-value. R-value of around 3.5 to 4.5 per inch, depending on density and radiation absorbing additives. ***Variant: EPS***

Expansion Joint A structural separation between building elements that allows independent movement without damage to the assembly.

Exterior Insulation Finishing System
A combination of insulation and stucco system cladding for buildings. Based on rain control strategy, there are two types of EIFS available in the U.S.—barrier face-sealed systems and drainable systems. ***Variant: EIFS***

Extruded Polystyrene Insulation
A rigid cellular foamed-plastic insulation material manufactured by extrusion of polystyrene in the presence of a blowing agent. The blowing agent dissipates out of the closed cell structure over time, creating a structure that resists liquid water penetration and vapor diffusion. The manufacturing process for XPS insulation results in a smooth surface skin. Typical density of 2 lb/ft^3 and R-value of 5 per inch (0.029 W/mK). ***Variant: XPS***

Façade Visible exterior face of a building.

Fancoil A mechanical space conditioning device comprising an enclosure, a fan to move air, and a finned extended surface heat exchanger (a “coil”). The coil transfers heat from refrigerant (“DX” or “direct expansion”), water or fluid to the air stream to heat or cool the air passing through. In many cases, a fancoil might also contain a filter and duct connection flanges. Can be installed as an exposed device or hidden in floors or ceilings; *see also AHU.*

Face-seal A building enclosure rain control strategy that relies on the exterior face of the enclosure to act as a perfect barrier to rain penetration. This method typically relies on exposed sealants to provide rain tight joints, and hence is highly reliant on workmanship and maintenance to achieve performance. Failure is defined as water penetration of the face. It is a sub-set of perfect barrier rain control strategies; *see also drained, concealed barrier, and mass.*

Fasteners A general term describing a variety of screws, nails, rivets, etc., that are used for mechanically securing various components of a building.

FCU Fancoil Unit; *see fancoil.*

FF Framing factor

FFA Finished floor area

Fiberglass-faced gypsum sheathing Moisture resistant type of exterior gypsum sheathing. The gypsum core is silicone treated for water repellency and the glass mat applied to each face as reinforcement meeting the requirements of ASTM C1177.

Finish coat The coating applied to the base coat to finish the lamina of an EIFS. The finish coat provides color, texture, water protection, dirt resistance and ultra violet ray resistance.

Fishmouth A deficiency in the installation of membranes (roofing, self-adhering membranes, etc.) that results in a fold in the edge of the membrane, through which air and/or water can penetrate.

Flash set The early hardening of stiffness in the working characteristics of a Portland cement paste, usually with the evolution of considerable heat. Stiffness cannot be dispelled nor the plasticity regained by further mixing without addition of water; also known as "quick set."

Flashing A waterproof material used to redirect or shed drained water, or occasionally to act as a capillary break.

Floor Plan A horizontal section draw showing the basic layout of a building, which includes placement of walls, windows and doors.

Foundation, Water-managed Systems for at or below-grade enclosure assemblies where gravity (drainage) is used to move liquid water away from the structure, relieving hydrostatic water forces.

Framing member Studs, joists, plates (tracks), bridging, bracing, and related accessories manufactured or supplied for wood or light gauge steel framing.

Grade Beam A foundation wall that is cast at or just below the grade of the earth, most often associated with the deepened perimeter concrete section in slab-on-grade foundations.

Gum Lip A method of sealing a flashing to a wall surface whereby the top edge of the flashing is bent outwards to form a sealant-filled reveal (typically at the vertical termination of a waterproof membrane).

Habitable Space Building space intended for continual human occupancy. Such space generally includes areas used for living, sleeping, dining and cooking, but does not generally include bathrooms, toilets, hallways, storage areas, closets, or utility rooms.

Header Framing members over windows, doors, or other openings. A beam placed perpendicular to joists and to which joists are nailed in framing for chimney, stairway, or other opening. ***Variant: Wood lintel***

Heating Degree Days Heating degree days (HDD) and cooling degree days (CDD) are climatalogical metrics used to express the magnitude of the heating or cooling load in a given location. These metrics are expressed in terms of a "base temperature" (e.g., "Heating Degree Days Base 65 F"). HDD are calculated by taking the daily average temperature; if it is colder than the "base temperature," the difference is calculated. This difference is summed for the period (week, month, year) in question. ***Variant: HDD***

HERS Home Energy Rating System developed by RESNET

Housewrap Any of the numerous spun-fiber polyolefin rolled sheet goods, or perforated plastic films designed to function as drainage planes, a class of sheathing membranes. Some are also used as part of an air barrier system if installed with proper sealing and structural support.

HRV Heat Recovery Ventilator

A device, oten fan-powered, that supplies and exhausts ventilation air to a space, and transfers the heat energy in the outgoing stream of air to the incoming stream or vice-versa.

HSPF **Heating Seasonal Performance Factor**
An energy rating metric for air source heat pumps (rating their heating performance, as opposed to SEER, which measures cooling performance)

HUD U.S. Department of Housing and Urban Development

HVAC Heating, ventilation, and air conditioning

Hydrophobic Materials that do not attract liquid water, they will force liquid water to form beads on their surface, act as capillary breaks and are non-hygroscopic.

Hygroscopic Materials that interact with water vapor by adsorbing water vapor as a function of the relative humidity of the surrounding air.

Indoor Air Air in a conditioned space.

IAQ indoor air quality

IDEC Indirect-Direct Evaporative Cooler

IDP Integrated Design Process

IECC International Energy Conservation Code

IEQ Indoor Environmental Quality

IGU Insulated glazing unit. Double or triple panes of glass sealed together to provide insulation value. The still gas between the panes acts as the insulation; *see also window.*

Initial set A time-related set caused by the hydration process of cement-based mixtures, or chemical reaction time of polymers.

Insulating Sheathing Non-structural insulating board products with varying R-values and a wide variation in vapor permeability and drainage characteristics. Materials include expanded polystyrene (EPS), extruded polystyrene (XPS), polyisocyanurate (most often foil-faced), rigid fiberglass, and mineral wool.

Insulation Thermal: Any material which significantly slows down or retards the flow or transfer of heat. Building insulation types are classified according to form (e.g loose-fill, batt, flexible, rigid, reflective, and foamed-in-place) or material (mineral fiber, organic fiber, foam plastic). All types are rated according to their ability to resist heat flow (R-Value or RSI). ***Variant: Thermal Insulation***

Electrical: A non-conductive wire covering, often rubber, thermoplastic, or asbestos. The thickness of insulation varies with wire size and type of material, application or other code limitations.

Jamb The vertical side or edge of a doorway, window, or other rectangular penetration through a wall.

Joints An interface between elements. Joints may be needed to allow for movement of different parts of a building or assembly, or may be required to make construction sequences practical. In all cases, the functional requirements of the enclosure must be maintained the same as for the body of an enclosure element, although aesthetic requirements may be relaxed. A joint may pass through the entire enclosure assembly, in which case it is a building movement or assembly joint, often commonly (and imprecisely) referred to as an expansion joint. Control joints are surface cuts or intentional geometric features which control the location of shrinkage cracks. Construction joints are formed between successive building elements parts during construction work; *see also cold joint.*

Jump Duct A flexible, short, U-shaped duct (typically 10-inch diameter) that connects a room to a common space as a pressure balancing mechanism. Jump ducts serve the same function as transfer grilles. Used when return ducts are not located in every room.

Kiln-dried Lumber Any lumber placed in a heated chamber or “shed” to reduce its moisture content to a specified range or average under controlled conditions. For softwood framing lumber, the moisture content of KD lumber is somewhat based on regional conventions but is most often an average of 12% by weight. In comparison, the moisture content of thoroughly air-dried softwood framing lumber is 15% to 20%.

Latent Heat Change of enthalpy (energy content) during a change of phase. Often used to describe the energy held by moisture (humidity) in the air.

Lamina Composite layer in EIFS that is installed over the insulation, comprised of the reinforcement, base coat and finish coat.

Low-E Most often used in reference to a low-emissivity coating for high-performance windows, but sometimes used to measure the performance of a radiant barrier. A thin metallic oxide coating increases the U-value and/or decreases the SHGC of the window by reducing heat flow from a warm(er) surface to a cold(er) glazing surface.

Maintenance A regular process of inspection, cleaning and minor repairs of buildings elements and exterior systems. Cleaning is for normal activities for those items as required on a regular basis, such as leaves from gutters and drains in the fall, and cleaning lint from dryer vents. Minor repairs encompass small projects that reinstate failed elements such as areas of cracked caulking or peeling paint.

Mastic Heavy-consistency compound that may remain adhesive and pliable with age. Is typically an airtight and waterproof compound applied to exterior walls and roof surfaces, or to provide a robust and durable air seal to air duct distribution systems.

Mcf Million cubic feet (typically used as a metric for natural gas)

Mechanical Ventilation Controlled, purposeful introduction of outdoor air to the conditioned space and/or removal of polluted air from a space driven by mechanical means (e.g., fans).

MEF Modified energy factor

Membrane A generic term relating to a variety of sheet goods used to control water, vapor and/or air movement.

MERV **Minimum Efficiency Reporting Value**
A metric of air filter efficiency, in terms of its ability to remove air particulates.

Mock-Up Full-scale but limited extent demonstration of a finished assembly, such as a wall system, window in a wall, a roof parapet, or washroom plumbing arrangement.

Mold A type of fungus that is different from plants, animals and bacteria. Molds are decomposers of dead organic material such as leaves, wood and plants. Molds sometimes can infect living plants and animals. The spores and hair-like bodies of individual mold colonies are too small for us to see without a microscope. When a lot of mold is growing on a surface, it often appears black or green. The color of mold is influenced by the nutrient source and the age of the colony. Mold growing on fabric is called mildew. ***Variant: Mould***

NAECA National Appliance Energy Conservation Act

NAHB National Association of Home Builders

NASEO National Association of State Energy Officials

NATE North American Technician Excellence

NHQ National Housing Quality

NZEH Net zero energy house (or home)

OASys a brand name of specific indirect/direct evaporative cooler

oc **on center**
Used to designate the spacing of repeating elements, often framing, but also lights, fasteners, windows, etc. ***Variants: o.c., O.C.***

Occupiable Space Any enclosed space inside the pressure boundary and intended for human activities, including but not limited to, all habitable spaces, toilets, closets, halls, storage and utility areas, and laundry areas. ***Variant: occupied space***

OSB **oriented strand board**
A composite wood panel made of flakes of wood, often aspen, bonded together with a small amount of adhesive. Orienting the flakes at the face with their grain in the long direction of the panel product results in improved bending strength (and distinguishes OSB from the older flakeboard product).

Outdoor Air Air outside of the building. ***Variant: OA***

Ozone An oxygen chemical compound of the form 0_3 instead of 0_2 (common atmosphereic oxygen). This 3-atom molecule is an even more active oxidizing agent than its more common 2-atom relative. At ground level, ozone is a pollutant and in the upper atmosphere it is a solar shield (location, location, location). Touted for its ability to "clean" air in room or household ozone generators, this application actually does more harm than good-ozone's highly reactive nature tends to accelerate the breakdown of synthetic materials in homes such as paints, plastics, and ever-available volatile organic compounds, often with less-than-desirable results.

Pa **Pascal**
A metric unit of measurement of pressure, defined as a Newton per square meter. One inch of water column is equal to about 250 Pa.

Parapet Wall A low wall around the perimeter of a roof that projects above the level of the top of the adjoining roof level.

PATH Partnership for Advancing Technology in Housing

Penetration A hole passing through a building enclosure or interior partition in which ducts, pipes, wires, structural elements, and windows pass.

PEX **Cross-linked polyethylene (PE)**
Durable form of small diameter polymer tubing used for plumbing and conveying heating and cooling fluids. ***Variant: pex***

PITI Principal, interest, tax, and insurance

Plate Sill plate or "mud sill": a horizontal member anchored to a masonry wall. Sole plate: bottom horizontal member of a frame wall. Top plate: top horizontal member of a frame wall supporting ceiling joists, rafters, or other members. ***Variants: sill plate, sole plate, top plate, mud sill***

Plywood A wood product made of three or more layers of veneer joined with glue, and usually laid with the grain of adjoining plies at right angles.

Ponding A condition where water stands on a roof, slab, or any other nominally horizontally surface for prolonged periods due to poor drainage and/or deflection of the surface.

Pot life The duration of time that the wet state material remains workable after it has been mixed.

Pressure Boundary The primary air enclosure boundary separating conditioned air and unconditioned air. Typically defined by the air barrier system/air control layer.

Pressure-Equalized Rainscreen
A specific type of drained wall system that uses spatial compartmentalization, large cladding vent to enclosure air leakage ratios, and stiff components to encourage short-term equalization of the drained cavity (also called the air chamber) air pressures with exterior wind pressures with the intent of reducing the air pressure difference across the screen (or cladding). Such systems rely on effective drained to control rain, and if functional, merely reduce the rainwater that the drained system must accommodate; *see also drained.* ***Variant: PER***

Primer A coating intended to prepare the surface of the substrate for the subsequent application of a paint, adhesive, sealant, or adhered membrane.

PSC Permanent split-capacitor (motor)

Psychrometrics The study of air and its energy and water vapor contents.

Punched Window A window installed as a "punched" opening surrounded by cladding, as opposed to being arranged in vertical or horizontal strips.

PV Photovoltaic

Rainscreen A building enclosure rain control strategy that accepts that some water will penetrate the outer surface (the cladding, which "screens" rain) and removes this water back to the exterior by gravity drainage over a drainage plane, through a drainage gap, and exiting via flashing and weep holes. It always refers to a Drained system, however, some use the term only for systems that have larger drainage gaps (e.g., $^1/_2$") or just for systems that are also ventilated (a ventilated drained approach) or just for systems that attempt to pressure-equalize; *see also drained, drainscreen.*

Relative Humidity (RH) The ratio (expressed as a percentage) of the amount of moisture within the air to the maximum amount of moisture that the air could possibly contain at a specific temperature.

R-value Quantitative measure of resistance to heat flow, the reciprocal of the U-factor. The units for R-value are ft^2 °F hr/Btu (English) or m^2 °K hr/W (SI or metric). It represents heat flow as equivalent to conduction. As an example of the context in to which R-value should be placed, 25% to 40% of a typical buildings energy use can often be attributed to air infiltration, and air conditioning loads are often dominated by solar heat gain.

Radiation A form of energy in electro-magnetic radiation. At normal temperatures, radiation is in the infra-red region of the spectrum, whereas solar radiation is at visible wavelengths. Controlling radiation is important to both heat transfer through porous insulations and through windows, as well as playing a significant role in daylighting and solar heating.

Rain Penetration Control Layer
The rain penetration control layer is the continuous layer (comprised of one of several materials and formed into planes to form a three dimensional boundary) in an enclosure assembly that is designed, installed, or acts to form the rainwater boundary. In face-sealed systems, this is the exterior-most face of the enclosure. In concealed barrier systems it is a plane concealed behind the exterior face. In drained systems the rain control plane is the drainage plane behind the drainage gap or drainage layer. In storage / mass/ reservoir systems the rain penetration control layer is typically the innermost storage mass layer.

Rainwater Boundary The rainwater boundary (comprised of one or several materials and formed into planes to create a three dimensional boundary) is the boundary beyond which rainwater is not intended to penetrate by design or beyond which damage can likely be expected to occur if penetration were to occur.

R.A.P. **Return-air pathway**
A commercial product that provides the same function as a transfer grille or jump duct, but also adds some light and sound transmission resistance.

RESNET Residential Energy Service Network

RH Relative humidity

Rigid Insulation Rigid board material that provides thermal resistance. Foam plastic such as EPS, XPS, and polyisocyanurate are commonly used, polyurethane and may be found.

Rough Opening The opening in a wall into which a door, window, or other enclosure component is to be installed.

SA Supply air

Saddle The transition of small horizontal surfaces, such as the top of a balcony guardrail or parapet wall, with a vertical surface, such as a wall.

Sealant A flexible, polymer-based elastomeric material installed wet and used in the assembly of the building enclosure to seal gaps, seams or joints and to provide a clean finish, or waterproof, or airtighten the joint.

SEER **Seasonal energy efficiency ratio**
An energy rating metric for cooling systems (air conditioners) that reports the cooling capacity (in Btu/hr) per Watt of electricity used in the system (fans, compressor, controls) when tested under conditions of 80°F indoor temperature with 50% indoor RH, and an 82°F outdoor temperature. Alternate measurement protocols allow for a binned approach with a range of outdoor temperatures.

Semi-rigid Insulation Formed board material comprised of mineral fibers that provides thermal resistance and supports its own weight. Semi-rigid mineral fiber insulation is normally used for its non-combustible properties and is typically comprised of glass or rockwool.

Sheathing A material used to provide structural stiffness to the wall framing and/or to provide structural backing for the cladding and air/water/vapor control layers. Typical materials are OSB (oriented strand board), plywood, or various forms of gypsum board.

Sheathing Membrane A generic term for a membrane layer that prevents the resists the passage of liquid water (and possibly air and vapor) through vertical, drained surfaces. Asphalt-impregnated building papers and felts and polymeric housewraps are the most common products available, but peel and stick air-water-vapor barrier membranes, trowel applied air and water barriers, etc. are currently available. See also housewrap, drainage plane.

SHGC **Solar heat gain coefficient**
A metric of a glazing or windows ability to reduce solar heat gain, it is the ratio of the amount of heat transmitted to the interior to the total solar radiation incident on the exterior plane of the glazing or window. Exterior temperature has a small influence on the result, and large angles of incidence will result in lower SHGC than more normal angles.

SHW Solar hot water

Siding The finish covering of the outside wall, whether made of horizontal weatherboards, vertical boards with battens, shingles, or other material.

SIP Structural Insulated Panels

SIPA Structural Insulated Panel Association

SIPS Structural Insulated Panel System

SLA Specific leakage area

Sourced Drained A term for a rain control strategy that use the face-sealed approach in body, and the drained approach at the joints or penetrations assumed to be the point at which water enters an assembly. Many windows, some EIFS, precast sandwich panels, and foamed metal panels use this approach; *see also drained.*

Spall A fragment of material, such as concrete or masonry, detached from a larger mass by a physical blow, freeze-thaw, high levels of compression, or subfluorescence.

Spectrally Selective Glazing A type of glazing that reduces solar gain while still providing daylighting; they have low solar heat gain coefficients (SHGC), high visible light transmittance (VT) and usually low U-values.

Stack Effect Air movement caused by warmer air rising and cold air falling. This generates small but steady pressures in direct relation to the size of the temperature difference and the height of the column of air. The resulting pressure differences can lead to air leakage and generate unplanned air flows within buildings that result in indoor air quality problems.

Stucco An exterior cladding formed in place on the wall made of inorganically-bonded sand and small aggregate. Typically Portland cement-based, but with additives of lime, surfactants, water repellents, etc.

Stud One of a series of wood or light steel vertical structural members placed as supporting elements in walls and partitions.

Subfluorescence A potentially harmful accumulation of water-soluble salts that re-crystalize just beneath the masonry, stucco, or concrete surface as moisture in the wall evaporates leaving the salts behind. See also efflorescence.

Surfactant An agent (e.g., detergent) that, when mixed with water, breaks the surface tension of water drops, thus enabling easier absorption of water through a material. Without surfactants, water would have a greater tendency to remain as drops on the surface of a given material. Widely present in water-based coatings and paints.

TAB Testing, adjusting, and balancing

TMY2 Typical Meteorological Year weather data (TMY2 or TMY3 reflects different data sets). ***Variant: TMY3***

Thermal Bridge A material with higher thermal conductivity transferring heat through an assembly with substantially lower thermal conductivity. For example, a steel stud in a wall will transfer more heat than the surrounding insulation, reducing the overall thermal control of the system.

Thermal Boundary The transition within a building enclosure between indoor temperature and outdoor temperature, intended to be located at the thermal control layer. It may, but does not have to align with the pressure boundary or rain water control boundary.

Thermal Control Layer The layer in a building enclosure (comprised of one or several materials and formed into planes to create a three dimensional boundary) that is designed, installed, and/or acts to form the thermal boundary in an enclosure assembly. It may be partially penetrated by thermally conductive elements (see thermal bridge).

Through-Wall Flashing Flashing that extends completely through a wall system. Designed and applied in combination with counter-flashings, to prevent water which may enter the wall above from proceeding downward in the wall or into a roof deck or roofing system. ***Variant: Thru-Wall Flashing***

TOU Time of use

UA Heat loss coefficient (U value multiplied by area)

U-factor Quantitative measure of heat flow or conductivity, the reciprocal of R-value. While building scientists will use R-values for measures of the resistance to heat flow for individual building materials, U-factor is usually used as a summary metric for the ease of heat transfer through building assemblies.

UL Underwriter's Laboratory

UV Ultraviolet radiation (from the sun), which has a degrading effect on many membranes, coatings, and sealants. Artificially generated UV may be used with HVAC systems to control or limit the growth of biologicals.

U-Value The thermal transmittance of a material, incorporating the thermal conductance of the structure along with heat transfer resulting from convection and radiation.

Vapor Barrier A material that has a permeance of 0.1 perm or less. A vapor barrier is a material that is essentially vapor impermeable. A vapor barrier is a Class I vapor control layer. The test procedure for classifying vapor barriers is ASTM E-96 Test Method A—the desiccant or dry cup method.

Vapor Control Layer The element (or elements) that is (or are) designed and installed in an assembly to control the movement of water by vapor diffusion.Language use example: "A Class III vapor control layer is used to manage vapor flow in this assembly."

Vapor Impermeable Materials with a permeance of 0.1 perm or less (e.g., rubber membranes, polyethylene film, glass, aluminum foil). A Class I vapor control layer.

Vapor Permeance A layer property that describes the ease with which vapor molecules diffuse through it. It is defined as the quantity of vapor flow across a unit area that will flow through a unit thickness under a unit vapor pressure difference. It is to vapor diffusion what conductance is to heat transfer. The unit of measurement is typically the "perm" ($gr/h{\cdot}ft^2{\cdot}in.\ Hg$); in metric/SI units, it is stated in terms of $ng/(s{\cdot}m^2{\cdot}Pa)$.

Vapor Permeable Materials with a permeance of greater than 10 perms (housewraps, building papers).

Vapor Permeance Classes

Test procedure for determining vapor permeance class is ASTM E-96 Test Method A (the desiccant or dry cup method).

Class I: Materials that have a permeance of 0.1 perm or less (***Note:*** This is the definition of a "vapor barrier".)

Class II: Materials that have a permeance of 1.0 perm or less and greater than 0.1 perm (***Note:*** This is the definition of a "vapor retarder".)

Class III: Materials that have a permeance of 10 perms or less and greater than 1.0 perm

Class IV: Materials that have a permeance of greater than 10 perms

Vapor Retarder A vapor retarder is a material that has a permeance of 1.0 perm or less and greater than 0.1 perm. A vapor retarder is a material that is vapor semi-impermeable. A vapor retarder is a Class II vapor control layer. The test procedure for classifying vapor retarders is ASTM E-96 Test Method A—the desiccant or dry cup method.

Vapor Semi-Impermeable

Materials with a permeance of 1.0 perm or less and greater than 0.1 perm (oil-based paints, most vinyl coverings).

Vapor Semi-Permeable Materials with a permeance of 10 perms or less and greater than 1.0 perm (plywood, OSB, most latex-based paints).

Ventilation The intentional flow of air into occupied spaces or behind cladding/roofing to move heat and moisture in a desirable manner.

VOC Volatile organic compounds

WRB Water resistant barrier

Water Resistant Barrier A sheet, spray- or trowel-applied membrane or material layer that prevents the passage of liquid water even after long or continuous exposure to moisture. ***Variant: Drainage plane***

Weep Hole An opening placed in a wall or window assembly to permit the escape of liquid water from within the assembly. Weep holes can also act as vents.

Wind Washing The phenomenon of air movement driven by wind pressures wind passing through or behind the thermal insulation within enclosures, causing significant loss of heat flow control and potentially causing condensation. Typicall occurs are exposed building edges, such as at the outside corners and roof eaves because of the large pressure gradients at these locations. This can be thought of as the "wind blowing through the insulating sweater" effect.

Window A manufactured assembly of a frame, sash, glazing and necessary hardware, made to fit an opening in a wall. Windows are comprised of a condensation track, frame, glazing, head, jamb, mullion, rail and sill.

Window Condensation Track A channel at the interior sill level of the window intended to intercept small amounts of water condensing on the interior surface of the glass. Condesnation tracks are sometimes part of curtainwalls and even precast.

Window Frame The stationary part of a window unit; the window sash fits into the window frame.

Window Glazing The glass portion of the window.

Window Head The horizontal window member at the top of a window opening.

Window Jamb Either of the vertical window members at the sides of a window opening.

Window Mullion The vertical window member between the glazed units.

Window Rail The horizontal window member between the glazed units.

Window Sill The horizontal window member at the base of a window opening.

WUFI Wärme Und Feuchte Instationär: a modeling program for simulating one-dimensional dynamic heat and moisture transfer through building enclosure.

Xeriscaping Climate-tuned landscaping that minimizes outdoor water use while maintaining soil integrity and building aesthetics. Typically includes emphasis on native plantings, mulching, and no or limited drip/subsurface irrigation.

XPS Extruded polystyrene; *see extruded polystyrene insulation.*

Zero Energy House or Building (ZNEH, ZNEB) Any house or building that averages out to net zero energy consumption. A zero energy home produces as much energy over a year as it consumes, typically using photovoltaics, or combined heat and power to generate electricity on site. In a zero energy home, efficiencies in the building enclosure and HVAC are great enough that electrical plug loads tend to dominate; therefore, these homes must have the added focus of high efficiency appliances and lighting. Several incompatible definitions are commonly used, including net-zero site energy, net-zero source energy, net-zero energy costs, and net-zero energy emissions. Therefore, terms should be clarified when this name is applied.

ZNE Zero net energy; *see Zero Energy House or Building.*

Index